Lecture Notes in Control and Information Sciences

Edited by M. Thoma and A. Wyner

Vol. 62: Analysis and Optimization of Systems
Proceedings of the Sixth International Conference on Analysis and Optimization of Systems
Nice, June 19–22, 1984
Edited by A. Bensoussan, J. L. Lions
XIX, 591 pages. 1984.

Vol. 63: Analysis and Optimization of Systems
Proceedings of the Sixth International Conference on Analysis and Optimization of Systems
Nice, June 19–22, 1984
Edited by A. Bensoussan, J. L. Lions
XIX, 700 pages. 1984.

Vol. 64: Arunabha Bagchi
Stackelberg Differential Games in Economic Models
VIII, 203 pages, 1984

Vol. 65: Yaakov Yavin
Numerical Studies in Nonlinear Filtering
VIII, 273 pages, 1985.

Vol. 66: Systems and Optimization
Proceedings of the Twente Workshop
Enschede, The Netherlands, April 16–18, 1984
Edited by A. Bagchi, H. Th. Jongen
X, 206 pages, 1985.

Vol. 67: Real Time Control of Large Scale Systems
Proceedings of the First European Workshop
University of Patras, Greece, Juli 9–12, 1984
Edited by G. Schmidt, M. Singh, A. Titli, S. Tzafestas
XI, 650 pages, 1985.

Vol. 68: T. Kaczorek
Two-Dimensional Linear Systems
IX, 397 pages, 1985.

Vol. 69: Stochastic Differential Systems – Filtering and Control
Proceedings of the IFIP-WG 7/1 Working Conference
Marseille-Luminy, France, March 12-17, 1984
Edited by M. Metivier, E. Pardoux
X, 310 pages, 1985.

Vol. 70: Uncertainty and Control
Proceedings of a DFVLR International Colloquium
Bonn, Germany, March, 1985
Edited by J. Ackermann
IV, 236 pages, 1985.

Vol. 71: N. Baba
New Topics in Learning Automata Theory and Applications
VII, 231 pages, 1985.

Vol. 72: A. Isidori
Nonlinear Control Systems: An Introduction
VI, 297 pages, 1985.

Vol. 73: J. Zarzycki
Nonlinear Prediction Ladder-Filters for Higher-Order Stochastic Sequences
V, 132 pages, 1985.

Vol. 74: K. Ichikawa
Control System Design based on Exact Model Matching Techniques
VII, 129 pages, 1985.

Vol. 75: Distributed Parameter Systems
Proceedings of the 2nd International Conference, Vorau, Austria 1984
Edited by F. Kappel, K. Kunisch, W. Schappacher
VIII, 460 pages, 1985.

Vol. 76: Stochastic Programming
Edited by F. Archetti, G. Di Pillo, M. Lucertini
V, 285 pages, 1986.

Vol. 77: Detection of Abrupt Changes in Signals and Dynamical Systems
Edited by M. Basseville, A. Benveniste
X, 373 pages, 1986.

Vol. 78: Stochastic Differential Systems
Proceedings of the 3rd Bad Honnef Conference, June 3–7, 1985
Edited by N. Christopeit, K. Helmes, M. Kohlmann
V, 372 pages, 1986.

Vol. 79: Signal Processing for Control
Edited by K. Godfrey, P. Jones
XVIII, 413 pages, 1986.

Vol. 80: Artificial Intelligence and Man-Machine Systems
Edited by H. Winter
IV, 211 pages, 1986.

For information about Vols. 1–61 please contact your bookseller or Springer-Verlag.

Lecture Notes in Control and Information Sciences

Edited by M. Thoma and A. Wyner

127

C. Heij

Deterministic Identification of Dynamical Systems

Springer-Verlag Berlin Heidelberg GmbH

Author
Dr. C. Heij
Department of Econometrics
Erasmus University Rotterdam
P. O. Box 1738
3000 DR Rotterdam
The Netherlands

ISBN 978-3-540-51323-0 ISBN 978-3-540-46196-8 (eBook)
DOI 10.1007/978-3-540-46196-8

Originally published by Springer-Verlag Berlin Heidelberg in New York 1989.

Offsetprinting: Mercedes-Druck, Berlin

2161/3020-543210

PREFACE

This monograph describes a deterministic approach to identification of linear dynamical systems. This subject is related with systems theory, statistics, time series analysis, econometrics and signal processing. The exposition is of an introductory nature. The main ideas and methods originate in linear systems theory and linear algebra.

Identification concerns the choice of a model for representing available data. The quality of a model depends upon the modelling objectives, on the model complexity and on the fit between model and data. The aim is to determine a simple model which is supported by the data. In general no simple relationships are satisfied exactly by the data. This discrepancy between observed data and simple relationships is often modelled by introducing stochastics. However, instead of stochastic uncertainty it is in our opinion primarily the complexity of reality which often prevents existence of simple exact models. In this case model errors do not reflect chance, but arise because a simple model can only give an approximate representation of complex systems. Therefore we will make no statistical assumptions. As we moreover pay special attention to data consisting of observed time series, the topic of this monograph is deterministic time series analysis and identification of dynamical systems.

A detailed overview of the contents of this monograph and a summary of the main results are given in section I.2.

Chapter I provides a brief introduction to modelling and identification. Chapter II contains an exposition of a deterministic approach to identification and a description and analysis of the class of finite dimensional, linear, time invariant dynamical systems. The material presented in this chapter forms the basis for chapter III on exact modelling, chapter IV on model reduction, and chapter V on approximate modelling. These three chapters can be read independently.

Conclusions are given at the end of each chapter and at the end of the main text. Proofs are collected in the appendix.

Acknowledgements

The approach and results presented in this monograph are inspired by the work of prof.dr.ir. J.C. Willems of the Mathematics Institute of the University of Groningen, The Netherlands. His view on applied mathematics and in particular his ideas on modelling have been a strong impetus for my research. I thank him for his stimulating ideas and for the inspiring working atmosphere which he creates.

I would also like to thank dr. J.W. Nieuwenhuis of the Econometrics Institute of the University of Groningen for our fruitful discussions and for his detailed comments on drafts of this monograph.

The text was typed mainly by Gineke Alberts and partly by Tamara Brünner and Ineke Kruizinga. I thank them for their skilful work.

CONTENTS

CHAPTER I

INTRODUCTION

1. Modelling

1.1. Modelling: specification and identification

Modelling is ubiquitous in scientific as well as in other human activities. A model is a condensed representation of relevant information. We construct and use models in order to describe aspects of experience, to predict future developments and in particular the effects of possible actions, to influence and control what concerns us.

A primary requirement for *scientific* modelling is the explicit description of the information and criteria on the basis of which a model is moulded. Some of the essential factors which play a role in scientific modelling are depicted in figure 1. Two of the main aspects are *specification* of the modelling problem and, subsequently, *identification* of the model.

In general terms, the identification of a model amounts to constructing a *model* on the basis of *data*. It is assumed that relevent data are available and that the class of candidate models, i.e., the model class, has been specified. For the identification of a model the quality of candidate models with respect to the data has to be assessed. This assessment, by means of a *criterion*, depends on the *objectives* underlying the modelling problem. An *identification procedure* describes the way in which a model is chosen (identified) from the model class for given data. The aim is to construct the procedure in such a way that the identified models are of good quality with respect to the data as measured by the criterion.

In modelling problems it is in general not known beforehand which data

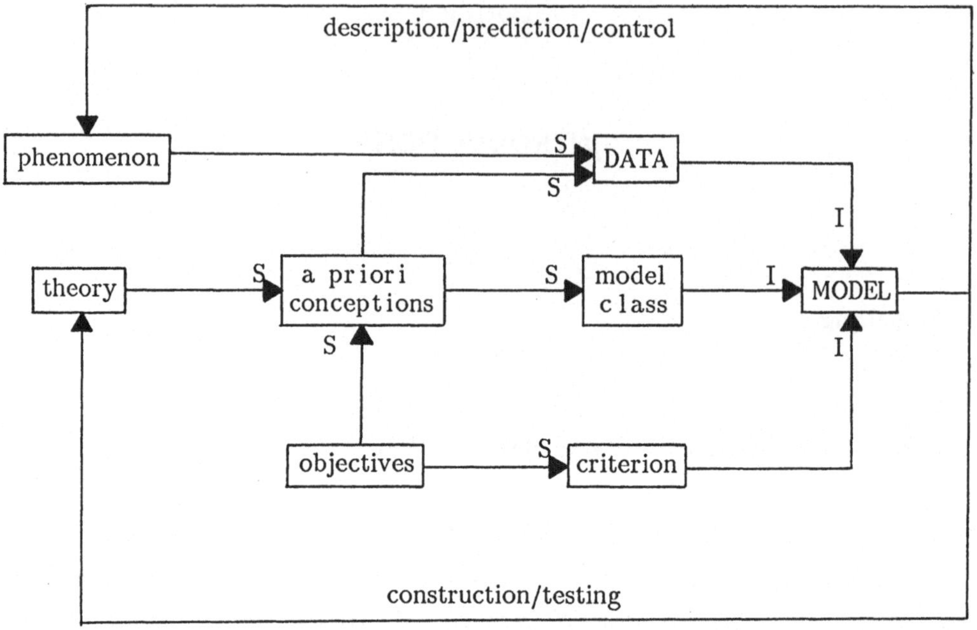

figure 1: modelling (S: specification; I: identification)

will be included for identification of a model. Moreover, in order to investigate the identification aspect of the modelling problem it is necessary to specify the model class and the objectives. This forms the specification aspect of modelling.

Often, the primary objective of constructing a model is not to model the data, but to model a phenomenon. It then is supposed that the data somehow reflect this phenomenon. The phenomenon is considered as a system which produces the data.

In the specification of the modelling problem one can incorporate prior knowledge concerning the phenomenon. This prior knowledge partly can be given by a *theory* concerning the phenomenon. Moreover, one will impose restrictions partly based upon the objectives of modelling and partly for convenience. This leads to a collection of *a priori conceptions*, on the basis of which one decides, e.g., which variables will be included in the model and what models will be considered. A final element of the specification aspect is the representation of the modelling objectives by means of a criterion.

Some of the main objectives of modelling are given in figure 1. An objective could be to model the phenomenon. One can think of description, prediction, or control of the phenomenon. Another objective could be to construct or validate theories concerning the phenomenon.

In section II.1.2 we give some simple examples illustrating the specification and identification aspects of modelling.

1.2. Specification

In the practice of modelling the specification aspect often is considered as being part of the relevant scientific discipline and the identification aspect as being a problem of construction of mathematical procedures and numerical algorithms.

In the sequel we will nearly exclusively be concerned with identification. The central problem then is to choose a model from a given model class on the basis of given data and a criterion assessing the quality of candidate models with respect to the data available, i.e., to transform data into a model of good quality.

Concerning the specification aspect we will restrict attention in the sequel to the specification of criteria which reflect the modelling objectives. It is beyond our scope to treat some of the other fundamental problems in specification. Hence, to mention just a few topics, we will not discuss the relationship between a scientific theory and the choice of data and a model class, the relationship between the phenomenon and the data, prior knowledge and its incorporation in the specification of the model class, or practical problems of data collection. Moreover, we will restrict attention to the objectives of *description* and *prediction*. So we will not consider the objective of control of a phenomenon or modelling with the explicit purpose of building and testing scientific theories.

Besides the specification of criteria, which we will consider in the sequel, we would like to comment on the specification of the model class. This specification implies prior conceptions of a mathematical nature. The choice between *deterministic* and *stochastic* models forms a particular example. This choice sometimes is based on a relevant scientific theory. In case the data consists of a random sample from a well-defined population the choice of stochastic models is based on the statistical aspect of sampling.

In the sequel we will pay special attention to the problem of identification in case the specification of the model class can *neither* be based on a *scientific* theory *nor* on *statistical* considerations like sampling from a population. A typical example is time series analysis for complex dynamical phenomena which lack the possibility of repetitive observation of different trajectories over time, like macro economics or industrial processes. Then the choice of the model class is mainly a matter of convenience. The current practice seems to be to take the models to be stochastic.

In modelling one generally is faced with the problem that the data do not satisfy exactly any simple deterministic relationship. This problem arises because simple models only can give an *approximation* of complex and partly unknown phenomena and also because only a small subset of all possible explanatory variables can usually be included in any specification.

A crucial step in the specification of a stochastic model class is the introduction of stochastics to explain this discrepancy between the data and simple, deterministic relationships. This can be done in various ways. It can be supposed that the data consist of noisy observations of nonobservable, deterministic variables which exactly satisfy simple relationships. This is the errors-in-variables approach. Another method is to assume that the exact relationships are disturbed by random shocks or disturbances. This is the errors-in-equations approach. In both cases it is assumed that there is some hidden, simple structure connecting the variables of interest. That this simple structure is not exactly put into evidence by the data is ascribed to noise and disturbances.

In our approach to the identification problem we make no stochastic assumptions. Instead we follow a completely *deterministic* approach. This does not mean that we want to identify a model which is exactly satisfied by the data. Instead we primarily aim for identification of *simple*, *accurate*, *approximate* relationships in the data. For complex phenomena the fact that the data do not exactly satisfy simple deterministic relationships is often not due to random disturbances or observational noise. Often the phenomena are simply too complex to be modelled exactly within the model class. The models even deliberately are chosen to be simple. Both for human understanding and for practical implementation a simple, slightly inaccurate model of a phenomenon often is preferred above a complex, more accurate one. Then the central issue is approximation. In this case the model errors have no existence as noise or disturbances in a reality outside of modelling, but are due to imposing a simple model on a complex phenomenon. Stated otherwise, errors are a result of modelling and not the other way round, i.e., model inaccuracy is not due to disturbances.

In section II.1.3 we elaborate the foregoing comments on the specification of the model class and motivate our choice of deterministic models.

1.3. Identification

As stated before we will be primarily concerned with the identification aspect of modelling. As we focus on situations where stochastic assumptions are not particularly relevant and as we consider the question of model identification as an approximation issue, we deal with *deterministic identification*.

It is assumed that the variables to be included in the model have been specified. The purpose of identification is to detect relationships between these variables. A set of variables together with the relationships interconnecting them is called a *system*. The interconnecting relationships are called the laws of the system. In identification the aim is to extract laws from the data. In this respect it is irrelevant whether the data is viewed as an entity itself or as a manifestation of an underlying phenomenon or a (meta–) physical system. It is equally irrelevant whether the model is viewed as a compressed representation of the data or as an efficient description of a data generating system.

We will pay special attention to the case of systems evolving over time, i.e., to *dynamical systems*. The data then are given in the form of a time series of observations of the variables included in the system. Time series analysis amounts to extracting dynamical laws contained in these data. In this case we deal with identification of dynamical systems.

Summarizing, we will present a deterministic approach to time series analysis, i.e., our theme is *deterministic identification of dynamical systems.*

2. Overview and summary

2.1. Problem statement

It is our primary aim to describe a deterministic approach to the problem of identification of dynamical systems. In order to do this we will develop a language to state and analyse deterministic identification problems. In the case of dynamical systems the data consists of a time series of observations of the relevant variables. In order to discuss identification of dynamical systems it then remains to specify a model class and, especially, to formulate criteria to evaluate the quality of candidate models with respect to the available data.

The main topics are the following. First we will put forward a modelling methodology for deterministic identification and specify a class of deterministic dynamical systems which will serve as our model class. This is done in chapter II. The remaining chapters are devoted to three main versions of the problem of deterministic identification of dynamical systems. In chapter III we consider *exact modelling.* In this case the aim is to identify a simple model for the data under the restriction that the data exactly satisfy the laws of the identified system. In chapter V we analyse *approximate modelling.* In this case the aim is to establish a compromise between simplicity and accuracy of candidate models. A decrease in accuracy with respect to the data is then allowed, provided that the gain in increasing simplicity of the model is large enough. In chapter IV we consider *model approximation.* This problem arises from exact modelling and is connected with approximate modelling. In this case the data consists of a model and the aim is to reduce the complexity of the model while loosing as little accuracy as possible.

In the next sections we give a brief summary of the contents of the next chapters and conclude with a summary of the main results and with some references.

2.2. Deterministic modelling

Chapter II starts with some simple examples illustrating the modelling approach as presented in section 1.1. Subsequently we formulate a modelling methodology for deterministic identification. According to this methodology an identification procedure identifies a deterministic dynamical system which for the given data is optimal with respect to a criterion of *utility* of models. This utility depends on the objectives of modelling. It is expressed in terms of a measure of *complexity* of models and a measure of *fit* between data and models. The utility function reflects a compromise between the simplicity and the goodness of fit of models. We present two particular utilities which will play a dominant role in the sequel. These utilities are illustrated by means of examples from econometrics, information theory, and speech processing.

The chapter is concluded by defining and investigating a class of deterministic dynamical systems. We consider *parametrization* of this class of models by means of *autoregressive equations* and define two *canonical forms* which will be used in chapter V. Further we summarize some results on *state space realizations* which play a central role in chapter IV. Finally we define a class of *finite time systems* and present some representation properties which will be used in chapter III.

2.3. Exact modelling

As a first instance of deterministic identification we consider exact modelling in chapter III. In this case we want to model the data by a model of least complexity under the restriction that the data satisfy all identified laws *exactly*. An example of this modelling problem is the following. Suppose that the data consists of a certain parametric description of a system and that the model class consists of another representation of systems. The exact modelling problem in this case amounts to finding an equivalent description of the system in terms of the model class. The identification problem then is a question of representation or realization.

First we give some examples of exact modelling. Next we consider the question of finding an exact model for a given time series of infinite length. Finally we develop procedures for exact modelling of a *finite* time series. A central issue here is to specify in which cases we have reason to accept laws which are exactly satisfied by the available data. We are only inclined to accept laws if they are somehow *corroborated* by the data. We define a concept of corroboration and formulate some desirable properties of exact deterministic identification procedures. These properties are investigated for the so-called partial realization procedure. An alternative procedure with optimal properties is constructed for the univariate case.

2.4. Model approximation

The problem of approximating a model by one of less complexity is considered in chapter IV. In this case the data consists of a model. The original, complex model could be the result of exact modelling of data, of approximate modelling with high accuracy, or of the interconnection of many subsystems. The aim is to approximate this model by a model in the model class. The approximate model should be simpler than the original model while the loss in accuracy should be as small as possible.

Two main elements in model approximation are the definition of a measure of *complexity* of models and the definition of measure of *distance* between models. We give a definition of complexity of dynamical systems. We introduce a quite natural distance measure for a certain class of dynamical systems, i.e., we take the gap metric for a class of l_2-systems. This distance can be calculated explicitly in terms of special representations for this class of systems. These representations are related to *scattering theory* and closely resemble innovation representations of stochastic processes. We present a new *balancing* method of model approximation and illustrate this method by means of some simple numerical simulations. In our exposition we use various representations of dynamical systems, especially in terms of *state* variables and *driving* variables.

2.5. Approximate modelling

Finally in chapter V we present a deterministic approach to approximate modelling. This problem of identification by approximation is of crucial interest in, e.g., statistics, econometrics, systems theory, and engineering.

It includes problems as structure identification, estimation, and model validation.

In order to describe approximate procedures for deterministic time series analysis we first present deterministic procedures for modelling *static* data. Both for the purpose of description and for that of prediction we formulate model utilities. These utilities are defined in terms of a measure of complexity of models and a measure of (descriptive or predictive) misfit of models with respect to data. We derive explicit algorithms for procedures corresponding to these utilities by using the *singular value decomposition.*

A main issue in time series analysis as well as in other areas of identification is that of *parametrization* and *identifiability.* Especially the numerical expression for the misfit of a model with respect to data raises problems in case of non–unique parametrizations. We define the misfit of a dynamical system in a way which does not involve parametrization. The misfit and corresponding utility of models can be numerically expressed in terms of special canonical parametrizations of dynamical systems. These *canonical forms* are in close correspondence with the objectives of description or prediction.

We describe four *procedures for deterministic time series analysis.* Two of these procedures correspond to the objective of description, the other two to the objective of prediction. Either the complexity of the model is minimized under the restriction that the misfit remains below a maximal tolerated level or the misfit of the model is minimized under the restriction that the complexity remains below a maximal tolerated level. We present numerical *algorithms* for these procedures. The algorithms are fairly simple and essentially consist of a recursive implementation of the static modelling procedures.

In contrast to current stochastic methods for time series analysis these deterministic procedures are not subject to problems of parameter identifiability or structure (order) estimation. A model is identified in terms of a canonical representation of dynamical systems which is directly related to the objective of modelling. The order of the identified model is determined directly by the data and the utility which represents the objective of modelling.

The procedures have a clear optimality property as data modelling procedures. The identified model represents the data in a way which is optimal with respect to a utility reflecting the purpose of modelling. One of the ways to evaluate if a procedure also has a satisfactory performance as a method of modelling phenomena is to check whether it is consistent. A *consistent* procedure identifies nearly optimal models of the phenomenon if the number of observations generated by the phenomenon is sufficiently large. Then in the limit the procedure would identify an optimal model of the phenomenon. Consistency is investigated for a class of deterministic generating systems and also for a class of stochastic generating systems. In the latter case the model class does not coincide with the class of generating systems.

The procedures and algorithms are illustrated by means of some simple numerical simulations.

2.6. Summary

We present a deterministic approach to identification of dynamical systems. We formulate this approach in terms of a general modelling methodology. Many of the existing modelling and identification procedures can be described in terms of this methodology.

For the case of exact modelling we present a new procedure which has optimal properties and which is inspired by general requirements of simplicity and corroboration. We define a new distance measure for dynamical systems and give a new solution for the problem of model approximation. Finally we describe procedures and algorithms for deterministic time series analysis. We define and investigate consistency of the procedures and show that the identified models are in some sense robust with respect to variations in the data.

2.7. Organization

Chapter II contains an exposition of the modelling objectives and the model class which we will consider in the sequel. The ideas and results presented in this chapter form the basis for the analysis of three instances of deterministic identification of dynamical systems in the next three chapters. In chapter III we use the results from sections II.2.3 and II.3.4, in chapter

IV those from sections II.2.2 and II.3.3, and in chapter V those from sections II.2 and II.3.2.

Chapters III, IV and V can be read independently, with the exception of section V.3.3 in which we use the concept of complexity of dynamical systems as defined in section IV.2.

The main text is devoted to the exposition of ideas, concepts and results. Proofs are collected in the appendix. This appendix is followed by a list of references and by a symbol index and a subject index.

In the text we explicitly denote definitions, lemmas, propositions, theorems and also remarks, notation and interpretation. The remarks contain material which can be skipped without impairing the continuity of the exposition. The notation parts contain notation and some minor concepts and definitions. The interpretation parts elucidate definitions and results. The end of remarks, examples, notation and interpretation is denoted by the symbol □.

The denotation of definitions and results is as follows. In each section the definitions and results are numbered in the order in which they are stated. For reference within a chapter we only give the relevant number, for reference to another chapter we give the chapter number followed by the relevant number.

Remark. Here we give no explicit description of computer programs for the algorithms of sections IV.5.3 and V.4. These programs are collected on diskette as a simple package which makes use of the program PC–MATLAB. □

2.8. References

First and for all, the approach and results presented here are dominantly and generously inspired by the work of Willems [73]. Our contribution should be seen as an offspring of this seminal work.

Most of the material presented in the sequel has been published elsewhere. The main parts of chapters II and V are contained in Heij and Willems [30]. The approach for descriptive modelling was presented in Willems [73]. The material of chapter III is extracted from Heij [28]. The analysis and the main results of chapter IV can be found in Willems and Heij [76]. Some preliminary results were presented in Heij [26], [27], Heij and Willems [29], and Willems and Heij [75].

The literature on identification is abundant. We just mention some of the main references which are related to our exposition in the next chapters.

For stochastic time series analysis we refer to Anderson [4], Box and Jenkins [5], Brillinger [6], and Hannan [22]. Some main text books on stochastic identification in econometrics are Fomby, Hill and Johnson [14], Koopmans [42], Malinvaud [53], and Theil [69].

System theoretic approaches to modelling are described in Caines [7], Kalman [35], [36], [37], Ljung [49], [50], Rissanen [60], and Willems [71], [73], [74]. Some related contributions on stochastic systems and identification are Anderson and Moore [3], Davis and Vinter [11], Finesso and Picci [13], Kalman and Bucy [38], Kumar and Varaiya [44], Ljung and Söderström [52], and the publications collected in Sorensen [65]. A statistical approach to dynamic systems identification is given by Akaike [1], Hannan and Deistler [23], Hannan and Kavalieris [24], Ljung [48], [50], Ljung and Caines [51], Shibata [62], as well as in the references mentioned for stochastic time series analysis and econometrics.

For an introduction into systems theory we refer to Chen [8], Kailath [33], Kalman, Falb and Arbib [39], Rosenbrock [61], and Wolovich [77]. A standard reference for statistics is Kendall and Stuart [41].

Some references for parametrization and related identification problems are Corrêa and Glover [9], Gevers and Wertz [16], Glover and Willems [18], Guidorzi [21], Hannan and Deistler [23], Hannan and Kavalieris [24], Hazewinkel and Kalman [25], the contributions in Hinrichsen and Willems [31], Nieuwenhuis and Willems [55], and Willems [73].

Methods of modelling inspired by information theory can be found in Akaike [1], Kullback [43], Rissanen [59], [60], and the publications collected in Slepian [64]. For modelling in speech processing we refer to Jayant and Noll [32].

Realization theory is exposed e.g. in Lindquist and Pavon [46], Lindquist and Picci [47], Silverman [63], Willems [72], [74], partial realization theory in Kalman [34] and Tether [68].

Some contributions on model reduction are contained in Glover [17], Moore [54], and Pernebo and Silverman [58].

We will extensively use results from linear algebra and matrix theory. We refer to Davis and Kahan [10], Gantmacher [15], Golub and Van Loan [19], [20], and Stewart [66], [67]. Results on the algebraic Riccati equation can be found in Payne and Silverman [57] and Willems [70]. Scattering theory was presented by Lax and Phillips [45]. For some results from functional analysis we refer to Akhiezer and Glazman [2] and Kato [40]. Finally we refer to Federer [12] and Northcott [56].

CHAPTER II

DETERMINISTIC MODELLING

1. Introduction and examples

1.1. Introduction

In section 1 of this chapter we illustrate the description of modelling as given in section I.1 by means of some simple examples. We especially pay attention to the various considerations which can play a role in specifying the model class. We slightly elaborate our preliminary exposition in section I.1.2 on this topic and motivate our choice of deterministic models.

In section 2 we formulate the identification problem in terms of some general modelling principles and define the concept of utility of a model with respect to given data. This utility is expressed in terms of a measure of complexity of models and a measure of misfit of models with respect to data. We define two particular utility functions which will play a dominant role in the sequel. The first utility corresponds to minimizing misfit under a complexity constraint and the second one to minimizing complexity under a misfit constraint. Model approximation as discussed in chapter IV is a special case of the first utility, exact modelling as discussed in chapter III of the second one. The procedures for deterministic time series analysis presented in chapter V consist of descriptive and predictive versions of these two utilities. We illustrate the utilities by means of examples from econometrics, information theory, and speech processing.

Finally in section 3 we introduce a class of deterministic models which throughout we use as our model class. We give a definition of these models in terms of (external) properties of dynamical systems. We describe results on representations of these models which will be extensively used in chapters

III, IV and V.

In section 3.2 we parametrize the models by means of autoregressive equations. This representation is used in chapter V. We investigate the non–uniqueness of this parametrization and comment on corresponding identification problems. The identification procedures of chapter V are based on model utilities which do not involve parametrization. For numerical implementation of the procedures it is a crucial result that these utility functions can be numerically expressed in terms of special parametrizations, which we call the canonical descriptive form and the canonical predictive form.

In section 3.3 we represent the models by means of state space realizations which are used in chapter IV.

We conclude the chapter by investigating finite time systems in section 3.4 and by giving some representation results which are used in chapter III.

Remark. Some parts of section 3 are rather technical. The concepts, definitions and results presented there form the basis for the procedures for deterministic identification of dynamical systems as described in the next chapters. □

1.2. Examples

We illustrate the modelling methodology described in section I.1 by means of some simple examples. We pay special attention to the relationship between scientific theory and the specification of the model class.

Example 1: a resistor

Suppose one wants to describe a resistor. On the basis of physical theory ("Ohm's law") one postulates a linear relationship between the voltage (V) across and the electrical current (I) through the resistor, i.e., $V=I.R$ with $R\geq 0$ the resistance. A resistor is then described by a parameter R. So the model class is parametrized by $\mathbb{R}_+$, i.e., the set of nonnegative real numbers. To identify R, suppose one performs a number (n) of experiments with resulting voltage and current measurements $(\tilde{V}_i,\tilde{I}_i)$, $i=1,\ldots,n$. See figure 2.

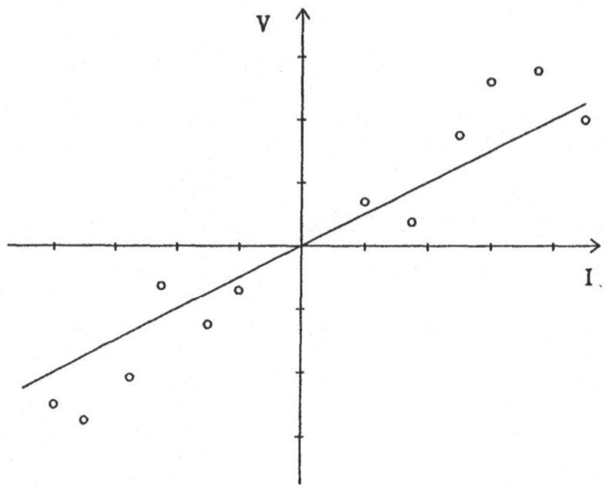

figure 2

The identification problem consists of choosing R on the basis of these data. In general there will exist no R such that $\tilde{V}_i = \tilde{I}_i . R$ for all $i=1,...,n$. This can be due to inaccurate measurements and to the fact that the linear relationship is an idealization – though it may be an accurate one. A reasonable criterion could be, for example, total least squares.

So in this case, in order to describe the resistor, one uses physical theory to specify the model class and the data to be collected.

Example 2: eye colour

Suppose one wants to predict the colour of the eyes of a person. On the basis of biological theory (genetics) one postulates a specific probabilistic relationship between this colour and the colour of the eyes of the ancestors. Assume that the colour is either brown (1) or blue (0), and that brown is dominant over blue. As model class one could take [0,1], where a particular model $p \in [0,1]$ means that p is the probability that the person has brown eyes. Suppose the data consist of the colour of the eyes of the parents and grandparents, as given in figure 3. In this particular case one can identify p by means of elementary probabilistic calculations. In general, identification of p also involves the colour of the eyes of the other ancestors.

One could now make a prediction for example by maximum likelihood, i.e., predicting the colour to be brown if and only if $p > \frac{1}{2}$.

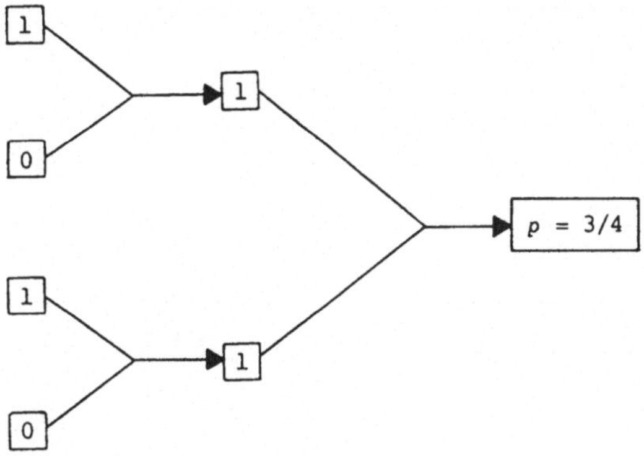

figure 3

So in this case, in order to predict the eye colour, one uses biological theory to specify the identification and prediction problem.

Example 3: consumption

Suppose one wants to predict the national consumption C_{t_0+1} for the coming year. On the basis of an economic theory one postulates that the dominant factor determining C_{t_0+1} is the national income Y_{t_0} in the current year. Suppose data for consumption and income, $(\tilde{C}_t, \tilde{Y}_t)$, $t=s,s+1,\ldots,t_0$, are available. For convenience one could postulate an affine relationship between consumption in a year and income in the preceding year. The model class for example could be parametrized by $\mathbb{R}_+^2$, where the parameter (a,b) with $a,b \geq 0$ describes the postulated relationship $C_{t+1}=a+b.Y_t$. In order to identify a model one could use the data to estimate a and b, for example, by means of ordinary least squares. If the resulting estimates $\hat{a},\hat{b}$ indeed are nonnegative, one could predict C_{t_0+1} by means of $\hat{a}+\hat{b}.\tilde{Y}_{t_0}$. See figure 4.

So in this case, in order to predict consumption, one uses economic theory to specify which data are relevant. The choice of the model class is mainly a matter of convenience. If the estimated values $\hat{a}$, $\hat{b}$ are not accepted as a reasonable description of consumptive behaviour one is ready to specify a different class of models.

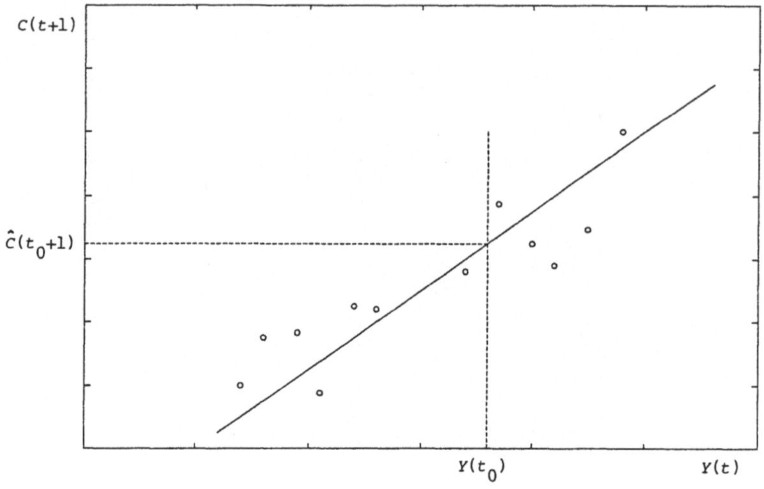

figure 4

Example 4: rainfall

Suppose one wants to regulate the water supply from a reservoir. The water of the reservoir is supplied to customers and replenished by rain. Suppose that one can construct a reasonable control strategy, once the rainfall is modelled.

As model class one could consider the class of (possibly multivariate) Gaussian ARMA processes. Suppose that rainfall data $\{\tilde{r}(t);\ t_1 \leq t \leq t_2\}$ are available. To identify a model on the basis of these data one could consider the objective of simultaneous prediction of the rainfall for a number of periods in the future.

So in this case, in order to formulate the water supply problem in terms of only the rainfall, one has used prior knowledge, e.g., of the demand pattern for water and of (stochastic) control theory. It is assumed that the rainfall can be modelled as an ergodic stationary stochastic process. This assumption can be supported in case the mechanism producing the rainfall is rather stable, i.e., if the climatological conditions are stable. This for example means that, although the rainfall is uncertain, some time averages of the rainfall are less uncertain.

Example 5: realization

Suppose one wants to interpolate n points $(x_i,y_i)\in\mathbb{R}^2$, $i=1,...,n$, by means of a polynomial p of lowest possible degree. So the data consists of n points in $\mathbb{R}^2$ and the model class consists of polynomials. As a criterion to choose p one requires that $y_i=p(x_i)$, $i=1,...,n$, and that the degree of p is minimal.

So in this case the objective is to give an exact description of the data in a most simple way. This is an example of exact modelling or realization. The concepts of phenomenon and theory do not play a role in the specification of the modelling problem. The choice of the model class is inspired by aesthetics or the desire to give a compact representation of the data.

Example 6: model approximation

Let the model class consist of the class of polynomials. Suppose that a model is available, e.g. identified by means of the realization procedure of example 5, and that one wants to construct an optimal linear approximation of this model.

Let the complexity of a polynomial be defined as its degree. Suppose that for the modelling problem at hand a relevant distance measure is given by $d(p_1,p_2):=\max\{|p_1(x)-p_2(x)|;\ x\in[0,1]\}$. Then the problem of optimal linear approximation is a model approximation problem, i.e., approximating a given model by one of less complexity.

So in this case, in order to get a simple model, one uses prior conceptions with respect to the use of models in order to define complexity and distance measures and a resulting model approximation problem.

1.3. Specification of the model class

The foregoing examples especially are intended to illustrate various considerations which can play a role in specifying the model class. In examples 1 and 2 well-established theories are used to choose the model class, one deterministic and the other probabilistic. In examples 5 and 6 the choice is inspired by aesthetics or convenience. In examples 3 and 4 the choice of the model class reflects an aim of simplicity.

A crucial question in the specification of the model class is whether it should consist of *stochastic* or of *deterministic* models. In examples 1 and 2 the choice is based on a relevant scientific theory. In examples 3 and 4, like in many modelling problems outside of the natural sciences, the choice is inspired by convenience.

In section I.1.2 we remarked that it seems current practice to model the discrepancy between data and simple relations by introducing ***stochastic*** models and by postulating that the data are generated by a hidden simple structure under stochastic distortions caused by *noise* or *disturbances*. For identification purposes it is nearly invariably assumed that the noise or disturbances are subject to fixed probabilistic laws. Based on this assumption, the quality of proposed identification procedures is assessed on the basis of statistical criteria like unbiasedness, consistency and efficiency.

No doubt this approach has led to a powerful modelling methodology and technology. Moreover, in case of sampling from a well-defined population a stochastic specification is a natural choice in order to reflect the statistical aspect of sampling.

However, especially in modelling complex dynamical phenomena which lack the possibility of repetitive observation, the stochastic specification raises two problems. First, it often seems difficult to defend the assumption of a fixed distortion generating mechanism. The model errors are mainly due to imposing a simple model structure on a complex phenomenon and have little to do with distortions of a stable nature. Second, it often seems difficult to give an interpretation of modelling a dynamical phenomenon by means of a stochastic process. It is not clear what it means to assign probabilities to sets of trajectories for the variables in case just one trajectory is observed.

We illustrate the foregoing by means of example 3 of section 1.2. Here the problem consists of modelling the relationship between income in one year and consumption in the next one by means of a model of the form $C_{t+1}=a+b{\cdot}Y_t$, $(a,b)\in\mathbb{R}_+^2$. In general there exists no $(a,b)\in\mathbb{R}_+^2$ such that $\tilde{C}_{t+1}=a+b{\cdot}\tilde{Y}_t$ for all available data $(\tilde{C}_t,\tilde{Y}_t)$.

In case one would like to make a stochastic specification of the model class one could draw attention to observation noise in order to explain the discrepancy between the data and the model class. However, it seems reasonable not to reproach only the data for this lack of fit. It might be true that the data only imperfectly measure the economic quantities consumption and income, but generally the model errors are too large to be imputed only to the data. Instead of this errors–in–variables specification it seems more reasonable to specify an errors–in–equations model class of the form $C_{t+1}=a+b{\cdot}Y_t+\varepsilon_t$. In the stochastic specification it is often assumed that the auxiliary variables (ε_t) form a stationary process. This assumption gives many additional means for evaluating the quality of modelling procedures.

In our approach the model errors are considered as being an *effect* of modelling. The model errors are due to imposing the simple model structure and (ε_t) represents the effect of neglected variables and nonlinearities. In this case there is no reason to assume that these errors can be modelled by means of a fixed probabilistic law. The phenomenon is not intrinsically *stochastic*, it is intrinsically *complex*. In our opinion the role of uncertainty and stochastics is overly stressed in the current modelling practice, at the expense of the role of complexity, especially for the case of modelling dynamical pheonomena which lack the possibility of repetitive observation of different trajectories.

As a basic question of modelling we will henceforth consider the problem of finding a simple model for often complex, but not intrinsically uncertain phenomena. The model errors then are due to complexity and our ignorance of phenomena or our practical abhorrence of overly complex models. We deliberately allow model errors because we want to do things in a simple way.

Remark. As stated before we will restrict attention in the sequel to the identification aspect of the modelling problem and to the specification of modelling objectives by means of a criterion. In section 2 we expose a deterministic identification methodology and define some general modelling objectives. A particular specification of these objectives by means of a criterion then leads to a particular deterministic identification problem. The chapters III, IV and V correspond to various possible specifications of the objectives. As model class we take the class of deterministic dynamical systems defined in section 3. □

2. Modelling objectives

2.1. Complexity, misfit, and utility

As described in section I.1.1 the identification aspect of modelling involves the objectives of modelling, denoted by π, a set of conceivable data, denoted by $\mathcal{D}$, and a model class, denoted by $\mathbb{M}$. For identification we need to construct data modelling procedures which for given data assign a model or a set of models in the prespecified class $\mathbb{M}$.

Definition 2-1 A data modelling *procedure* is a map $P:\mathcal{D}\to 2^{\mathbb{M}}$.

Interpretation. A procedure associates with any conceivable data a set of models. Usually $P(d)$ will be a singleton, but it need not be. □

The aim is to construct procedures which are optimal with respect to the objectives π. This means that for given data $d\in\mathcal{D}$ the identified model(s) $P(d)$ should, within the model class $\mathbb{M}$, reflect the data in a way which is optimal with respect to the objectives π.

A general objective is to construct models which are both simple and accurate. On the one hand we want to infer from the data as much structure as possible. A model is considered to be simple if it claims many laws, i.e., if it imposes many restrictions on the variables under consideration. So a simpler model is more easily falsified. We call this striving for structure a *simplicity* or *falsifiability* principle. On the other hand, although we want to infer many laws from the data, we want to accept laws only if there is sufficient evidence for them from the data. We should have some reason for claiming laws on the basis of data. We call this requirement of "reason" or "evidence" a *corroboration* principle. Often the evidence for a law is measured by a criterion of *fit* of the law with respect to the data.

We assume that the objectives π can be specified by a *complexity map* $c:\mathbb{M}\to C$ and a *misfit map* $\varepsilon:\mathcal{D}\times\mathbb{M}\to E$. We assume the spaces C and E to be partially ordered. It is desirable to have models for which both the complexity and the misfit are "small". In this case we get simple models which are corroborated by the data. However, these desires in general are competitive. We therefore assume that π can be expressed by means of a *utility function*, i.e., a map $u:C\times E\to U$, where U is a partially ordered set. The aim then is to choose a model for which the complexity and misfit are such that the corresponding utility is maximal.

Notation. For a partial ordering $\leq$ on U, $m\in U'\subset U$ is said to be a maximal element of U' if $\{u'\in U',\ m\leq u'\}\Rightarrow\{u'\leq m\}$. □

Definition 2-2 The procedure $P_u:\mathcal{D}\to 2^{\mathbb{M}}$ *corresponding* to the utility $u:C\times E\to U$ is defined by $P_u(d):=\text{argmax}\{u(c(M),\varepsilon(d,M));\ M\in\mathbb{M}\}$ for $d\in\mathcal{D}$.

Interpretation. P_u assigns to data the set of models for which the utility is maximal. □

Remark. This clearly raises questions of existence and unicity of maximal elements. □

Remark. In the remainder of this section we consider two instances of this specification of the modelling objectives which will play a dominant role in the next chapters. In fact many of the classical identification procedures can be described in this context. □

2.2. Modelling under a complexity constraint

2.2.1. Procedure

Suppose that the complexity space C and the misfit space E both are totally ordered. A possible reconciliation between the objectives of low complexity and of low misfit is to specify a *maximal tolerated complexity* and to minimize the misfit under this constraint.

Notation. Given $c_{tol}\in C$, we define the utility $u_{c_{tol}}$ as follows. Let $\underline{u}\notin C\times E$ and $U:=(C\times E)\cup\{\underline{u}\}$. For $c>c_{tol}$ let $u_{c_{tol}}(c,\varepsilon):=\underline{u}$, and for $c\leq c_{tol}$ define $u_{c_{tol}}(c,\varepsilon):=(c,\varepsilon)$. Denote the orderings on C and E by $\leq$. On U we impose the following total ordering: $\underline{u}<(c,\varepsilon)$ for all $(c,\varepsilon)\in C\times E$, and $(c_1,\varepsilon_1)<(c_2,\varepsilon_2)$ if $\varepsilon_1>\varepsilon_2$ or if $\varepsilon_1=\varepsilon_2$ and $c_1>c_2$. So a complexity above c_{tol} is not allowed. Further, models of low misfit are preferred, and for models of equal misfit low complexity is preferred. The procedure $P_{c_{tol}}$ now is defined as the procedure corresponding to $u_{c_{tol}}$. □

Definition 2-3 $P_{c_{tol}}(d):=\text{argmax}\{u(c(M),\varepsilon(d,M));\ M\in\mathbb{M}\}$, where $\{u(c_1,\varepsilon_1)=u(c_2,\varepsilon_2)\}:\Leftrightarrow\{c_1,c_2>c_{tol}$ or $(c_1,\varepsilon_1)=(c_2,\varepsilon_2)\}$ and $\{u(c_1,\varepsilon_1)<u(c_2,\varepsilon_2)\}:\Leftrightarrow\{c_1>c_{tol}\geq c_2$, or $c_1,c_2\leq c_{tol}$, $\varepsilon_1>\varepsilon_2$, or $c_1,c_2\leq c_{tol}$, $\varepsilon_1=\varepsilon_2$, $c_1>c_2\}$.

Interpretation. Given the complexity constraint, the procedure $P_{c_{tol}}$ determines the models of minimal misfit, and among models of minimal misfit it selects those of minimal complexity. □

Remark. Model approximation as discussed in chapter IV can be considered as a special case of this approach to modelling. Moreover, two of the identification procedures for deterministic time series analysis presented in chapter V are of this type. □

Example. We first illustrate the approach by a simple geometric example. Let $\mathcal{D}$ consist of the bounded convex subsets of $\mathbb{R}^2$ and $\mathbb{M}$ of the convex polyhedral subsets of $\mathbb{R}^2$. For $M\in\mathbb{M}$ define the complexity $c(M)$ as the number of extremal points of M. For $C\in\mathcal{D}$ and $M\in\mathbb{M}$ define the misfit $\varepsilon(C,M)$ as the Lebesgue measure of the symmetric difference $(C\backslash M)\cup(M\backslash C)$. Suppose c_{tol} is given. Then $P_{c_{tol}}$ models C by means of the convex hull of at most c_{tol} points in such a way that the measure of the resulting symmetric difference is minimal. Among solutions it chooses those with minimal number of extremal points. It can be shown that this last step in fact never will be invoked. See figure 5 for an illustration for the case $C:=\{(x,y)\in\mathbb{R}^2;\ x^2+y^2\leq 1,\ x\geq 0,\ y\geq 0\}$ and $c_{tol}:=4$. It can be shown that then $P_{c_{tol}}(C)$ is the convex hull of $(0,0)$, $(0,a)$, $(a,0)$ and (b,b), with $a:=2(\alpha^2+1)^{1/2}/(4\alpha^2+1)^{1/2}$ and $b:=\alpha a/(1+\alpha)$, where $\alpha:=1+\sqrt{2}$. □

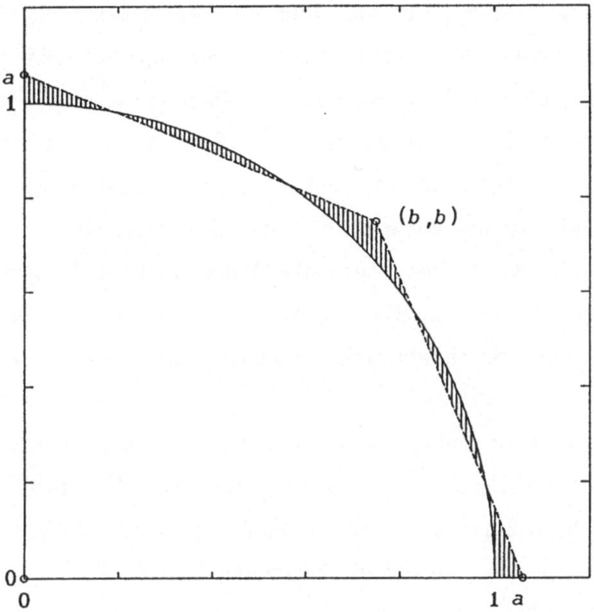

figure 5

Example. Another example is maximum likelihood identification in time series analysis, see, e.g., Box and Jenkins [5], or Hannan and Deistler [23]. For instance, let $\mathcal{D}$ consist of the set of univariate time series of finite length and $\mathbb{M}$ of the class of stochastic ARMA–models. For $M \in \mathbb{M}$ define the complexity $c(M)$ as, e.g., $\max\{d_1, d_2\}$, where d_1 denotes the degree of the autoregressive part and d_2 of the moving average part of M. For $w \in \mathcal{D}$ and $M \in \mathbb{M}$ define the misfit $\varepsilon(w,M)$ as the inverse of the likelihood of M for w. Suppose c_{tol} is given. Then $P_{c_{tol}}$ models the time series by means of an ARMA–model of maximum likelihood, under the restriction $\max\{d_1, d_2\} \leq c_{tol}$. Among optimal solutions it minimizes $\max\{d_1, d_2\}$. This last step generally need not be invoked, as for non–degenerate data the optimal solutions satisfy $d_1 = d_2 = c_{tol}$. □

Remark. In the next section we give another example of modelling with given tolerated complexity. □

2.2.2. Simultaneous equation models

We consider a modelling procedure which is sometimes followed in macro–econometrics and other disciplines dealing with complex dynamical phenomena. See, e.g., Fomby, Hill and Johnson [14, sections 19, 20 and 21].

Suppose one wants to describe the relationship between two groups of variables, one consisting of n_2 variables collected in $y\in\mathbb{R}^{n_2}$ and the other consisting of n_1 variables collected in $x\in\mathbb{R}^{n_1}$. For example, y could consist of the values of n_2 variables of interest at time t, and x of certain lagged values of these and possibly some other, auxiliary variables.

Suppose one decides to use stochastic linear models. In general no simple linear relationship will be exactly satisfied by the data. It is assumed that these deviations can be adequately modelled by means of a (Gaussian) disturbance term.

The model class of simultaneous equation models in this case can be parametrized by $\{(A,B,\Sigma);\ A\in\mathbb{R}^{n_2\times n_2}$ nonsingular, $B\in\mathbb{R}^{n_2\times n_1}$, $\Sigma\in\mathbb{R}^{n_2\times n_2}$, $\Sigma=\Sigma^T\geq 0\}$. The parameter (A,B,Σ) corresponds to the model $Ay+Bx=\varepsilon$, where ε is a Gaussian random variable with mean zero and covariance matrix Σ.

Let data $\{(\tilde{x}_i,\tilde{y}_i);\ i=1,...,n\}$ be available. One possible approach to identify a model on the basis of these data, i.e., to estimate (A,B,Σ), is the following. Suppose the data are generated by a stochastic system $A_0y_i+B_0x_i=\varepsilon_i$, $i=1,...n$, where the ε_i are independent identically distributed zero mean Gaussian random variables with covariance matrix Σ_0. First estimate $(-A_0^{-1}B_0,A_0^{-1}\Sigma_0(A_0^{-1})^T)$, e.g., by maximum likelihood. Denote the resulting estimates by $(\hat{\Pi},\hat{S})$. Impose restrictions on the parameter (A,B) such that the map $f:(A,B)\rightarrow -A^{-1}B$ becomes a bijection. The injectivity of f is called identifiability in the literature. In this case the model could be estimated as $(\hat{A},\hat{B}):=f^{-1}(\hat{\Pi})$ and $\hat{\Sigma}:=\hat{A}\hat{S}\hat{A}^T$. This is the so-called method of indirect least squares.

We state some of the essential elements in this approach.

First, identifiability often is obtained by imposing prior restrictions on A and B, declaring certain elements of these matrices to be zero. The interpretation is that every equation corresponds to a part of the phenomenon which only incorporates certain variables. These zero restrictions are often inspired by theory. Imposing the restrictions resembles fixing the tolerated complexity, interpreted as the number of non-zero coefficients.

Second, it is not so much the (least squares) misfit as the variance of the estimated parameters which determines the confidence in the model. In a strict sense, every observation fits any model for which $\Sigma>0$. However, inspection of the estimated variability of the parameter estimates corresponds to some intuitive concept of misfit.

Finally, both the complexity and the "confidence" are defined in terms of parametrizations of models. In particular, every equation is investigated independent of the other ones. For example, declaring a parameter in a particular equation to be zero does not imply the absence of a direct relationship between the corresponding variables, as such a relationship could be due to the other equations.

Remark. In chapter V we decribe four modelling procedures for modelling dynamical phenomena which do not make use of stochastic assumptions. This in particular avoids the assumption of a fixed distribution which generates the disturbances. The procedures are based on complexity and misfit measures which can be expressed in terms of canonical parametrizations of dynamical models. These canonical forms are directly inspired by the objectives of modelling and do not depend on a theory concerning the phenomenon. The resulting measures have an unambiguous interpretation in terms of model quality, as opposed to parameter quality. Moreover, the measures take the simultaneous nature of the model equations explicitly into account. □

2.3. Modelling under a misfit constraint

2.3.1. Procedure

Again suppose that both C and E are totally ordered. Another possible reconciliation between the objectives of low complexity and of low misfit is to specify a *maximal tolerated misfit* and to minimize the complexity under this constraint.

Notation. Given $\varepsilon_{tol}\in E$, we define the utility $u_{\varepsilon_{tol}}$ as follows. Let $\underline{u}\notin C\times E$ and $U:=(C\times E)\cup\{\underline{u}\}$. For $\varepsilon\geq\varepsilon_{tol}$ let $u_{\varepsilon_{tol}}(c,\varepsilon):=\underline{u}$, and for $\varepsilon<\varepsilon_{tol}$ $u_{\varepsilon_{tol}}(c,\varepsilon):=(c,\varepsilon)$. On U we impose the following total ordering: $\underline{u}<(c,\varepsilon)$ for all $(c,\varepsilon)\in C\times E$, and $(c_1,\varepsilon_1)<(c_2,\varepsilon_2)$ if $c_1>c_2$ or if $c_1=c_2$ and $\varepsilon_1>\varepsilon_2$. So misfits of ε_{tol} or larger are not allowed. Further, models of low complexity are preferred, and for models of equal complexity low misfit is preferred. The procedure $P_{\varepsilon_{tol}}$ now is defined as the procedure corresponding to $u_{\varepsilon_{tol}}$. □

Definition 2-4 $P_{\varepsilon_{tol}}(d)$:=argmax$\{u(c(M),\varepsilon(d,M));\ M\in\mathbb{M}\}$, where $\{u(c_1,\varepsilon_1)=u(c_2,\varepsilon_2)\}:\Leftrightarrow\{\varepsilon_1,\varepsilon_2\geq\varepsilon_{tol}$ or $(c_1,\varepsilon_1)=(c_2,\varepsilon_2)\}$ and $\{u(c_1,\varepsilon_1)<u(c_2,\varepsilon_2)\}:\Leftrightarrow\{\varepsilon_1\geq\varepsilon_{tol}>\varepsilon_2$, or $\varepsilon_1,\varepsilon_2<\varepsilon_{tol}$, $c_1>c_2$, or $\varepsilon_1,\varepsilon_2<\varepsilon_{tol}$, $c_1=c_2$, $\varepsilon_1>\varepsilon_2\}$.

Interpretation. Given the misfit constraint, the procedure $P_{\varepsilon_{tol}}$ determines the models of minimal complexity, and among models of minimal complexity it selects those of minimal misfit. □

Remark. Two of the identification procedures for deterministic time series analysis presented in chapter V are of this type. □

Remark. The procedure corresponding to the requirement $\varepsilon\leq\varepsilon_{tol}$ (instead of $\varepsilon<\varepsilon_{tol}$) will be denoted by $\bar{\bar{P}}_{\varepsilon_{tol}}$. Exact modelling as described in chapter III can be considered as a special case of this approach to modelling with $\varepsilon_{tol}=0$, i.e., allowing no misfit. □

Example. Returning to the geometric example of section 2.2.1, suppose that ε_{tol} is given. Then $P_{\varepsilon_{tol}}$ models a bounded convex set C by means of the convex hull of a minimal number of points under the misfit restriction, and chooses among solutions those with minimal misfit. □

Example. Concerning the maximum likelihood example of section 2.2.1, for given minimal tolerated likelihood $1/\varepsilon_{tol}$ the procedure $P_{\varepsilon_{tol}}$ minimizes $\max\{d_1,d_2\}$ under the likelihood constraint, and chooses among solutions those with maximal likelihood. This gives a solution to the so-called problem of model order selection. □

Remark. In the next sections we give two other examples of modelling with given misfit constraint, one for the case of exact modelling and one for the case of approximate modelling. □

2.3.2. Minimum description length principle

As an example of exact modelling we mention the minimum description length principle of Rissanen, see e.g. Rissanen [59]. In this case the data set $\mathcal{D}$ consists of finite sequences of (finite precision) real numbers. The model class $\mathbb{M}$ consists of finite sequences of binary digits. A model represents data exactly by means of an injective code $C:\mathcal{D}\to\mathbb{M}$. It is assumed that C codes the data d by means of an auxiliary (countable) class $\mathbb{P}=\{P_\theta;\ \theta\in\Theta\}$ of probability distributions on $\mathcal{D}$, in the following way. The binary sequence $C(d)$ consists of an initial part describing the parameter θ and a remaining part describing the data in a way which is optimal in P_θ (minimum mean description length code for P_θ).

The complexity of a model is defined as the length of the binary sequence. Given the class $\mathbb{P}$, the minimum description length principle corresponds to the procedure which consists of coding the data by means of the shortest possible binary sequence, i.e., by the model of least complexity. This minimum description length principle balances the desire for a small number of parameters (in θ) and a simple description of the data by means of P_θ (maximum likelihood).

Remark. It is interesting to note that this approach gives a deterministic interpretation, in terms of exact modelling, of, e.g., maximum likelihood estimation and modelling by means of minimizing prediction errors. □

2.3.3. Speech processing

An example of modelling under a misfit constraint is speech processing. Let S denote the set of binary sequences of finite length. The problem is to encode, transmit and decode a signal $s\in S$ in the simplest way possible, given a tolerated misfit and an auxiliary class of models $\mathbb{M}_{aux}\subset S$. An encoder is a map $f:S\to\mathbb{M}_{aux}\times S$ transforming a signal s into a transmitted signal $t\in S$. The signal t consists of an initial part describing the auxiliary model and a remaining part describing the signal s in an approximate way by means of the auxiliary model. A decoder is a map g: $\mathbb{M}_{aux}\times S\to S$ transforming a signal t into a decoded signal $\hat{s}$. See figure 6.

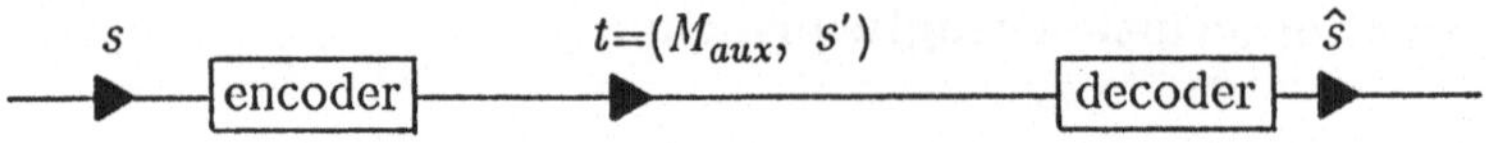

figure 6

For example, $\mathbb{M}_{aux}$ could be chosen to be the (set of parameters of the) class of autoregressive systems. The initial part of t then describes the order and the numerical values of the parameters of the auxiliary system. The remaining part of t could be used to describe some of the prediction errors of the estimates generated by the auxiliary system with respect to the signal s. The decoder could construct a signal $\hat{s}$ based upon the estimates generated by the auxiliary system and the transmitted prediction errors. See e.g. Jayant and Noll [32].

Here the set of conceivable data is $\mathcal{D}=S$ and the model class is $\mathbb{M}=\mathbb{M}_{aux}\times S$. Define the complexity of a model $t\in\mathbb{M}_{aux}\times S$ as the length of the string t. Let $\delta(s,\hat{s})$ denote a measure of the error of the decoded signal $\hat{s}$ with respect to the original signal s. Define the misfit of a model t with respect to data s by $\varepsilon(s,t):=\delta(s,\hat{s})$ where $\hat{s}:=g(t)$. So the misfit measures the error made in reconstructing the signal s from the transmitted signal t. Given a tolerated misfit, the aim is to minimize the complexity of the transmitted signal, i.e., of the model.

Remark. This approach resembles the minimum description length principle, though in speech processing it is not required that the data can be reconstructed exactly from the transmitted signal. □

3. Model class

3.1. Deterministic dynamical systems

3.1.1. Definition

Definition 3-1 A *deterministic dynamical system* is a triple $(T,W,\mathcal{B})$ with $T\subset\mathbb{R}$ the time set, W the signal set, and $\mathcal{B}\subset W^T$ the *behaviour* of the system.

Remark. The behaviour $\mathcal{B}$ we will sometimes call a system or a model. □

Interpretation. A deterministic dynamical system describes the relationships between variables of interest in the following way. Let W be the set in which the variables on every time instant take their values, and let T denote the time set under consideration. The behaviour $\mathcal{B}$ then consists of a set of time series $w:T\rightarrow W$ with the interpretation that time series $w\in\mathcal{B}$ are compatible with the laws of the system, while time series $w\notin\mathcal{B}$ are not compatible with these laws. This gives a deterministic description of the system. □

Remark. Note that in general a deterministic dynamical system $\mathcal{B}$ by no means uniquely describes a trajectory for the variables of interest. This would only be the case if $\mathcal{B}$ would consist of a singleton. Instead, $\mathcal{B}$ gives a partition of the set of all possible time series in a set of trajectories compatible with the laws of the system and a set of trajectories incompatible with these laws. Which of the trajectories in $\mathcal{B}$ actually will be realized will depend on factors which are not modelled by the system, i.e., on the initial conditions and on the environment with which the system interacts. The model $\mathcal{B}$ only describes the relationships between the variables, teared from their environment. So in a deterministic system the time evolution of the variables is *not* deterministic in the sense of being uniquely determined. Deterministic modelling does not imply modelling all relevant factors or perfect predictability of the variables. A deterministic model expresses uncertainty or *indeterminateness* by describing a set of possible trajectories, without assigning probabilities. The environment is not modelled as a stochastic mechanism, as in a stochastic model specification, but as an uncertain mechanism, i.e., the environment is not modelled by the system. □

Remark. For some illustrative examples we refer to Willems [73],[74]. □

3.1.2. Linear, time invariant, complete systems

In the sequel we nearly invariably will use as model class a special class of deterministic dynamical systems. We define this model class in this section and give representation properties in sections 3.2 and 3.3.

In the identification problem the model class reflects prior restrictions which we impose on the models. The structure of these models is accepted, in principle independent from the data. The identification procedure infers additional model structure from the data.

Notation. We restrict attention to discrete time systems with time set $T=\mathbb{Z}$, i.e., the integers, and with signal set $\mathbb{R}^q$ for some integer $q\geq 1$. So there are q variables of interest which take on real values and which evolve in discrete time. A system then is a subset $\mathcal{B}\subset(\mathbb{R}^q)^{\mathbb{Z}}$, i.e., a set of q-dimensional time series. □

As model class we take a subclass of the systems in $(\mathbb{R}^q)^{\mathbb{Z}}$, namely those which are *linear*, *time invariant*, and *complete*.

A system $\mathcal{B}\subset(\mathbb{R}^q)^{\mathbb{Z}}$ is called *linear* if it is a linear subspace of $(\mathbb{R}^q)^{\mathbb{Z}}$.

Notation. In order to define time invariance, let σ denote the left shift, i.e., if $w:\mathbb{Z}\to\mathbb{R}^q$ then $\sigma w:\mathbb{Z}\to\mathbb{R}^q$ is defined by $(\sigma w)(t):=w(t+1)$, $t\in\mathbb{Z}$. □

A system $\mathcal{B}$ is called *time invariant* if $\sigma\mathcal{B}=\mathcal{B}$, i.e., shifted time series of the system also satisfy the laws of the sytem. This has the interpretation that the laws of the system are time invariant.

Remark. Time invariance of $\mathcal{B}$ by no means implies that trajectories in $\mathcal{B}$ do not vary over time. For example, trajectories which consist of a finite number of periodic and exponential components can be modelled perfectly by time invariant systems. Seasonal fluctuations and trends can be modelled, e.g., by including time variables in the signal set W. Particular time varying trajectories can be modelled by the time varying characteristics of so-called inputs, cf. the interpretation of definition 3-24 in section 3.3 in the sequel. □

A system $\mathcal{B}$ is called *complete* if $\{w\in\mathcal{B}\}\Leftrightarrow\{w|_{[t_0,t_1]}\in\mathcal{B}|_{[t_0,t_1]}$ for all $-\infty<t_0\le t_1<+\infty\}$, where $[t_0,t_1]:=\{t\in\mathbb{Z};\ t_0\le t\le t_1\}$. This means that, in order to check whether a time series $w\in(\mathbb{R}^q)^{\mathbb{Z}}$ belongs to $\mathcal{B}$ or not, it suffices to consider the model restrictions for all finite time intervals. Moreover, it can be shown that if $\mathcal{B}$ is linear and time invariant, then $\mathcal{B}$ is complete if and only if there exists a $\Delta\ge 0$ such that $\{w\in\mathcal{B}\}\Leftrightarrow\{w|_{[t,t+\Delta]}\in\mathcal{B}|_{[0,\Delta]}$ for all $t\in\mathbb{Z}\}$. So in this case the laws which are imposed by $\mathcal{B}$ are local in time, as it suffices to consider only time intervals of length $\Delta+1$.

Definition 3-2 The model class $\mathbb{B}$ is defined by $\mathbb{B}:=\{\mathcal{B}\subset(\mathbb{R}^q)^{\mathbb{Z}};\ \mathcal{B}$ linear, time invariant, complete$\}$.

Interpretation. If for modelling a dynamical phenomenon we specify $\mathbb{B}$ as model class it is supposed that it is reasonable to model the phenomenon by means of a system which is linear, time invariant, and complete. The interpretation is that the model gives a description of the phenomenon which is *local*, both in space (linearity) and in time (time invariance and completeness). □

Remark. The choice of $\mathbb{B}$ as model class here is made a priori, i.e., independent of the data. We will not discuss validation of this choice by means of the data. The choice is supported by the parametrization result of section 3.2. □

Remark. We mention that the model class $\mathbb{B}$ consists exactly of those subsets $\mathcal{B}\subset(\mathbb{R}^q)^{\mathbb{Z}}$ which are linear, shift invariant, and *closed* in the topology of pointwise convergence in $(\mathbb{R}^q)^{\mathbb{Z}}$. We use this characterization in some parts of chapters III and IV. □

We summarize the properties of $\mathbb{B}$ which were stated before. We refer to Willems [73, proposition 4 and theorem 5], and Willems [74, section 1.4.2].

Proposition 3-3 If $\mathcal{B}\subset(\mathbb{R}^q)^{\mathbb{Z}}$ is linear and time invariant, then $\{\mathcal{B}\in\mathbb{B}\}\Leftrightarrow$ $\{\exists\Delta\geq0$ such that $\{w\in\mathcal{B}\}\Leftrightarrow\{w|_{[t,t+\Delta]}\in\mathcal{B}|_{[0,\Delta]}$ for all $t\in\mathbb{Z}\}\}\Leftrightarrow\{\mathcal{B}$ closed in topology of pointwise convergence$\}$.

3.2. Autoregressive parametrizations

Remark. This section is related *only* to chapter V. □

In this section we discuss a convenient parametrization for the model class $\mathbb{B}$ of linear, time invariant, complete systems. It is shown that $\mathbb{B}$ coincides with the class of systems describable by a finite number of autoregressive equations.

The content of the section is as follows. We define a class of autoregressive systems which provides a parametrization of $\mathbb{B}$ and we analyse some properties of this parametrization. We comment on identification problems in case of non-unique parametrization and define two special forms of parametrization which we call the canonical descriptive form and the canonical predictive form. These parametrizations are used in the algorithms for deterministic time series analysis in section V.4. We conclude the section by giving an example illustrating the canonical forms.

3.2.1. Autoregressive systems

Notation. As before we consider dynamical systems with time set $T=\mathbb{Z}$ and with signal set $W=\mathbb{R}^q$.

Let $\mathbb{N}:=\{1,2,3,...\}$. For $g\in\mathbb{N}$ let $\mathbb{R}^{g\times q}[s,s^{-1}]$ denote the set of finite Laurent series in s with coefficients in $\mathbb{R}^{g\times q}$, i.e., $R\in\mathbb{R}^{g\times q}[s,s^{-1}]$ if there exist $d_1,d_2\in\mathbb{Z}$ with $d_1\leq d_2$ and matrices $R_k\in\mathbb{R}^{g\times q}$, $k=d_1,d_1+1,...,d_2$, such that $R=\Sigma_{k=d_1}^{d_2}R_ks^k$. By a slight abuse of language we call R a polynomial matrix in s and s^{-1}. As before let σ denote left shift. By σ^{-1} we denote the inverse of σ. The autoregressive system $\mathcal{B}(R)$ now is defined as $\ker(R(\sigma,\sigma^{-1}))$, i.e., $\mathcal{B}(R)$ is the set of those time series $w:\mathbb{Z}\to\mathbb{R}^q$ for which $R(\sigma,\sigma^{-1})w=0$, i.e.,

$\Sigma_{k=d_1}^{d_2} R_k w(t+k)=0$ for all $t\in\mathbb{Z}$. □

Definition 3-4 Let $R\in\mathbb{R}^{g\times q}[s,s^{-1}]$. Then the *autoregressive system* (*AR-system*) $\mathcal{B}(R)$ is defined by $\mathcal{B}(R):=\{w\in(\mathbb{R}^q)^{\mathbb{Z}};\ R(\sigma,\sigma^{-1})w=0\}$. The rows of R are called *laws* for $\mathcal{B}(R)$.

Interpretation. If R has rows $r_i\in\mathbb{R}^{1\times q}[s,s^{-1}]$, $i=1,\ldots,g$, then $\mathcal{B}(R)$ consists of those time series for which all the autoregressive laws r_i are satisfied, $i=1,\ldots,g$. □

Notation. We denote the class of all AR-systems by $\mathbb{B}(AR)$, i.e., $\mathbb{B}(AR):=\{\mathcal{B}\subset(\mathbb{R}^q)^{\mathbb{Z}};\ \exists g\ \exists R\in\mathbb{R}^{g\times q}[s,s^{-1}]$ such that $\mathcal{B}=\mathcal{B}(R)\}$. □

Remark. This class of systems is interesting for a number of reasons. First, it forms a class of models often used in practical modelling situations where one wants to describe linear relationships between the variables and their lagged values as, e.g., in econometrics, signal processing, and linear control. Second, this class of systems includes some widely used systems as, for example, linear input/output systems with finite dimensional state space, see section 3.3. Third, there exists a nice interpretation of AR-systems on the behavioural level of sets of time series. The following result states that AR-systems are exactly those systems which are linear, time invariant, and complete. □

Theorem 3-5 $\mathbb{B}=\mathbb{B}(AR)$.

Proof. See the appendix.

Interpretation. This result provides another motivation for taking $\mathbb{B}$ as model class. Every linear, time invariant, complete system $\mathcal{B}$ has an *AR-representation*, i.e., there exist $g\in\mathbb{N}$ and $R\in\mathbb{R}^{g\times q}[s,s^{-1}]$ such that $\mathcal{B}=\mathcal{B}(R)$. We call R a parametrization of $\mathcal{B}$. □

Remark. For identification it is of crucial importance to have such a numerical description of systems in terms of a finite number of parameters.

However, given a system $\mathcal{B}$ the polynomial matrix R such that $\mathcal{B}=\mathcal{B}(R)$ is not unique. In section 3.2.2 we characterize the set of all AR–parametrizations for a given system $\mathcal{B}\in\mathbb{B}$. In section 3.2.4 we define a class of parsimonious parametrizations. These are used in sections 3.2.5 and 3.2.6 to define canonical forms which play an important role in the numerical algorithms for deterministic time series analysis given in section V.4. □

3.2.2. Equivalent AR-parametrizations

As before let $\mathbb{B}$ denote the class of systems $\mathcal{B}\subset(\mathbb{R}^q)^{\mathbb{Z}}$ which are linear, time invariant, and complete. According to theorem 3–5 every system $\mathcal{B}\in\mathbb{B}$ admits an autoregressive parametrization. We will now define and investigate equivalence of parametrizations.

Notation. We use the following terminology and notation. R_1 is called *equivalent* to R_2, notation $R_1\sim R_2$, if $\mathcal{B}(R_1)=\mathcal{B}(R_2)$. For $\mathcal{B}\in\mathbb{B}$ let $\mathcal{B}^{\perp}$ denote the family of laws which are satisfied by the behaviour $\mathcal{B}$, i.e., $\mathcal{B}^{\perp}:=\{r\in\mathbb{R}^{1\times q}[s,s^{-1}];\ r(\sigma,\sigma^{-1})w=0$ for all $w\in\mathcal{B}\}$. Let $R\in\mathbb{R}^{g\times q}[s,s^{-1}]$ have rows $r_i\in\mathbb{R}^{1\times q}[s,s^{-1}]$, $i=1,\ldots,g$, notation $R=\mathrm{col}(r_1,\ldots,r_g)$, then the polynomial module generated by $r_1,\ldots,r_g$ is denoted by $M(R):=\{r\in\mathbb{R}^{1\times q}[s,s^{-1}];\ \exists p_i\in\mathbb{R}[s,s^{-1}]$, $i=1,\ldots,g$, such that $r=\Sigma_{i=1}^{g}p_ir_i\}$. Let $\mathbb{B}^{\perp}$ denote the class of these (finitely generated) submodules of $\mathbb{R}^{1\times q}[s,s^{-1}]$. By $\dim(M)$ we denote the dimension of $M\in\mathbb{B}^{\perp}$ as a module, i.e., $\dim(M)$ is the minimal number of elements of M which generate M. Finally, $U\in\mathbb{R}^{g\times g}[s,s^{-1}]$ is called unimodular if it is invertible in $\mathbb{R}^{g\times g}[s,s^{-1}]$. □

The next proposition summarizes some results on AR–representations of models in $\mathbb{B}$.

Proposition 3-6 (*i*) For every $\mathcal{B}\in\mathbb{B}$, $\mathcal{B}^{\perp}\in\mathbb{B}^{\perp}$; the map $f:\mathbb{B}\to\mathbb{B}^{\perp}:\mathcal{B}\to\mathcal{B}^{\perp}$ is a *bijection* of $\mathbb{B}$ onto $\mathbb{B}^{\perp}$, and $\{\mathcal{B}=\mathcal{B}(R)\}\Leftrightarrow\{\mathcal{B}^{\perp}=M(R)\}$;

(*ii*) if $\dim(\mathcal{B}^{\perp})=p$, then there exists an $R\in\mathbb{R}^{p\times q}[s,s^{-1}]$ with $\mathcal{B}=\mathcal{B}(R)$; moreover, this R is unique up to left multiplication by a unimodular matrix.

Proof. See the appendix.

Interpretation. The equivalence class of AR–parametrizations of a given model $\mathcal{B}\in\mathbb{B}$ consists of those polynomial matrices $R\in\mathbb{R}^{g\times q}[s,s^{-1}]$, for some $g\in\mathbb{N}$, for which the rows generate $\mathcal{B}^{\perp}$. Then the (autoregressive) laws which are satisfied for any time series in $\mathcal{B}$ consist of the rows of R and (polynomial) combinations of them. □

3.2.3. Identification and parametrization

In chapter V we consider approximate modelling of an observed time series by means of a model in the model class $\mathbb{B}$. Hence it is supposed that it is reasonable to identify a model which is linear, time invariant, and complete. The interpretation is that the model gives a local description.

The aim is to formulate reasonable identification procedures which can be implemented algorithmically and for which efficient numerical solution procedures can be constructed. An identification procedure corresponds to a utility on $\mathbb{B}$. For *algorithmic* and *numerical* implementation it is crucial that this utility can be expressed in terms of a finite number of parameters. If we consider AR–representations of systems in $\mathbb{B}$ then the main problem is that defining a utility in terms of these representations need not automatically be compatible with a utility on $\mathbb{B}$. This problem arises because the representation of a system by means of autoregressive laws is highly non–unique, as described in proposition 3–6.

We next define the concept of canonical form and comment on its use in identification.

Notation. Let $\mathbb{A}:=\bigcup\{\mathbb{R}^{g\times q}[s,s^{-1}];\ g\in\mathbb{N}\}$ denote the class of AR–parametrizations, i.e., the set of all polynomial matrices with q columns and an arbitrary finite number of rows. In section 3.2.2 we introduced the natural equivalence relation $\sim$ on $\mathbb{A}$, declaring R_1 and R_2 to be equivalent if they parametrize the same model, i.e., $\{R_1\sim R_2\}:\Leftrightarrow\{\mathcal{B}(R_1)=\mathcal{B}(R_2)\}$. We define C to be a *canonical form* for $\mathbb{A}$ if $C\subset\mathbb{A}$ and C contains at least one element of every equivalence class in $\mathbb{A}$, i.e., for every $R\in\mathbb{A}$ there exists an $R_c\in C$ such that $R\sim R_c$. We call C *minimal* if it contains exactly one element of every class, i.e., if in addition

$\{R_1, R_2 \in C, R_1 \sim R_2\} \Rightarrow \{R_1 = R_2\}$. □

A general approach to identification in case of non–unique parametrization is to express a utility of models directly in terms of a canonical form. A model then is identified by optimization of a criterion which is expressed in terms of canonical parameters.

Remark. We refer to Corrêa and Glover [9], Gevers and Wertz [16], Glover and Willems [18], Guidorzi [21] and Hannan and Deistler [23] for some contributions on canonical forms, minimality, i.e., parameter identifiability, and identification. □

Remark. It is crucial that the criterion in terms of parameters corresponds to a sensible utility of models. Moreover, it is reasonable to require that the criterion is constant on equivalence classes in case the canonical form is not minimal. □

In chapter V we define utilities on $\mathbb{B}$ corresponding to the objectives of description or prediction. It is shown that these utilities can be numerically expressed in terms of two canonical forms. These canonical forms are not minimal.

Remark. This minimality is rather intrinsic, i.e., forcing a reduction of the canonical forms such that the resulting forms are minimal would require arguments which are not related to the objectives of modelling. We refer to propositions 3–14 and 3–18. □

Remark. The canonical forms lead to simple algorithms for determining models which are optimal with respect to the corresponding utilities on $\mathbb{B}$. This is described in sections V.3 and V.4. In the next sections we introduce a kind of minimal parametrization and define the canonical forms. □

3.2.4. Shortest lag and tightest equation representations

In sections 3.2.5 and 3.2.6 we introduce canonical forms for the

AR–parametrization of linear, time invariant, complete systems. These canonical forms consist of a selection from a class of parsimonious parametrizations which we define in this section and which we denote by shortest lag (or tightest equation) representations.

Notation. For $r\in\mathbb{R}^{1\times q}[s,s^{-1}]$, $r=\Sigma_{k=-\infty}^{\infty}r_ks^k$, $r_k\in\mathbb{R}^{1\times q}$, define the *order* of r by $d(r):=\max\{k;r_k\neq0\}-\min\{k;r_k\neq0\}$. As before let $R=\text{col}(r_1,\ldots,r_g)\in\mathbb{R}^{g\times q}[s,s^{-1}]$ denote the polynomial matrix with rows $r_1,\ldots,r_g$, then the order of R is defined as $d(R):=\max\{d(r_i);\ i=1,\ldots,g\}$. Suppose $r_i=\Sigma_{k=d_i'}^{d_i''}r_k^{(i)}s^k$, with $d_i''\geq d_i'$, $r_{d_i'}^{(i)}\neq0\neq r_{d_i''}^{(i)}$, so $d(r_i)=d_i''-d_i'$. Let $L_+:=\text{col}(r_{d_i''}^{(i)};\ i=1,\ldots,g)$ and $L_-:=\text{col}(r_{d_i'}^{(i)};\ i=1,\ldots,g)$ be the leading and trailing coefficient matrices of R respectively. Then R is called *bilaterally row proper* if L_+ and L_- both have full row rank g. Let $R=\text{col}(r_1,\ldots,r_g)\in\mathbb{R}^{g\times q}[s,s^{-1}]$, where it is assumed that the rows are ordered such that $d(r_1)\leq\ldots\leq d(r_g)$, then $(d(r_1),\ldots,d(r_g))$ is called the lag structure of R. □

In the sequel we will make use of the equation structure of R, which is defined in terms of the lag structure, as follows.

Definition 3-7 If $R\in\mathbb{R}^{g\times q}[s,s^{-1}]$ has lag structure $(d_1,\ldots,d_g)$, then the *equation structure* of R is defined as $e(R):=(e_t;\ t\geq0)$, where $e_t:=\#\{i;\ d_i=t\}$ is the number of rows in R of order t.

Notation. For lag structures we define a total *ordering* by $\{(d_1',\ldots,d_{g'}')\leq(d_1'',\ldots,d_{g''}'')\}:\Leftrightarrow\{(d_1',\ldots,d_{g'}')=(d_1'',\ldots,d_{g''}'')$, or $g'<g''$, or there is a $g\leq g'=g''$ such that $d_g'<d_g''$ and $d_i'=d_i''$ for all $i<g\}$. So few equations and short lags are preferred. We order equation structures by $\{e'\leq e''\}:\Leftrightarrow\{e'=e''$, or $\Sigma_{t=0}^{\infty}e_t'<\Sigma_{t=0}^{\infty}e_t''$, or $\Sigma_{t=0}^{\infty}e_t'=\Sigma_{t=0}^{\infty}e_t''$ and there is a t_0 such that $e_{t_0}'>e_{t_0}''$ and $e_t'=e_t''$ for all $t<t_0\}$. For $\mathcal{B}\in\mathbb{B}$ we call R a *shortest lag* or *tightest equation* representation of $\mathcal{B}$ if $\mathcal{B}=\mathcal{B}(R)$ and the lag or equation structure respectively is minimal in the class of AR–representations of $\mathcal{B}$. □

Remark. Clearly, every system $\mathcal{B}\in\mathbb{B}$ has shortest lag and tightest equation representations. The following proposition characterizes these minimal representations. □

Proposition 3-8 If $\mathcal{B}=\mathcal{B}(R)$ then the following statements are equivalent:

(*i*) R is bilaterally row proper;

(*ii*) R is a tightest equation representation of $\mathcal{B}$;

(*iii*) R is a shortest lag representation of $\mathcal{B}$.

Remark. For the equivalence of (*i*) and (*iii*) we refer to Willems [73, theorem 6]. The equivalence of (*ii*) and (*iii*) is immediate as the orderings of lag structures and equation structures are equivalent. □

Next we characterize the class of all tightest equation representations for a given system $\mathcal{B}\in\mathbb{B}$. This result is used in sections 3.2.5 and 3.2.6 to express two canonical forms.

Notation. Let $\mathcal{B}\in\mathbb{B}$ and $\mathcal{B}^{\perp}:=\{r\in\mathbb{R}^{1\times q}[s,s^{-1}];\ r(\sigma,\sigma^{-1})w=0$ for all $w\in\mathcal{B}\}$. Let $\mathbb{R}_t^{1\times q}[s]:=\{r\in\mathbb{R}^{1\times q}[s];\ r=\Sigma_{k=-\infty}^{\infty}r_ks^k,\ r_k\in\mathbb{R}^{1\times q},\ r_k=0$ for $k<0$ and for $k>t\}$. Let $\mathcal{B}_t^{\perp}$ denote the class of polynomials in s of order at most t which are contained in $\mathcal{B}^{\perp}$, i.e., $\mathcal{B}_t^{\perp}:=\mathcal{B}^{\perp}\cap\mathbb{R}_t^{1\times q}[s]$. Then $\mathcal{B}_t^{\perp}$ describes the family of laws of order at most t which are satisfied by the behaviour $\mathcal{B}$. □

Definition 3-9 The bijection $v_t:\mathbb{R}_t^{1\times q}[s]\rightarrow(\mathbb{R}^{1\times q})^{t+1}$ is defined as follows. Let $r=\Sigma_{k=0}^{t}r_ks^k\in\mathbb{R}_t^{1\times q}[s]$, then $v_t(r)\in(\mathbb{R}^{1\times q})^{t+1}$ is defined by $v_t(r):=(r_0,r_1,\ldots,r_t)$.

Remark. Now we can identify $\mathcal{B}_t^{\perp}$ with the subspace $v_t(\mathcal{B}_t^{\perp})\subset(\mathbb{R}^{1\times q})^{t+1}$. It is easily seen that $v_t(\mathcal{B}_t^{\perp})$ is isomorphic with the (Euclidean) orthogonal complement in $(\mathbb{R}^q)^{t+1}$ of $\mathcal{B}_t:=\mathcal{B}|_{[-t,0]}=\mathcal{B}|_{[s,s+t]}$ for any $s\in\mathbb{Z}$, i.e., the behaviour on an interval of length $t+1$. □

Notation. We define spaces $L_t\subset\mathcal{B}^{\perp}$ as follows. Let $L_0:=\mathcal{B}_0^{\perp}$ consist of the zero order laws for $\mathcal{B}$. Define $V_0:=v_0(L_0)$. Observe that $\mathcal{B}_0^{\perp}+s\mathcal{B}_0^{\perp}\subset\mathcal{B}_1^{\perp}$. We will say that the first order laws in $\mathcal{B}_0^{\perp}+s\mathcal{B}_0^{\perp}$ are *implied* by zero order laws. *Truly* first order laws for $\mathcal{B}$, collected in $L_1\subset\mathcal{B}_1^{\perp}$, are required to be independent of those implied laws. Formally, let V_1 be a *complementary space* of $v_1(\mathcal{B}_0^{\perp}+s\mathcal{B}_0^{\perp})$ in $v_1(\mathcal{B}_1^{\perp})$, i.e., $V_1\cap v_1(\mathcal{B}_0^{\perp}+s\mathcal{B}_0^{\perp})=\{0\}$ and $V_1+v_1(\mathcal{B}_0^{\perp}+s\mathcal{B}_0^{\perp})=v_1(\mathcal{B}_1^{\perp})$. Then $L_1:=v_1^{-1}(V_1)$. Analogously, the t–th order laws in $\mathcal{B}_{t-1}^{\perp}+s\mathcal{B}_{t-1}^{\perp}\subset\mathcal{B}_t^{\perp}$ are implied by lower order laws. Truly t-th order laws are collected in $L_t \subset \mathcal{B}_t^\perp$, defined as $L_t := v_t^{-1}(V_t)$ for a complementary space V_t of $v_t(\mathcal{B}_{t-1}^\perp + s\mathcal{B}_{t-1}^\perp)$ in $v_t(\mathcal{B}_t^\perp)$, i.e., $V_t \cap v_t(\mathcal{B}_{t-1}^\perp + s\mathcal{B}_{t-1}^\perp) = \{0\}$ and $V_t + v_t(\mathcal{B}_{t-1}^\perp + s\mathcal{B}_{t-1}^\perp) = v_t(\mathcal{B}_t^\perp)$. Let $n_t := \dim(V_t)$. □

Remark. Note that the spaces L_t have the property that $\mathcal{B}_t^\perp = \mathcal{B}_{t-1}^\perp + s\mathcal{B}_{t-1}^\perp + L_t$. Moreover note that the spaces V_t and L_t in general are not uniquely defined. However, the numbers n_t are uniquely defined by $\mathcal{B}$. □

Remark. The following proposition establishes the relationship between the sets L_t and tightest equation representations of a model $\mathcal{B} \in \mathbb{B}$. It is a restatement of parts of Willems [73, theorems 6 and 8]. □

> **Proposition 3-10** Let $\mathcal{B} \in \mathbb{B}$. Then there exists a d such that $n_d \neq 0$ and $n_t = 0$ for all $t > d$. Any tightest equation representation R of $\mathcal{B}$ has equation structure $e(R) = (n_0, \ldots, n_d, 0, 0, \ldots)$. Moreover, R is a tightest equation representation of $\mathcal{B}$ if and only if there exists a choice of the complementary spaces V_t, of bases $\{v_i^{(t)};\ i=1,\ldots,n_t\}$, and of numbers $k_i(t) \in \mathbb{Z}$ for $i=1,\ldots,n_t$, $t=0,\ldots,d$, such that the rows of R consist of $\{\sigma^{k_i(t)} \cdot v_t^{-1}(v_i^{(t)});\ i=1,\ldots,n_t,\ t=0,\ldots,d\}$.

Remark. In the next two sections we introduce two canonical forms which correspond to a special choice of the complementary spaces V_t. This choice is inspired by the purpose of description or prediction. □

3.2.5. Canonical descriptive form

The descriptive procedures for deterministic time series analysis in chapter V correspond to utilities which can be expressed in terms of a canonical AR-parametrization which we call the canonical descriptive form. This form corresponds to a tightest equation representation of a special type. Note that proposition 3-10 characterizes the non-unicity of tightest equation representations in terms of the choice of the complementary spaces V_t and of bases of these spaces. The canonical descriptive form selects particular complementary spaces, but the choice of bases is left arbitrary. Hence the canonical descriptive form is not minimal.

In the canonical descriptive form we choose truly t–th order laws of $\mathcal{B}$ such that they are (Euclidean) orthogonal to the t–th order laws which are implied by lower order ones.

Notation. Formally, we define $L_t^D \subset \mathcal{B}_t^\perp$ as follows. $L_0^D := \mathcal{B}_0^\perp$, and $L_t^D := v_t^{-1}\{[v_t(\mathcal{B}_{t-1}^\perp + s\mathcal{B}_{t-1}^\perp)]^\perp \cap [v_t(\mathcal{B}_t^\perp)]\}$. So the laws $r \in L_t^D$ are orthogonal to those in $\mathcal{B}_{t-1}^\perp + s\mathcal{B}_{t-1}^\perp$. □

Interpretation. Orthogonality is imposed to ensure that the laws in L_t^D are "far" from being implied by laws of lower order. □

Remark. In some cases it could be reasonable to choose other inner products than the Euclidean one. □

Now R is defined to be in canonical descriptive form if it is itself a tightest equation representation of the corresponding behaviour $\mathcal{B}(R)$ and if the laws of truly order t are contained in L_t^D. We then say that laws of different order are orthogonal.

Definition 3–11 $R \in \mathbb{R}^{g \times q}[s]$ is in *canonical descriptive form* (CDF) if
- (*i*) R is a tightest equation representation of $\mathcal{B}(R)$;
- (*ii*) laws of different order are orthogonal.

Proposition 3–12 (CDF) is a canonical form.

Proof. See the appendix.

Remark. Note that for R in (CDF) $R \in \mathbb{R}^{g \times q}[s]$, i.e., R is a polynomial matrix in s. □

For numerical implementation of the descriptive procedures of chapter V we use a description of (CDF) in terms of matrices, as follows.

Notation. For $r = \Sigma_{k=0}^{\infty} r_k s^k \in \mathbb{R}^{1 \times q}[s]$ define the degree of r as $\max\{k;\ r_k \neq 0\}$. Let $R \in \mathbb{R}^{g \times q}[s]$ and let $R^{(t)} := \mathrm{col}(r_i^{(t)};\ i=1,\ldots,n_t)$ consist of the rows of R of degree t, $t\geq 0$, $n_t\geq 0$, $\Sigma_{t=0}^{\infty}n_t=g$. Let d be the maximal degree of rows of R. Define $N_t:=\text{col}(v_d(r_i^{(t)});\ i=1,\dots,n_t)\in\mathbb{R}^{n_t\times(d+1)q}$, say $N_t=[R_0^{(t)}\ \dots\ R_d^{(t)}]$, $R_i^{(t)}\in\mathbb{R}^{n_t\times q}$, $i=0,\dots,d$. Define $L_-:=\text{col}(R_0^{(0)},\dots,R_0^{(d)})\in\mathbb{R}^{g\times q}$ and $L_+:=\text{col}(R_0^{(0)},\dots,R_d^{(d)})\in\mathbb{R}^{g\times q}$. Define $s:\mathbb{R}^{1\times(d+1)q}\to\mathbb{R}^{1\times(d+1)q}$ as follows. If $v=(v_0,\dots,v_{d-1},v_d)$ with $v_i\in\mathbb{R}^{1\times q}$, $i=0,\dots,d$, then $s(v):=(0,v_0,\dots,v_{d-1})$. Let $\bar{V}_0:=N_0$ and define $\bar{V}_t$ for $t=1,\dots,d$ inductively by $\bar{V}_t:=\text{col}(\bar{V}_{t-1},s\bar{V}_{t-1},N_t)$. Finally, for matrices A_1 and A_2 let $A_1\perp A_2$ denote that every row of A_1 is orthogonal to any row of A_2, i.e., $A_1A_2^T=0$. □

Proposition 3-13 $R\in\mathbb{R}^{g\times q}[s]$ is in (CDF) if and only if

(i) L_+ and L_- have full row rank, and

(ii) $N_t\perp\text{col}(\bar{V}_{t-1},s\bar{V}_{t-1})$ for all $t=1,\dots,d$.

Proof. See the appendix.

Remark. An example is given in section 3.2.7. □

So, whether R is in (CDF) or not can be checked by means of proposition 3-13 in terms of matrices which can be easily calculated from R. These algebraic conditions play a role in the algorithms of section V.4.

The next proposition describes the non-unicity of (CDF) representations of systems $\mathcal{B}\in\mathbb{B}$.

Proposition 3-14 Let $\mathcal{B}\in\mathbb{B}$, $\mathcal{B}=\mathcal{B}(R)$ with R in (CDF), $d(R)=d$ and $e(R)=(n_0,\dots,n_d,0,0,\dots)$. Let the rows of R be ordered with increasing degree. Then $\mathcal{B}=\mathcal{B}(R')$ with R' in (CDF) if and only if there exists a permutation matrix Π and a blockdiagonal matrix $A=\text{diag}(A_{00},\dots,A_{dd})$ with $A_{tt}\in\mathbb{R}^{n_t\times n_t}$ nonsingular, $t=0,\dots,d$, such that $R'=\Pi AR$.

Proof. See the appendix.

Interpretation. So indeed (CDF) is not minimal. The non-uniqueness corresponds to a choice of bases for the spaces L_t^D of truly t-th order laws of $\mathcal{B}$ which are orthogonal to laws of order unequal to t. □

3.2.6. Canonical predictive form

The predictive procedures for deterministic time series analysis in chapter V also correspond to utilities which can be expressed in terms of a canonical AR–parametrization, which we call the canonical predictive form. This form also corresponds to a particular tightest equation representation of the AR–equations describing a behaviour. Again, the complementary spaces V_t of section 3.2.4 are chosen in a particular way and the choice of bases is left arbitrary. The spaces are intimately connected with the purpose of prediction and corresponding utilities which will be defined in chapter V.

To define the canonical predictive form, we consider the (forward) predictive interpretation of a law $r \in \mathbb{R}^{1\times q}[s]$. Let $r = \Sigma_{k=0}^{d} r_k s^k$ with $r_0 \neq 0 \neq r_d$. The law r corresponding to $r(\sigma)w=0$ predicts that, given $w(\tau)$ for $\tau=t-d,\ldots,t-1$, $w(t)$ will be such that $r_d w(t) = -\Sigma_{k=0}^{d-1} r_k w(t-d+k)$, $t\in\mathbb{Z}$. We call r a predictive law of order d, r_d a predicted functional of order d, and $-\Sigma_{k=0}^{d-1} r_k s^k$ a prediction polynomial of order d. Intuitively stated, we will choose the complementary spaces V_t such that the predicted functionals of different order are orthogonal and such that prediction polynomials of a certain order are orthogonal to predictive laws of lower order. This ensures that predictive laws of different order are "far" from each other.

Notation. Formally, for $\mathcal{B}\in\mathbb{B}$ define $L_t^P \subset \mathcal{B}_t^\perp$ as follows. Let $F_t := \{\tilde{r} \in \mathbb{R}^{1\times q};\ \exists r \in \mathcal{B}_t^\perp,\ r = \Sigma_{k=0}^{t} r_k s^k,$ such that $r_t = \tilde{r}\}$ denote the set of predicted functionals of order at most t. Then $L_0^P := \mathcal{B}_0^\perp$ and $L_t^P := v_t^{-1}\{[v_t(F_{t-1}.s^t) + v_t(\mathcal{B}_{t-1}^\perp)]^\perp \cap [v_t(\mathcal{B}_t^\perp)]\}$. □

R is said to be in canonical predictive form if it is itself a tightest equation representation of the corresponding behaviour $\mathcal{B}(R)$ and if the predictive laws of order t are contained in L_t^P. We will then say that predicted functionals of different order are orthogonal, corresponding to $v_t(L_t^P) \perp v_t(F_{t-1}.s^t)$, and that the prediction polynomials are orthogonal to predictive laws of lower order, corresponding to $v_t(L_t^P) \perp v_t(\mathcal{B}_{t-1}^\perp)$.

Definition 3-15 $R \in \mathbb{R}^{g\times q}[s]$ is in *canonical predictive form* (CPF) if

(*i*) R is a tightest equation representation of $\mathcal{B}(R)$;

(*ii*) predicted functionals of different order are orthogonal;

(*iii*) prediction polynomials are orthogonal to predictive laws of lower order.

Proposition 3–16 (CPF) is a canonical form.

Proof. See the appendix.

Using the notation of section 3.2.5, proposition 3–17 gives simple algebraic conditions for R to be in (CPF). These conditions are used in the algorithms of section V.4.

Proposition 3–17 $R\in\mathbb{R}^{g\times q}[s]$ is in (CPF) if and only if

(*i*) L_+ and L_- have full row rank;

(*ii*) $R_t^{(t)}\perp R_\tau^{(\tau)}$ for all $t\neq\tau$, $t,\tau=0,\ldots,d$;

(*iii*) $N_t\perp\bar{V}_{t-1}$ for all $t=1,\ldots,d$.

Proof. See the appendix.

Remark. The non–unicity of (CPF) representations is exactly of the same kind as described for (CDF) in proposition 3–14. The following proposition is proved by replacing L_t^D by L_t^P in the proof of proposition 3–14. □

Proposition 3–18 For $\mathcal{B}\in\mathbb{B}$ the representation (CPF) is unique up to a permutation of rows and a choice of bases in the spaces L_t^P.

3.2.7. Example

We conclude this section by giving a simple example illustrating the canonical forms (CDF) and (CPF).

Consider $\mathcal{B}\in\mathbb{B}$ defined by $\mathcal{B}:=\{w\in(\mathbb{R}^3)^{\mathbb{Z}};\ w_1(t)+w_2(t-1)=0,\ w_1(t)+w_3(t)+w_2(t-2)=0,\ t\in\mathbb{Z}\}$. Then $\mathcal{B}=\mathcal{B}(R)$ with $R:=\begin{pmatrix}0&1&0\\0&1&0\end{pmatrix}+\begin{pmatrix}1&0&0\\0&0&0\end{pmatrix}.s+\begin{pmatrix}0&0&0\\1&0&1\end{pmatrix}.s^2$. R is neither in (CDF) nor in (CPF). Let $U_1:=\begin{pmatrix}1&0\\-1/2&1\end{pmatrix}+\begin{pmatrix}0&0\\-1/2&0\end{pmatrix}.s$, $U_2:=\begin{pmatrix}1&0\\-1/2&1\end{pmatrix}+\begin{pmatrix}0&0\\-1&0\end{pmatrix}.s$, $R_1:=U_1\cdot R$ and $R_2:=U_2\cdot R$. Then $\mathcal{B}=\mathcal{B}(R_1)=\mathcal{B}(R_2)$, $R_1=\begin{pmatrix}0&1&0\\0&1/2&0\end{pmatrix}+\begin{pmatrix}1&0&0\\-1/2&-1/2&0\end{pmatrix}.s+\begin{pmatrix}0&0&0\\1/2&0&1\end{pmatrix}.s^2$ is in

(CDF) and $R_2=\begin{bmatrix} 0 & 1 & 0 \\ 0 & ½ & 0 \end{bmatrix}+\begin{bmatrix} 1 & 0 & 0 \\ -½ & -1 & 0 \end{bmatrix}.s+\begin{bmatrix} 0 & 0 & 0 \\ 0 & 0 & 1 \end{bmatrix}.s^2$ is in (CPF). This can be easily checked by means of propositions 3–13 and 3–17.

3.3. State space realizations

Remark. This section is related *only* to chapter IV. □

In this section we briefly discuss another parametrization of the model class $\mathbb{B}$, using the notion of state. The main result is that $\mathbb{B}$ coincides with the class of linear, time invariant systems which have a finite dimensional state space realization. These realizations are used in chapter IV where we consider model approximation.

Remark. For an intrinsic definition of state on a set theoretic level we refer to Willems [74, sections 1.4.4 and 2]. Here we only present some results which are relevant for chapter IV. □

We once again restrict attention to discrete time systems with $T=\mathbb{Z}$. State space systems are dynamical systems with a particular first order autoregressive representation.

Definition 3–19 A *finite dimensional* (forward) *state space system* is a system $(\mathbb{Z}, \mathbb{R}^m\times\mathbb{R}^n\times\mathbb{R}^q, \mathcal{B}_s)$, where for some matrices $(A,B,C,D)\in\mathbb{R}^{n\times n}\times\mathbb{R}^{n\times m}\times\mathbb{R}^{q\times n}\times\mathbb{R}^{q\times m}$ $\mathcal{B}_s$ has a representation $\mathcal{B}_s=\mathcal{B}_s(A,B,C,D):=\{(v,x,w)\in(\mathbb{R}^m\times\mathbb{R}^n\times\mathbb{R}^q)^{\mathbb{Z}};\ \begin{bmatrix} \sigma x \\ w \end{bmatrix}=\begin{bmatrix} A & B \\ C & D \end{bmatrix}\begin{bmatrix} x \\ v \end{bmatrix}\}$.

Definition 3–20 $\mathcal{B}_s$ is called a (forward) *realization* of $\mathcal{B}\subset(\mathbb{R}^q)^{\mathbb{Z}}$ if $\mathcal{B}=\{w;\ \exists(v,x)$ such that $(v,x,w)\in\mathcal{B}_s\}$. It is called a *minimal* realization if *both* m and n are individually as small as possible.

Remark. Let $\mathcal{B}$ have realization $\mathcal{B}_s$. In $\mathcal{B}_s$ we call w an *external* variable, v and x internal variables, v a *driving* variable and x a *state* variable. Note that v is a free variable in $\mathcal{B}_s$ and that $x(t)$ has the property of making $\mathcal{B}|_{(-\infty,t-1]}$ and $\mathcal{B}|_{[t,\infty)}$ conditionally independent in a set theoretic sense. By this we mean the following. Suppose that $(v_i,x_i,w_i)\in\mathcal{B}_s$, $i=1,2$, and that for some $t\in\mathbb{Z}$

$x_1(t)=x_2(t)$. Then $w \in \mathcal{B}$, where w is defined by $w|_{(-\infty,t-1]}:=w_1|_{(-\infty,t-1]}$ and $w|_{[t,\infty)}:=w_2|_{[t,\infty)}$. Hence if at time t the state for two trajectories is the same, then the future of one trajectory is compatible with the past of the other one. This corresponds to an intuitive notion of state. □

Notation. By $\mathbb{B}_s$ we denote the class of dynamical systems which have a finite dimensional realization. □

Remark. The following result states that this class coincides with the class of linear, time invariant, complete systems. For a proof we refer to Willems [73, theorems 1 and 3]. □

Theorem 3-21 $\mathbb{B}=\mathbb{B}_s$.

Next we give a parametric characterization of minimality.

Remark. Note that it is not evident that minimal realizations exist, as we require both m and n to be individually as small as possible. □

Notation. We call $(A,B,C,D) \in \mathbb{R}^{n \times n} \times \mathbb{R}^{n \times m} \times \mathbb{R}^{q \times n} \times \mathbb{R}^{q \times m}$ minimal if $\mathcal{B}_s(A,B,C,D)$ is a minimal realization of its external behaviour $\mathcal{B}:=\{w;\ \exists(v,x)$ such that $(v,x,w) \in \mathcal{B}_s(A,B,C,D)\}$. We call (A,B,C,D) perfectly observable if $\{(v,x,w) \in \mathcal{B}_s(A,B,C,D),\ w|_{[0,n-1]}=0\} \Rightarrow \{x(0)=0\}$. □

Proposition 3-22 (*i*) Every $\mathcal{B} \in \mathbb{B}$ has a minimal realization;
(*ii*) $\{(A,B,C,D)$ is minimal$\} \Leftrightarrow \{(A,B,C,D)$ is perfectly observable, $(A\ \ B) \in \mathbb{R}^{n \times (n+m)}$ is surjective, and D is injective$\}$.

Proof. See the appendix.

From this result we immediately get the following corollary. See also Willems [74, section 4.8.1].

Corollary 3-23 For any $\mathcal{B} \in \mathbb{B}$ there exists a number $m \in \{0,1,2,\dots,q\}$ and a permutation matrix $\Pi \in \mathbb{R}^{q \times q}$ such that $\Pi\mathcal{B}=\{(u,y) \in (\mathbb{R}^m \times \mathbb{R}^{q-m})^{\mathbb{Z}};\ \exists x \in (\mathbb{R}^n)^{\mathbb{Z}}$ such that $\begin{pmatrix} \sigma x \\ y \end{pmatrix} = \begin{pmatrix} \tilde{A} & \tilde{B} \\ \tilde{C} & \tilde{D} \end{pmatrix} \begin{pmatrix} x \\ u \end{pmatrix}\}$ for matrices $(\tilde{A},\tilde{B},\tilde{C},\tilde{D})$ of appropriate dimensions.

Remark. For $\mathcal{B}\in\mathbb{B}$ the number m is uniquely defined, but Π need not be unique. □

Definition 3-24 $\mathcal{B}_{i/s/o}(\tilde{A},\tilde{B},\tilde{C},\tilde{D}) := \{(u,x,y)\in(\mathbb{R}^m\times\mathbb{R}^n\times\mathbb{R}^{q-m})^{\mathbb{Z}};\ \begin{pmatrix} \sigma x \\ y \end{pmatrix} = \begin{pmatrix} \tilde{A} & \tilde{B} \\ \tilde{C} & \tilde{D} \end{pmatrix} \begin{pmatrix} x \\ u \end{pmatrix}\}$ is called an *input/state/output realization* of $\mathcal{B}$ if there exists a permutation matrix Π such that $\Pi\mathcal{B}=\{(u,y);\ \exists x$ such that $(u,x,y)\in\mathcal{B}_{i/s/o}\}$. It is called minimal if n is as small as possible.

Interpretation. So any $\mathcal{B}\in\mathbb{B}$ has a minimal input/state/output realization for which the external variables can be split in two parts, $\Pi w = \begin{pmatrix} u \\ y \end{pmatrix}$, such that u plays the role of (external) driving forces, called *inputs*, and y that of the external variables restricted by $\mathcal{B}$, called *outputs*. This clearly illustrates the fact that a deterministic system does not describe a unique trajectory for the variables of interest. One of the possible interpretations of the model $\mathcal{B}$ is that it describes the evolution of the outputs conditionally, i.e., dependent on the inputs and a finite number of "initial" conditions. The inputs and initial conditions are free in the sense that they can be chosen arbitrarily. □

Remark. In case $m=0$ the sytem $\mathcal{B}$ is called autonomous, in case $m=q$ the system is $(\mathbb{R}^q)^{\mathbb{Z}}$, i.e., all variables are free. □

In the next proposition we describe the class of all minimal realizations of a given system.

Proposition 3-25 If $\mathcal{B}_s(A,B,C,D)$ is a minimal realization of $\mathcal{B}$, then all minimal realizations of $\mathcal{B}$ are obtained by $\mathcal{B}_s(S(A+BF)S^{-1}, SBR, (C+DF)S^{-1}, DR)$, where $S\in\mathbb{R}^{n\times n}$ and $R\in\mathbb{R}^{m\times m}$ are nonsingular and $F\in\mathbb{R}^{m\times n}$ is arbitrary.

Proof. See the appendix.

Remark. Note that minimal realizations are obtained from each other by means of state feedback and change of coordinates on the state space and the space of driving variables, i.e., by means of the transformation $\begin{bmatrix} S & 0 \\ -R^{-1}F & R^{-1} \end{bmatrix}$ on $\mathbb{R}^n\times\mathbb{R}^m$. Stated otherwise, let (A_i,B_i,C_i,D_i), i=1,2, be called equivalent if $\mathcal{B}_s(A_i,B_i,C_i,D_i)$ realize the same behaviour, i=1,2. The set of equivalence classes then consists of the equivalence classes for the so–called feedback group. □

In chapter IV we also need backward state space systems.

Definition 3–26 A finite dimensional *backward* state space system is defined by behaviour ${}^R\mathcal{B}_s(A,B,C,D):=\{(v,x,w)\in(\mathbb{R}^m\times\mathbb{R}^n\times\mathbb{R}^q)^{\mathbb{Z}};\ \begin{bmatrix} \sigma^{-1}x \\ w \end{bmatrix} = \begin{bmatrix} A & B \\ C & D \end{bmatrix}\begin{bmatrix} x \\ v \end{bmatrix}\}$.

Notation. The time reverse operator $\mathcal{R}$ on $(\mathbb{R}^d)^{\mathbb{Z}}$ is defined by $(\mathcal{R}z)(t):=z(-t)$, $t\in\mathbb{Z}$, $z\in(\mathbb{R}^d)^{\mathbb{Z}}$. □

Remark. It is easily seen that $\mathcal{R}\mathbb{B}=\mathbb{B}$ and that $\mathcal{B}$ has a realization $\mathcal{B}_s(A,B,C,D)$ if and only if ${}^R\mathcal{B}_s(A,B,C,D)$ is a realization of $\mathcal{R}\mathcal{B}$. Hence $\mathcal{B}\in\mathbb{B}$ if and only if it has a finite dimensional backward realization ${}^R\mathcal{B}_s$. □

Remark. The concepts of minimality for backward realizations and of backward input/state/output realizations are defined in an obvious way. The propositions on the class of all minimal backward realizations and the characterization of minimaltiy are exactly analogous to those for forward realizations. This is obvious by considering $\mathcal{R}\mathbb{B}$. □

Remark. There exists a close connection between state trajectories of minimal forward and backward realizations. The following result can be obtained from abstract realization theory by noting that $x(t)$ of a forward realization and $\tilde{x}(t-1)$ of a backward realization of $\mathcal{B}\in\mathbb{B}$ both make $\mathcal{B}|_{(-\infty,t-1]}$ and $\mathcal{B}|_{[t,\infty)}$ conditionally independent, cf. the remark following definition 3–20. We refer to Willems [74, section 4.7.5]. □

Proposition 3–27 Let $\mathcal{B}\in\mathbb{B}$ have minimal realizations $\mathcal{B}_s$ and ${}^R\mathcal{B}_s$. Let $w\in\mathcal{B}$ and $(v,x,w)\in\mathcal{B}_s$, $(\tilde{v},\tilde{x},w)\in{}^R\mathcal{B}_s$. Then $\tilde{x}=\sigma x$ (up to an isomorphism on $\mathbb{R}^n$).

Remark. A special case is obtained if in $\mathcal{B}_s(A,B,C,D)$ A is invertible. One then easily shows $\{(v,x,w)\in\mathcal{B}_s(A,B,C,D)\} \Leftrightarrow \{(v,\sigma x,w)\in^R\mathcal{B}_s(A^{-1},-A^{-1}B,CA^{-1},D-CA^{-1}B)\}$.□

3.4. Finite time systems

Remark. This section is related *only* to chapter III. □

3.4.1. Introduction

The main part of chapter III on exact modelling deals with exact modelling of a finite time series. To investigate this problem we use some results on finite time systems which we describe in this section.

Definition 3-28 A (discrete) *finite time* system is a dynamical system $(T,W,\mathcal{B})$ where T is a finite subset of $\mathbb{Z}$.

Notation. As before, let $\mathbb{N}:=\{1,2,3,...\}$ and for $t_1,t_2\in\mathbb{N}$, $t_1\leq t_2$, let $[t_1,t_2]:=\{t\in\mathbb{N};\ t_1\leq t\leq t_2\}$. We throughout assume that the finite time set can be represented by $[1,T]$ for some $T\in\mathbb{N}$ and that $W=\mathbb{R}^q$. Then a behaviour $\mathcal{B}\subset(\mathbb{R}^q)^{[1,T]}$ is a set of sequences in $\mathbb{R}^q$ of length T. We identify $(\mathbb{R}^q)^{[1,T]}$ with $(\mathbb{R}^q)^T$. □

We again restrict attention to behaviours which are linear and shift invariant. In this section we define these properties, in the next one we consider parametrization by means of autoregressive equations.

Let $\mathcal{B}\subset(\mathbb{R}^q)^T$. It is called *linear* if it is a linear subspace of $(\mathbb{R}^q)^T$. It is called *shift invariant* if $\mathcal{B}|_{[2,T]}\subset\mathcal{B}|_{[1,T-1]}$, *translation invariant* if $\mathcal{B}|_{[2,T]}=\mathcal{B}|_{[1,T-1]}$.

Notation. Let $\sigma\mathcal{B}:=\mathcal{B}|_{[2,T]}$. Here σ is considered as the left shift defined on $\cup\{(\mathbb{R}^q)^T;\ T=2,3,4,...\}$, where for $w\in(\mathbb{R}^q)^T$, $T\in\{2,3,4,...\}$, $\sigma w\in(\mathbb{R}^q)^{T-1}$ is defined by $\sigma w:=w|_{[2,T]}$. □

Interpretation. The interpretation of shift invariance is as follows. Let $\mathcal{B}$ be shift invariant and $w\in\mathcal{B}$. Then there exists $f\in W$ such that $\tilde{w}\in\mathcal{B}$, where $\tilde{w}|_{[1,T-1]}:=\sigma w$ and $\tilde{w}(T):=f$. Described intuitively this means that there exists a future realization f which follows on w and which is compatible with the behaviour $\mathcal{B}$. If $\mathcal{B}$ is shift invariant but not translation invariant this means that there exist $w \in \mathcal{B}$ for which there is no past realization compatible with $\mathcal{B}$. If $\mathcal{B}$ is translation invariant then for every $w \in \mathcal{B}$ there exist a past and a future of w compatible with $\mathcal{B}$. □

Definition 3-29 The class $\mathbb{B}_T$ ($\tilde{\mathbb{B}}_T$) of linear, shift (translation) invariant, finite (T) time systems is given by $\mathbb{B}_T := \{\mathcal{B} \subset (\mathbb{R}^q)^T;\ \mathcal{B} \text{ linear}, \sigma\mathcal{B} \subset \mathcal{B}|_{[1,T-1]}\}$, $\tilde{\mathbb{B}}_T := \{\mathcal{B} \subset (\mathbb{R}^q)^T;\ \mathcal{B} \text{ linear}, \sigma\mathcal{B} = \mathcal{B}|_{[1,T-1]}\}$.

Remark. It is clear that $\tilde{\mathbb{B}}_T \subset \mathbb{B}_T$ and that both classes are closed under addition. This resembles the situation for the class $\mathbb{B}$ of linear, time invariant, complete systems in $(\mathbb{R}^q)^{\mathbb{Z}}$. Note that $\mathbb{B}$ is closed under intersection, as $\mathcal{B}(R_1) \cap \mathcal{B}(R_2) = \mathcal{B}\begin{bmatrix} R_1 \\ R_2 \end{bmatrix}$. However, $\tilde{\mathbb{B}}_T$ and $\mathbb{B}_T$ are *not* closed under intersection. This can be illustrated by taking $q=1$, $T=4$, $\mathcal{B}_1 := \{w \in \mathbb{R}^4;\ w(4) = w(1)\}$ and $\mathcal{B}_2 := \{w \in \mathbb{R}^4;\ w(3) = -w(1);\ w(4) = -w(2)\}$. Then $\mathcal{B}_i \in \tilde{\mathbb{B}}_4$, $i=1,2$, but $\mathcal{B} := \mathcal{B}_1 \cap \mathcal{B}_2 = \{w \in \mathbb{R}^4;\ \exists \alpha \in \mathbb{R} \text{ such that } w = \alpha(1,-1,-1,1)\}$ and $\mathcal{B} \notin \mathbb{B}_4$. □

In chapter III we use the following proposition, which proof is immediate.

Proposition 3-30 (*i*) If $\mathcal{B}$ is shift invariant, $\Delta \in \mathbb{N}$, $1 \le t_1 \le t_1 + \Delta \le t_2 + \Delta \le T$, then $\mathcal{B}|_{[t_1+\Delta, t_2+\Delta]} \subset \mathcal{B}|_{[t_1,t_2]}$;
(*ii*) if $\mathcal{B}$ is translation invariant, $\Delta \in \mathbb{Z}$, $1 \le t_1 \le t_2 \le T$, $1 \le t_1 + \Delta \le t_2 + \Delta \le T$, then $\mathcal{B}|_{[t_1+\Delta, t_2+\Delta]} = \mathcal{B}|_{[t_1,t_2]}$.

3.4.2. Autoregressive parametrizations

In this section we show that systems in $\mathbb{B}_T$ can be described by a finite number of autoregressive equations. However, not every finite number of autoregressive equations defines a system in $\mathbb{B}_T$. This in contrast to the case $T=\mathbb{Z}$, cf. theorem 3–5.

Notation. For $r \in \mathbb{R}^{1\times q}[s]$ we denote the *degree* of r by $\bar{d}(r)$, i.e., if $r = \Sigma_{k=0}^{\infty} r_k s^k \in \mathbb{R}^{1\times q}[s]$ then $\bar{d}(r) := \max\{k;\ r_k \neq 0\}$. We define $\bar{d}(0) := -\infty$. For $R \in \mathbb{R}^{g\times q}[s]$ with rows $r_i \in \mathbb{R}^{1\times q}[s]$, $i \in [1,g]$, we define $\bar{d}(R) := \max\{\bar{d}(r_i);\ i \in [1,g]\}$. By $\mathcal{B}_T(R)$ we denote the behaviour in $(\mathbb{R}^q)^T$ where the autoregressive equations $r_i(\sigma)w=0$, $i\in[1,g]$, are all satisfied. □

Remark. By definition we consider r with $\bar{d}(r)\geq T$ as imposing *no* restriction on the behaviour on $[1,T]$. Note that for r with $\bar{d}(r)\geq T$ $r(\sigma)w$ for $w\in(\mathbb{R}^q)^T$ is undefined. As an example, take $q=1$, $T=3$, $r=s^3$, $w\in\mathbb{R}^3$, then $\sigma^2 w=w(3)$ has length 1, hence $\sigma^3 w$ is undefined. □

Definition 3-31 Let $R\in\mathbb{R}^{g\times q}[s]$ have rows r_i and let $I_T(R):=\{i\in[1,g];\ \bar{d}(r)\leq T-1\}$. Then the *finite time autoregressive system* $\mathcal{B}_T(R)$ is defined by $\mathcal{B}_T(R):=\{w\in(\mathbb{R}^q)^T;\ [r_i(\sigma)w](t)=0,\ t\in[1,T-\bar{d}(r_i)],\ i\in I_T(R)\}$.

Notation. We denote the class of all finite (T) time autoregressive systems by $\mathbb{B}_T(AR)$., i.e., $\mathbb{B}_T(AR):=\{\mathcal{B}\subset(\mathbb{R}^q)^T;\ \exists g\ \exists R\in\mathbb{R}^{g\times q}[s]$ such that $\mathcal{B}=\mathcal{B}_T(R)\}$. □

Proposition 3-32 $\mathbb{B}_T\subset\mathbb{B}_T(AR)$.

Proof. See the appendix.

Remark. There holds $\mathbb{B}_T\neq\mathbb{B}_T(AR)$. As an example we consider the case $T=4$, $R:=\text{col}(s^3-1,s^2+1)$. Then $\mathcal{B}_4(R)=\{w\in\mathbb{R}^4;\ \exists\alpha\in\mathbb{R}$ such that $w=\alpha(1,-1,-1,1)\}\notin\mathbb{B}_4$. Hence any linear shift invariant finite time system is a finite time AR–system and $\mathbb{B}_T$ consists of a strict subclass of the systems described by a finite number of autoregressive equations. □

Notation. In order to describe the subclass of the AR–systems which coincides with $\mathbb{B}_T$ we use the following terminology. Let $0\neq R\in\mathbb{R}^{g\times q}[s]$ have rows $r_i\in\mathbb{R}^{1\times q}[s]$, $i\in[1,g]$. Let $r_i(s)=\sum_{k=d_i'}^{d_i''} r_k^{(i)}s^k$ with $\bar{d}(r_i)=d_i''\geq d_i'$ and $r_{d_i'}^{(i)}\neq 0\neq r_{d_i''}^{(i)}$. Let $L_+,L_-\in\mathbb{R}^{g\times q}$ denote the matrices with rows $r_{d_i''}^{(i)}$ and $r_{d_i'}^{(i)}$ respectively. Recall that R is called bilaterally row proper if $\text{rank}(L_+)=\text{rank}(L_-)=g$. We call R row proper if $\text{rank}(L_+)=g$, zero order bilaterally row proper if R is bilaterally row proper and if in addition $d_i'=0$, $\forall i\in[1,g]$. We define 0 to be zero order bilaterally row proper. □

The next proposition describes subclasses of $\mathbb{B}_T(AR)$ which give a parametrization of $\mathbb{B}_T$ and $\tilde{\mathbb{B}}_T$.

Proposition 3-33 (*i*) $\{\mathbb{B}\in\mathbb{B}_T\} \Leftrightarrow \{\exists g,\ \exists R\in\mathbb{R}^{g\times q}[s],\ \bar{d}(R)\leq T-1,\ R$ row proper, such that $\mathcal{B}=\mathcal{B}_T(R)\}$;

(*ii*) $\{\mathcal{B}\in\tilde{\mathbb{B}}_T\} \Leftrightarrow \{\exists g,\ \exists R\in\mathbb{R}^{g\times q}[s],\ \bar{d}(R)\leq T-1,\ R$ zero order bilaterally row proper, such that $\mathcal{B}=\mathcal{B}_T(R)\}$.

Proof. See the appendix.

Remark. Henceforth, if we write $\mathcal{B}_T(R)$ we *throughout assume* that R is row proper, and zero order bilaterally row proper only if we explicitly state that $\mathcal{B}_T(R)\in\tilde{\mathbb{B}}_T$. □

Remark. Note that $\mathcal{B}_T(R)$ need not belong to $\tilde{\mathbb{B}}_T$ if R is bilaterally row proper. A simple counterexample is $R=\begin{bmatrix} s & 0 \\ 0 & s \end{bmatrix}$ for $T=2$. □

Remark. In accordance with the terminology of definition 3-4, if $\mathcal{B}=\mathcal{B}_T(R)$ where R has rows r_i, $i\in[1,g]$, then $\{r_i;\ i\in I_T(R)\}$ are called *laws* governing the behaviour $\mathcal{B}$. □

Next we derive a representation of the sum of two systems in $\mathbb{B}_T$ and of the largest system in $\mathbb{B}_T$ contained in the intersection of two systems in $\mathbb{B}_T$. The representation results are used in chapter III.

Remark. Recall that the intersection itself need not belong to $\mathbb{B}_T$. However, for every set in $(\mathbb{R}^q)^T$ there exists a largest system in $\mathbb{B}_T$ contained in this set, because $\mathbb{B}_T$ is closed under addition. □

Lemma 3-34 Let $R_i\in\mathbb{R}^{g_i\times q}[s]$ be row proper with $\bar{d}(R_i)\leq T-1$, $i=1,2$. Then $\{\mathcal{B}_T(R_1)\subset\mathcal{B}_T(R_2)\} \Leftrightarrow \{\exists F\in\mathbb{R}^{g_2\times g_1}[s]$ such that $R_2=FR_1\}$.

Proof. See the appendix.

Remark. If we drop the condition $\bar{d}(R_i)\leq T-1$, $i=1,2$, then $(\Rightarrow)$ no longer holds true, as can be seen, e.g., by taking $T=3$, $R_1=s-1$, $R_2=s^3$. □

Notation. In describing the intersection or sum of systems in $\mathbb{B}_T$ we use the following concepts. Let $R_i \in \mathbb{R}^{g_i \times q}[s]$, $i=1,2$. Then R is called a least common left multiple of R_1 and R_2, notation $R \in \mathrm{LCLM}(R_1,R_2)$, if (1) $\exists F_i$ with $R=F_iR_i$, $i=1,2$, and (2) $\{\exists \tilde{F}_i$ with $\tilde{R}=\tilde{F}_iR_i$, $i=1,2\} \Rightarrow \{\exists F$ with $\tilde{R}=FR\}$. R is called a greatest common right divisor of R_1 and R_2, notation $R \in \mathrm{GCRD}(R_1,R_2)$, if (1) $\exists F_i$ with $R_i=F_iR$, $i=1,2$, and (2) $\{\exists \tilde{F}_i$ with $R_i=\tilde{F}_i\tilde{R}$, $i=1,2\} \Rightarrow \{\exists F$ with $R=F\tilde{R}\}$. □

Lemma 3-35 (*i*) For any row proper R_1, R_2, $\mathrm{LCLM}(R_1,R_2)$ and $\mathrm{GCRD}(R_1,R_2)$ are nonempty and contain row proper elements; (*ii*) if $R',R'' \in \mathrm{LCLM}(R_1,R_2)$ are both row proper, then there exists a unimodular U such that $R''=UR'$, and the same holds true for GCRD.

Proof. See the appendix.

Proposition 3-36 (*i*) $\mathcal{B}_T(R_1)+\mathcal{B}_T(R_2)=\mathcal{B}_T(R)$ for $R \in \mathrm{LCLM}(R_1,R_2)$ row proper;
(*ii*) $\mathcal{B}_T(R)=\Sigma\{\mathcal{B}_T(\tilde{R});\ \mathcal{B}_T(\tilde{R}) \subset \mathcal{B}_T(R_1) \cap \mathcal{B}_T(R_2)\}$ for $R \in \mathrm{GCRD}(R_1,R_2)$ row proper, if $\bar{d}(R_i) \le T-1$, $i=1,2$.

Proof. See the appendix.

Interpretation. A row proper least common left multiple of R_1 and R_2 gives the laws of the sum $\mathcal{B}_T(R_1)+\mathcal{B}_T(R_2)$, and a row proper greatest common right divisor of R_1 and R_2 gives the laws of the largest system in $\mathbb{B}_T$ contained in the intersection $\mathcal{B}_T(R_1) \cap \mathcal{B}_T(R_2)$. □

Remark. The implication $\{\mathcal{B}_T(R)=\mathcal{B}_T(R_1)+\mathcal{B}_T(R_2)\} \Rightarrow \{R \in \mathrm{LCLM}(R_1,R_2)\}$ does not hold true. For example, take $q=g_1=g_2=1$, $T=3$, $R_1=\sigma^2$, $R_2=\sigma+1$, then $\mathcal{B}_T(R_1)+\mathcal{B}_T(R_2)=\mathbb{R}^3=\mathcal{B}_T(0)$ but $0 \notin \mathrm{LCLM}(R_1,R_2)$. Moreover, (*ii*) does not hold true generally if $\bar{d}(R_i) \le T-1$, $i=1,2$, is not satisfied. For example, take $q=g_1=g_2=1$, $T=3$, $R_1=\sigma^2(\sigma-1)$, $R_2=(\sigma-1)^3$, then $\mathrm{GCRD}(R_1,R_2)=\sigma-1$ but $\mathcal{B}_3(\sigma-1) \neq \mathbb{R}^3=\mathcal{B}_3(R_1) \cap \mathcal{B}_3(R_2)$. □

4. Conclusion

In this chapter we motivated and illustrated a deterministic approach to system identification. In this approach a model is identified by optimizing a utility of models which is expressed in terms of complexity and misfit. Two utilities which play a dominant role in the sequel correspond to minimizing misfit under a complexity constraint and minimizing complexity under a misfit constraint. The deterministic procedures for time series analysis in chapter V are based on these utilities. Exact modelling, described in chapter III, is based on the second utility, and model approximation, described in chapter IV, is based on the first one.

We defined and analyzed a class of deterministic dynamical systems which will be our model class in identification. This model class is defined in terms of external properties of the behaviour. The models can be represented by means of autoregressive equations. We derived two canonical forms which play a dominant role in the approximate identification procedures of chapter V. Further we summarized some results on state space realizations which are used in chapter IV. Finally we described finite time systems which are used in chapter III.

CHAPTER III

EXACT MODELLING

1. Introduction and examples

1.1. Introduction

As a first instance of deterministic identification we consider exact modelling in this chapter. Exact modelling is a special case of modelling under a misfit constraint. We want to model the data by means of a model of least complexity under the restriction that the data exactly satisfy all identified laws. This corresponds to the procedure $\bar{\bar{P}}_{\varepsilon_{tol}}$ defined in section II.2.3.1, with $\varepsilon_{tol}=0$.

In section 1 we give some examples of exact modelling procedures which play a role in the sequel. We briefly describe the minimal realization problem in systems theory. In this case the data consists of a certain parametric description of a system, called the impulse response. The model class consists of another representation of systems, i.e., input/state/output systems as described in section II.3.3. The complexity of a model is defined as the dimension of the state space. The minimal realization problem consists of finding an input/state/output system of least complexity which is compatible with the given impulse response.

In section 2 we consider exact modelling of a given time series of infinite length by means of a linear, time invariant, complete system. Finally in section 3 we develop procedures for exact modelling of a finite time series. In this case a central question is when we have reason to accept laws which are exactly satisfied by the available data. We only have reason to accept laws if they are somehow corroborated by the data. We give a definition of corroboration and of some desirable properties of exact modelling procedures. These properties are investigated for two simple procedures. We conclude by presenting an alternative procedure with optimal properties.

1.2. Undominated unfalsified modelling

Let S be a set, and let the set of conceivable data consist of finite numbers of observations in S, i.e. $\mathcal{D}:=\{V\subset S;\ \#(V)<\infty\}$, where $\#(V)$ denotes the number of elements in V. Let a model M consist of a subset $M\subset S$ and let $\mathbb{M}\subset 2^S$ denote a class of models.

A model M is called *unfalsified* by a measurement $d\in\mathcal{D}$ if $d\subset M$.

We call a model M_1 less complex than a model M_2 if $M_1\subset M_2$, cf. the falsifiability principle stated in section II.2.1. This induces a partial ordering on $\mathbb{M}$. We call a model M *undominated* unfalsified in $\mathbb{M}$ for d if $d\subset M\in\mathbb{M}$ and $\{d\subset M'\in\mathbb{M},\ M'\subset M\}\Rightarrow\{M'=M\}$, i.e., if M is unfalsified and there exist no unfalsified models $M'\neq M$ of less complexity. We define $P^u(d)$ as the collection of undominated unfalsified models in $\mathbb{M}$ for d. So P^u identifies the models which are unfalsified by the data and which are as small as possible in the sense of set inclusion. We call P^u the procedure of *undominated unfalsified modelling*.

Remark. This procedure could be expressed by means of a utility, as follows. Define a misfit ε by $\varepsilon(d,M):=1$ if $d\not\subset M$, $\varepsilon(d,M):=0$ if $d\subset M$, and define a complexity c by $c(M):=M$. Let $\underline{u}\notin\mathbb{M}$, $U:=\mathbb{M}\cup\{\underline{u}\}$, and define a utility uu by $uu(M,1):=\underline{u}$ and $uu(M,0):=M$. Define a partial ordering $\leq$ on U as follows: $\underline{u}\leq M$ for all $M\in\mathbb{M}$, and for M_1, $M_2\in\mathbb{M}$, $M_1\leq M_2$ if and only if $M_1\supset M_2$. Then P^u coincides with the procedure P_{uu} corresponding to the utility uu, cf. definition II.2–2. □

Remark. In sections 2 and 3 we consider a special case of this procedure, i.e., for modelling an observed time series by means of a dynamical system. □

1.3. The minimal realization problem

The *minimal realization problem* of linear systems theory is a problem of exact modelling. Both the data and the model class consist of representations of dynamical systems. Given a representation of a system, the problem is to construct a representation in the model class which parametrizes the same system.

Remark. For a thorough discussion and solution of this problem we refer to Kalman, Falb and Arbib [39], and Silverman [63]. Here we only describe the data, the model class, and the modelling problem. □

Notation. Let $\mathbb{Z}_+:=\mathbb{N}\cup\{0\}=\{0,1,2,...\}$. Let $(\mathbb{R}^d)_0^{\mathbb{Z}}:=\{z\in(\mathbb{R}^d)^{\mathbb{Z}};\ \exists t_0\in\mathbb{Z}$ such that $z(t)=0\ \forall t<t_0\}$, i.e., the set of sequences in $\mathbb{R}^d$ with left bounded support. For given $p,m\in\mathbb{N}$ and $G:=(G_k;\ k\in\mathbb{Z}_+)$, $G_k\in\mathbb{R}^{p\times m}$, $k\in\mathbb{Z}_+$, let the system $S(G)$ be defined by $S(G):=\{(u,y)\in(\mathbb{R}^m\times\mathbb{R}^p)_0^{\mathbb{Z}};\ y(t)=\ \Sigma_{k=0}^{\infty}G_ku(t-k),\ t\in\mathbb{Z}\}$. Note that for every $u\in(\mathbb{R}^m)_0^{\mathbb{Z}}$ $y\in(\mathbb{R}^p)_0^{\mathbb{Z}}$ is well-defined in this way. Further, for given $n\in\mathbb{Z}_+$ and $(A,B,C,D)\in\mathbb{R}^{n\times n}\times\mathbb{R}^{n\times m}\times\mathbb{R}^{p\times n}\times\mathbb{R}^{p\times m}$ let the system $S(A,B,C,D)$ be defined in a way analogous to definition II.3-24, i.e., $S(A,B,C,D):=\{(u,y)\in(\mathbb{R}^m\times\mathbb{R}^p)_0^{\mathbb{Z}};\ \exists x\in(\mathbb{R}^n)_0^{\mathbb{Z}}$ such that $\begin{pmatrix}\sigma x\\ y\end{pmatrix}=\begin{pmatrix}A & B\\ C & D\end{pmatrix}\begin{pmatrix}x\\ u\end{pmatrix}\}$. □

The data set is $\mathcal{D}:=(\mathbb{R}^{p\times m})^{\mathbb{Z}_+}$ for some fixed $p,m\in\mathbb{N}$. The data $G\in\mathcal{D}$ consists of a so-called *impulse response* sequence. This impulse response represents the dynamical input/output system $S(G)$. The model class is $\mathbb{M}:=\bigcup\{\mathbb{R}^{n\times n}\times\mathbb{R}^{n\times m}\times\mathbb{R}^{p\times n}\times\mathbb{R}^{p\times m};\ n\in\mathbb{Z}_+\}$. A model (A,B,C,D) represents the dynamical input/output system $S(A,B,C,D)$. So n is the number of state variables in $S(A,B,C,D)$. The complexity of a model is defined as the corresponding number n of state variables.

A model (A,B,C,D) is called a *realization* of an impulse response G if $S(G)=S(A,B,C,D)$. It is called a *minimal* realization if in addition n is as small as possible among realizations of G. The minimal realization problem consists of identifying a minimal realization for a given impulse response, i.e., of identifying an exact model of least possible complexity.

Remark. For $G\in\mathcal{D}$ and $M:=(A,B,C,D)\in\mathbb{M}$ define the misfit by $\varepsilon(G,M):=0$ if M is a realization of G and $\varepsilon(G,M):=1$ otherwise. Let $U:=\{0,-1,-2,-3,...\}\cup\{-\infty\}$ and define a utility by $u(n,1):=-\infty$, $u(n,0):=-n$, $n\in\mathbb{Z}_+$. The procedure corresponding to this utility solves the minimal realization problem. In case a solution exists, it is unique up to a choice of basis in the state space. We refer to Kalman, Falb and Arbib [39, theorem 10.6.9]. □

Remark. This minimal realization problem can be considered as a special case of exact modelling of an infinite time series, see section 2. □

Remark. Note that the model class $\mathbb{M}$ differs from the class $\mathbb{B}$ defined in section II.3.1.2, cf. also corollary II.3–23. In $M\in\mathbb{M}$ the sequences have left bounded support, in $\mathcal{B}\in\mathbb{B}$ this is not required. Moreover, any model $M\in\mathbb{M}$ contains at least one input and one output, i.e., $m\geq 1$ and $p\geq 1$. This does not necessarily hold true on $\mathbb{B}$. □

1.4. The partial realization problem

The *partial realization problem* of linear systems theory is a problem of minimal realization in case the data consists of only a *finite part* of the impulse response.

The model set is as in section 1.3. The data set is $\mathcal{D}:=\bigcup\{(\mathbb{R}^{p\times m})^T;\ T\in\mathbb{N}\}$, where T denotes the number of elements of the impulse response sequence which are available. In general it is assumed that the data consists of the initial part of the impulse response.

A model (A,B,C,D) is called a realization of the partial impulse response $G=(G_t;\ t\in[0,T-1])$ if there exists an impulse response G^e such that $G^e|_{[0,T-1]}=G$ and $S(G^e)=S(A,B,C,D)$. It is called a minimal realization if in addition the number n of state variablels in $S(A,B,C,D)$ is minimal among all realizations of G.

Remark. This problem always has a solution. However, it no longer need be unique up to a choice of basis in the state space. We refer to Kalman [34] and Tether [68]. □

Remark. The partial realization problem can be considered as a special case of exact modelling of a finite time series which we consider in section 3. We refer to section 3.3.5. □

2. Exact modelling of an infinite time series

In this section we consider exact modelling of a given q–dimensional time series of *infinite* length by means of a linear, time invariant, complete system. So the data set is $\mathcal{D}:=(\mathbb{R}^q)^{\mathbb{Z}}$ and the model class is $\mathbb{M}:=\mathbb{B}$ as defined in section II.3.1.2.

We call a model $\mathcal{B}_1\in\mathbb{B}$ less complex than a model $\mathcal{B}_2\in\mathbb{B}$ if $\mathcal{B}_1\subset\mathcal{B}_2$. This is in accordance with the falsifiability principle discussed in section II.2.1.

The procedure of undominated unfalsified modelling P^u defined in section 1.2 corresponds to modelling a given time series $w\in(\mathbb{R}^q)^{\mathbb{Z}}$ by means of those models in $\mathbb{B}$ which are not falsified by w and which are of minimal complexity.

Notation. For $w\in(\mathbb{R}^q)^{\mathbb{Z}}$ let $\mathbb{B}(w):=\{\mathcal{B}\in\mathbb{B};\ w\in\mathcal{B}\}$ denote the class of unfalsified models for w and let $\mathcal{B}^*(w):=\bigcap\{\mathcal{B};\ \mathcal{B}\in\mathbb{B}(w)\}$. □

Remark. $\mathbb{B}(w)\neq\emptyset$ as $(\mathbb{R}^q)^{\mathbb{Z}}\in\mathbb{B}(w)$ for any $w\in(\mathbb{R}^q)^{\mathbb{Z}}$. From the property of closedness stated in proposition II.3–3 it easily follows that $\mathcal{B}^*(w)\in\mathbb{B}$. Hence $w\in\mathcal{B}^*(w)\in\mathbb{B}$ and $\{w\in\mathcal{B}\in\mathbb{B}\}\Rightarrow\{\mathcal{B}^*(w)\subset\mathcal{B}\}$. □

In this identification problem there always exists a unique undominated unfalsified model, i.e., $P^u(w)=\mathcal{B}^*(w)$. This model is less complex than any other unfalsified model. The model $\mathcal{B}^*(w)$ is called the *most powerful unfalsified model* for w.

Remark. Similar results hold true in case the data consists of a finite number of infinite time series. The minimal realization problem forms a particular example. The impulse response $G\in(\mathbb{R}^{p\times m})^{\mathbb{Z}_+}$ can be considered as consisting of m time series, where for $i=1,\ldots,m$ the i–th column of G describes the output on $\mathbb{Z}_+$ corresponding to a unit pulse on time $t=0$ in the i–th input channel. The input and output on $\{\ldots,-2,-1\}$ are defined to be zero. Constructing the most powerful unfalsified model for these data results in a solution of the minimal realization problem. For a thorough discussion we refer to Willems [73, sections 15 and 17], where also algorithms are given to determine a minimal realization (A,B,C,D) of the most powerful unfalsified model for m observed time series. □

Remark. In section 3.1 we investigate some properties of the procedure P^u for modelling a finite time series $w \in (\mathbb{R}^q)^T$, $T \in \mathbb{N}$, by means of models in $\mathbb{B}_T$. It turns out that in general there does not exist a unique undominated unfalsified model, in which case there is no most powerful unfalsified model in the class $\mathbb{B}_T$. This is due to the fact that $\mathbb{B}_T$ is not closed under intersection, see section II.3.4.1. □

Remark. In the sequel we do not consider exact modelling of an infinite time series, but exact modelling of a finite time series in the remainder of this chapter, approximate modelling of a finite time series in chapter V, and model approximation in chapter IV. □

3. Exact modelling of a finite time series

3.1. Properties of identification procedures

3.1.1. Simplicity, corroboration, and prudence

We start by briefly restating the identification problem as described in section I.1.1 and the deterministic modelling methodology as described in section II.2.1 for the case of exact modelling of a finite time series.

Suppose that we want to find an exact model for data concerning q real-valued variables of interest, where the data consists of a (finite) time series of length T, $w := (w(t);\ t \in [1,T]) \in (\mathbb{R}^q)^T$. As model class we take the class of linear, shift invariant systems $\mathbb{B}_T$ as defined in section II.3.4.1.

Interpretation. By taking $\mathbb{B}_T$ as model class we a priori only want to detect autoregressive laws for the data which are such that *future* observations $w(t)$, $t > T$, are *conceivable* which satisfy these laws, see section II.3.4.1. Alternatively, in case we only want to accept laws if these also conceivably hold true for the past, we take as model class the class of linear, translation invariant models $\tilde{\mathbb{B}}_T$. □

The aim is to construct a data modelling procedure P which assigns to data $w \in (\mathbb{R}^q)^T$ models in $\mathbb{B}_T$ ($\tilde{\mathbb{B}}_T$), $T \in \mathbb{N}$. We will follow the methodology of section II.2.1 and require that the procedure is based on principles of simplicity and corroboration.

A model is considered to be simple if it is small in some set-theoretic sense. We want to infer from the data as much structure as possible in order to get many laws.

Remark. Note that if both R and $R' := \begin{bmatrix} R \\ r \end{bmatrix}$ are row proper, then $\mathcal{B}_T(R') \subset \mathcal{B}_T(R)$, see lemma II.3–34. So a more simple model is more easily falsifiable, as it imposes more restrictions. □

Remark. Note that simplicity here is connected with systems as sets of trajectories and not with the simplicity of the laws of systems. For example, $(\mathbb{R}^q)^T$ is the most complex model in $\mathbb{B}_T$, although with respect to laws it is very simple as there are no laws at all. A simple model is one which claims many laws. □

As a measure of complexity of a model in $\mathbb{B}_T$ we here simply take its dimension.

Definition 3–1 The *complexity* of $\mathcal{B} \in \mathbb{B}_T$ is defined as $c(\mathcal{B}) := \dim(\mathcal{B})$.

Although we want to get models of low complexity, i.e., with much structure, on the other hand we want to accept structure only if there is some evidence for it from the data. Note that data independent structure is reflected in the a priori specified model class and that the data modelling procedure identifies the additional structure which is accepted after observing the data. We should have some reason for claiming structure on the basis of data. Laws should be *corroborated* by the data.

Notation. Let the data set be $\mathcal{D} := \bigcup\{(\mathbb{R}^q)^T;\ T \in \mathbb{N}\}$ and the model class $\mathbb{M} := \bigcup\{\mathbb{B}_T;\ T \in \mathbb{N}\}$. A modelling procedure is a map $P : \mathcal{D} \rightarrow 2^{\mathbb{M}}$ such that for $w \in (\mathbb{R}^q)^T$

$P(w)\subset\mathbb{B}_T$. By P_T we denote $P|_{(\mathbb{R}^q)^T}$, i.e., $P_T:(\mathbb{R}^q)^T\rightarrow 2^{\mathbb{B}_T}$. Analogously, for $\tilde{\mathbb{M}}:=\bigcup\{\tilde{\mathbb{B}}_T;\ T\in\mathbb{N}\}$ and $P:\mathcal{D}\rightarrow 2^{\tilde{\mathbb{M}}}$ we also denote $P_T:=P|_{(\mathbb{R}^q)^T}:(\mathbb{R}^q)^T\rightarrow 2^{\tilde{\mathbb{B}}_T}$. □

We state in definition 3–2 what we mean by corroboration and formulate in definition 3–3 the concept of prudence for modelling procedures. Intuitively stated, a prudential procedure only accepts laws for which there is reason to be accepted, i.e., it only accepts laws which are corroborated by that procedure.

Remark. In order to illustrate the concepts of corroboration and prudence as well as the properties defined in section 3.1.2 we consider the procedure P^u of undominated unfalsified modelling, cf. section 1.2. Recall that $\mathcal{B}\in\mathbb{B}_T$ is called unfalsified by $w\in(\mathbb{R}^q)^T$ if $w\in\mathcal{B}$ and that it is called undominated unfalsified if in addition $\{w\in\mathcal{B}'\in\mathbb{B}_T,\ \mathcal{B}'\subset\mathcal{B}\}\Rightarrow\{\mathcal{B}'=\mathcal{B}\}$. The procedure P^u assigns to $w\in(\mathbb{R}^q)^T$ the set of undominated unfalsified models in $\mathbb{B}_T$. □

Example. Let $q=1$, $T=4$. Using an "incomplete Hankel array", described in section 3.2.1, it can be derived that $P^u_4(0,1,0,1)=\mathcal{B}_4(\sigma^2-1)$. For $w=(1,1,0,1)$ the class of unfalsified models is $\{\mathcal{B}_4(\sigma^2+\sigma-1\}$, $\mathcal{B}_4(\sigma^3+\alpha\sigma^2+\beta\sigma-1-\beta)$, $\alpha,\beta\in\mathbb{R}\}$ and $P^u_4(1,1,0,1)=\{\mathcal{B}_4(\sigma^2+\sigma-1)$, $\mathcal{B}_4(\sigma^3+\alpha\sigma^2+\beta\sigma-1-\beta)$, $\alpha-\beta\neq 2\}$. So in this case there exists more than one undominated unfalsified model. This is essentially due to the fact that $\mathbb{B}_T$ is not closed under intersection. □

Notation. Let $L\subset\mathbb{R}^n$ be a linear subspace of dimension d with a basis $\{b_1,\ldots,b_d\}$, and for $x\in L$ let $(x_1,\ldots,x_d)\in\mathbb{R}^d$ be defined by $x=\Sigma_{i=1}^d x_i b_i$. A mapping $p:L\rightarrow\mathbb{R}$ is called a polynomial (on L) if $p(x)$ is a polynomial in the d variables $(x_1,\ldots,x_d)$.

A subset $V\subset L$ is called an algebraic variety in L if $V=p^{-1}(0)$ for some polynomial p on L. It is called a proper algebraic variety if in addition $V\neq L$, i.e., $p\neq 0$. We call a set $\pi\subset L$ *generic* in L if there is a proper algebraic variety V in L such that $\pi\supset(L\backslash V)$. Moreover we call a property for points in L generic in L (or: generic in $x\in L$) if the set of points which have the property is a generic set in L.

It is easily seen that these concepts are independent from the choice of basis of L. □

Remark. This algebraic concept of genericity is stronger than the topological concept of genericity. A set is called topologically generic if it contains an open and dense subset. An (algebraically) generic set $\pi \subset L$ even contains an open set of full Lebesgue measure in L. We refer to section V.4.4. □

Definition 3-2 Let $\mathcal{B} \in \mathbb{B}_T$, $P_T:(\mathbb{R}^q)^T \to 2^{\mathbb{B}_T}$. Then

(*i*) $\{\mathcal{B}$ is *weakly corroborable* by $P_T\}: \Leftrightarrow \{\mathcal{B} \notin P_T w$ is not generic in $w \in \mathcal{B}\}$;

(*ii*) $\{\mathcal{B}$ is *strongly corroborable* by $P_T\}: \Leftrightarrow \{P_T w = \mathcal{B}$ is generic in $w \in \mathcal{B}\}$.

Interpretation. If a system $\mathcal{B}$ is strongly corroborable by a modelling procedure P_T then for generic observations from $\mathcal{B}$ P_T assigns exactly the right model, i.e., $\mathcal{B}$ is (strongly) identifiable by P_T. On the other hand, if $\mathcal{B}$ is not weakly corroborable by P_T then for generic observations from $\mathcal{B}$ P_T does not identify $\mathcal{B}$ as a possible generating system. □

Example. Let $q=1$, $T=4$ and consider P^u. Now $\mathbb{R}^4$ is not even weakly corroborable by P_4^u, as $P_4^u w=\mathbb{R}^4$ if and only if $w=\alpha(0,0,0,1)$ for some $\alpha \neq 0$. It can be shown that every other model in $\mathbb{B}_4$ is weakly corroborable by P_4^u. $\mathcal{B}_4(\sigma^2-1)$ is weakly, but not strongly corroborable by P_4^u. $\mathcal{B}_4(\sigma-1)$ is strongly corroborable by P_4^u. In fact, with $c:\mathbb{B}_4 \to [0,4]$, $c(\mathcal{B}):=\dim(\mathcal{B})$, it can be shown that the class of models which are strongly corroborable by P_4^u is given by $c^{-1}(\{0,1\})$ and $\mathcal{B}_4(\sigma^2)$. □

Remark. Clearly P^u puts all stress on simplicity, with the ordering corresponding to set inclusion, while P^u does not take corroboration into account. □

Notation. To define the concept of prudence we use the notation $\mathbb{B}^w_{P_T}:=\{\mathcal{B} \in \mathbb{B}_T;\ \mathcal{B}$ is weakly corroborable by $P_T\}$, $\mathbb{B}^s_{P_T}:=\{\mathcal{B} \in \mathbb{B}_T;\ \mathcal{B}$ is strongly corroborable by $P_T\}$, and $\mathrm{im}(P_T):=\{\mathcal{B} \in \mathbb{B}_T;\ \exists w \in (\mathbb{R}^q)^T$ such that $\mathcal{B} \in P_T w\}$. □

Definition 3-3 (*i*) P is called a *weakly prudential* procedure if $\mathrm{im}(P_T) \subset \mathbb{B}^w_{P_T}$ for all $T \in \mathbb{N}$;

(*ii*) P is called a *strongly prudential* procedure if $\mathrm{im}(P_T) \subset \mathbb{B}^s_{P_T}$ for all $T \in \mathbb{N}$.

Interpretation. A weakly (strongly) prudential procedure only identifies models which are weakly (strongly) corroborable by that procedure. So a prudential procedure only accepts laws which it can "*verify*" or *corroborate* itself.

If a model is not weakly corroborable by a procedure it is natural to require that this model will never be assigned by the procedure. For suppose that $\mathcal{B}\in P_T w$ while $\mathcal{B}\notin \mathbb{B}^w_{P_T}$. Then generically on $\mathcal{B}$ P_T assigns only wrong models. If $\mathcal{B}$ is assigned on the basis of w this means that we have to suppose that we were extraordinarily lucky to observe precisely this w in order to identify $\mathcal{B}$, as for generic data generated by $\mathcal{B}$ the procedure P_T would not have identified the generating system. Assigning $\mathcal{B}$ is reckless. So weak prudence seems a minimum requirement.

On the other hand, strongly prudential procedures are very cautious. A model $\mathcal{B}$ only is assigned if this model also generically will be identified on the basis of observations of the system $\mathcal{B}$. So if identifying the right model is not guaranteed for generic observations from a system $\mathcal{B}$, then a strongly prudential procedure does not dare to make this assignment $\mathcal{B}$. □

Remark. The relaxation that conditions are satisfied generically instead of universally is essential in the definition of corroborability and prudence as well as in many definitions which follow. This has to do with the fact that a model $\mathcal{B}_T(R)\in\mathbb{B}_T$ contains submodels with more structure, i.e., if R' is such that there exists an F with $FR'=R$ then $\mathcal{B}_T(R')\subset\mathcal{B}_T(R)$ and hence exceptional observations from $\mathcal{B}_T(R)$ may exhibit the stronger structure of $\mathcal{B}_T(R')$. We mention that for $\mathcal{B}_1,\mathcal{B}_2\in\mathbb{B}_T$ with $\mathcal{B}_2\subset\mathcal{B}_1$ and $\mathcal{B}_2\neq\mathcal{B}_1$ the set $\mathcal{B}_1\backslash\mathcal{B}_2$ is generic in $\mathcal{B}_1$. □

Example. Consider P^u for $q=1$, $T=4$. From the foregoing example it follows that P^u_4 is not strongly prudential and not even weakly prudential. However, every observation $w\in\mathbb{R}^4$ with $w\notin\{\alpha(0,0,0,1),\ 0\neq\alpha\in\mathbb{R}\}$ is modelled in the class $\mathbb{B}^w_{P^u_4}$. □

The aim is to construct modelling procedures for modelling a finite time series which take account of the principles of simplicity and corroboration. In the next section we formulate some other desirable properties of procedures. In section 3.2 we consider a well-known modelling procedure which strives for maximal simplicity without taking into account corroboration. In section 3.3 we present an alternative procedure which not only is strongly prudential but which also has other desirable properties described in the next section.

3.1.2. Some other properties

In this section we define some possible properties of procedures and comment on their interpretation. We illustrate these properties by the procedure P^u of undominated unfalsified modelling. In sections 3.2 and 3.3 we define some other procedures and investigate their properties.

Definition 3-4 P is called *exact* if it only assigns unfalsified models, i.e., $\{\mathcal{B}\in P_T w\} \Rightarrow \{w\in\mathcal{B}\}$ for all $w\in(\mathbb{R}^q)^T$, $T\in\mathbb{N}$.

Example. P^u is exact. □

Remark. Clearly this requirement of exact modelling is a strong one. It is motivated by purposes like exact data representation and transformation of model representations as discussed in sections 1.3 and 1.4. The case of approximate modelling in which it is not required that $w\in\mathcal{B}$ but that w is "near to" $\mathcal{B}$ has often more appeal in applications. Even if the data is exact, i.e., there are no errors of observation, one generally is more interested in approximate simple structure than in exact more complex structure. This is discussed in chapter IV. If the data is corrupted by noise then exact modelling is not a reasonable requirement. In applications generally the phenomenon under observation will not correspond to any model in the a priori class of models and also the observations will contain inaccuracies. Approximate modelling is discussed in chapter V. □

Definition 3-5 (*i*) P is called *monotone* on $\mathcal{B}\in\mathbb{B}_T$ if generically in $w\in\mathcal{B}$ the following holds true for all $t\in[2,T]$: $\{\mathcal{B}_{t-1}\in P_{t-1}(w|_{[1,t-1]}),\ \mathcal{B}_t\in P_t(w|_{[1,t]})\}\Rightarrow \{\mathcal{B}_t|_{[1,t-1]}\subset\mathcal{B}_{t-1}\}$;

(*ii*) P is called monotone if it is monotone on every $\mathcal{B}\in\mathbb{B}_T$, for all $T\in\mathbb{N}$;

(*iii*) P is called bilaterally monotone if on every $\mathcal{B}\in\tilde{\mathbb{B}}_T$ it is monotone and if in addition generically in $w \in \mathcal{B}$ the following holds true for all $t \in [2,T]$: $\{\mathcal{B}_{T-t+1} \in P_{T-t+1}(w|_{[t,T]}),\ \ \mathcal{B}_{T-t+2} \in P_{T-t+2}(w|_{[t-1,T]})\} \Rightarrow \{\sigma\mathcal{B}_{T-t+2} \subset \mathcal{B}_{T-t+1}\}$.

Interpretation. If P is monotone then on getting a new observation $w(t)$, the structure assigned on the basis of $(w(1),...,w(t-1))$ generically is verified or even made more tight. A monotone procedure assigns more structure if the observation period gets longer. This does not only hold true generically on $(\mathbb{R}^q)^T$ but even generically on every system $\mathcal{B} \in \mathbb{B}_T$, $T \in \mathbb{N}$. □

Example. For $q=1$ P^u is only monotone on $\mathbb{B}_2$ or, for $T>2$, on $c^{-1}(\{0,1\})$, i.e., on $\{0\}$ and on $\mathcal{B}_T(\sigma-\alpha)$, $\alpha \in \mathbb{R}$. Consider, e.g., $T=3$, $\mathcal{B}=\mathcal{B}_3(\sigma^2)=\{a,b,0);\ a,b \in \mathbb{R}\}$, and let $w=(a,b,0)$ with $a \neq 0 \neq b$. Then $P_2^u(w|_{[1,2]})=\mathcal{B}_2(\sigma-(b/a))$ while $\mathcal{B}_3(\sigma^2) \in P_3^u w$, and $\mathbb{R}^2=\mathcal{B}_3(\sigma^2)|_{[1,2]} \not\subset \mathcal{B}_2(\sigma-(b/a))$. □

Next we define two properties of procedures which are connected with the prior restriction that we only assign models in the class $\mathbb{B}_T$ (or $\tilde{\mathbb{B}}_T$), i.e., shift (or translation) invariant, linear models. So we describe the phenomenon which generates the data as a shift (translation) invariant, linear system. This leads to some desirable properties of P.

Definition 3-6 (*i*) P is called *shift invariant* on $\mathcal{B} \in \mathbb{B}_T$ if generically in $w \in \mathcal{B}$ the following holds true for all $t \in [2,T]$: $\{\mathcal{B}_{t-1}(R) \in P_{t-1}(w|_{[2,t]}),\ \mathcal{B}' \in P_t(w|_{[1,t]})\} \Rightarrow \{\mathcal{B}' \subset \mathcal{B}_t(\sigma.R)\}$.

(*ii*) P is called shift invariant if it is shift invariant on every $\mathcal{B} \in \mathbb{B}_T$.

Interpretation. Assume that we only observe $w|_{[2,t]}$ and want to model the phenomenon on $[1,t]$. The prior assumption that assigned models have to be shift-invariant, i.e., $\mathcal{B}|_{[1,t-1]} \supset \mathcal{B}|_{[2,t]}$, does not imply any prior restriction on $w(1)$, given $w|_{[2,t]}$. It seems reasonable first to model $w|_{[2,t]}$ and to impose no restriction on $w(1)$, i.e., if $\mathcal{B}_{t-1}(R) \in P_{t-1}(w|_{[2,t]})$ then take $\mathcal{B}_t(\sigma R)$ as a model on $[1,t]$. This amounts to letting $w(1)$ be arbitrary and assigning the laws R on $[2,t]$. Now assume that we also observe $w(1)$, so we have more information concerning the phenomenon. It is reasonable to demand that we can make more accurate models, i.e., to require that P is shift invariant. □

Proposition 3-7 If P is bilaterally monotone on $\mathcal{B}\in\tilde{\mathbb{B}}_T$ then it is shift invariant on $\mathcal{B}$.

Proof. See the appendix.

Remark. Because of this proposition we do not define a concept of translation invariance for procedures, as it would be a concept very close to bilateral monotonicity. □

Example. For P^u take $q=1$, $T=3$. P^u is not shift invariant on $\mathcal{B}_3(\sigma^2-1)$, as for $w=(a,b,a)$, $a\neq b\neq 0$, $P_2^u(w|_{[2,3]})=\mathcal{B}_2(\sigma-(a/b))$ while $\mathcal{B}_3(\sigma^2-1)\in P_3^u(w)$, and $\mathcal{B}_3(\sigma^2-1)\not\subset\mathcal{B}_3(\sigma^2-(a/b)\sigma)$. □

Definition 3-8 P is called *linear* if for all $T\in\mathbb{N}$ there holds

(*i*) $P_T(\alpha w)=P_T w$ for all $w\in(\mathbb{R}^q)^T$, $0\neq\alpha\in\mathbb{R}$, and

(*ii*) for all $\mathcal{B}_1,\mathcal{B}_2\in\mathbb{B}_T$ there holds generically in $(w_1,w_2)\in\mathcal{B}_1\times\mathcal{B}_2$ that $\{\mathcal{B}\in P_T(w_1+w_2),\ \mathcal{B}'\in P_T w_1,\ \mathcal{B}''\in P_T w_2\}\Rightarrow\{\mathcal{B}'+\mathcal{B}''\subset\mathcal{B}\}$.

Interpretation. The phenomenon a priori is assumed to be linear, and (*i*) reflects that the observations w and αw, $\alpha\neq 0$, are in a sense equivalent. Concerning (*ii*), consider the situation of constructing a model for the signal w_1+w_2 either from observation of w_1+w_2 or from observation of both w_1 and w_2. In the latter case we have more information concerning the structure of the phenomenon which we want to model and hence it is reasonable to demand that we can make more accurate models. □

Example. P^u satisfies (*i*), but not (*ii*). Take $q=1$, $T=3$, $\mathcal{B}_i=\mathcal{B}_3(\sigma-i)$, $i=1,2$. Then generically in $w_i\in\mathcal{B}_i$ $P_3^u w_i=\mathcal{B}_i$, $i=1,2$. According to proposition II.3-36(*i*) $\mathcal{B}_1+\mathcal{B}_2=\mathcal{B}_3(\sigma^2-3\sigma+2)$. If $w=(a,b,3b-2a)\in\mathcal{B}_1+\mathcal{B}_2$, $a\neq 0$, $2a^2+b^2-3ab\neq 0$, then $\mathcal{B}_3(\sigma^2+(2a-3b)/a)\in P_3^u w$, and $\mathcal{B}_3(\sigma^2-3\sigma+2)\not\subset\mathcal{B}_3(\sigma^2+(2a-3b)/a)$. So (*ii*) is not satisfied generically on $\mathcal{B}_1\times\mathcal{B}_2$. □

In addition to the concept of prudence defined in the foregoing section we finally define another property reflecting the wish to accept laws only if there is some evidence for them.

Definition 3-9 (*i*) P_T is called *truthful* on $\mathcal{B}_0 \in \mathbb{B}_T$ if generically in $w \in \mathcal{B}_0$ there holds $\{\mathcal{B} \in P_T w\} \Rightarrow \{\mathcal{B}_0 \subset \mathcal{B}\}$;

(*ii*) P is called truthful if P_T is truthful on every $\mathcal{B}_0 \in \mathbb{B}_T$, for all $T \in \mathbb{N}$.

Interpretation. P is truthful if generically the laws which are accepted by P in fact are also satisfied for the phenomenon which generates the data. So the identified models contain the phenomenon as a subset. It is not required that all laws of the phenomenon are detected. □

Example. P^u is not truthful. Take $q=1$, $T=2$, $\mathcal{B}_0=\mathbb{R}^2$. Then generically in $w \in \mathcal{B}_0$ $\dim(P_2^u w)=1$. □

Summarizing, the procedure P^u which identifies undominated unfalsified models is exact, it is not monotone, not shift invariant, not linear, not truthful and not even weakly prudential. Moreover, P^u is a complex procedure in the sense that generally it identifies many models for each observation.

In the next section we investigate the partial realization procedure P^K for the case of a one-dimensional time series, i.e., $q=1$. This procedure is a refinement of P^u in the sense that, for all $T \in \mathbb{N}$ and $w \in \mathbb{R}^T$, $P_T^K w \subset P_T^u w$. In fact P_T^K chooses among the undominated unfalsified models those of minimal complexity. In section 3.3 we define procedures P^0 and a refinement P^* which take corroboration into account. The procedure P^* identifies the unfalsified model of minimal complexity in the class of models for which there is some evidence from the data. This procedure is exact, monotone, shift invariant, linear, truthful and strongly prudential, see theorem 3-28.

Remark. If $q=1$, $T=\infty$, then the procedures P_∞^u, P_∞^K, P_∞^0, P_∞^* for modelling a time series $w \in (\mathbb{R})^{\mathbb{N}}$ can be shown to be equivalent. So P_∞^u is exact, truthful, and even strongly prudential. This is implied by the fact that for every system $\mathcal{B}_\infty(R) := \{w \in (\mathbb{R})^{\mathbb{N}};\ [R(\sigma)w](t)=0 \text{ for all } t \in \mathbb{N}\}$ generically in $w \in \mathcal{B}_\infty(R)$ the only undominated unfalsified model is $\mathcal{B}_\infty(R)$ itself, so $P_\infty^u w=\mathcal{B}_\infty(R)$ generically in $w \in \mathcal{B}_\infty(R)$. We return to this in section 3.3.3. □

3.2. The partial realization procedure

3.2.1. Procedure

From now on in this chapter we throughout restrict attention to the univariate case, i.e., $q=1$.

Remark. For a multivariable exact modelling procedure with desirable properties we refer to section V.5.2.2. □

The "partial realization" procedure P^K is a *refinement* of P^u. It assigns to data w the least complex model(s) unfalsified by w. This implies that all assigned models are undominated unfalsified. So in the class of exact procedures P^K is the one which *maximizes* the *simplicity* of the model, i.e., it minimizes the dimension.

Remark. Stated otherwise, this procedure identifies the shortest lag AR-relations exactly satisfied by the data, as for $q=1$ and $0 \neq R \in \mathbb{R}[s]$ $\dim(\mathcal{B}_T(R)) = \min\{T, \bar{d}(R)\}$, where $\bar{d}(R)$ is the degree of R as defined in section II.3.4.2. □

> **Definition 3-10** The *partial realization procedure* P^K assigns least complex unfalsified models, i.e., for $T \in \mathbb{N}$ $P_T^K : \mathbb{R}^T \to 2^{\mathbb{B}_T}$ is defined by $\{\mathcal{B} \in P_T^K w\} :\Leftrightarrow \{w \in \mathcal{B} \in \mathbb{B}_T \text{ and } c(\mathcal{B}) = \min\{c(\mathcal{B}');\ w \in \mathcal{B}' \in \mathbb{B}_T\}\}$.

To determine $P_T^K w$ we use an algorithm for the partial realization problem as given, e.g., in Kalman [34].

Remark. As described in section 1.4, partial realization theory concerns the realization of a partial impulse response sequence in a minimal way, i.e., with minimal dimension of the state space. The procedures for modelling a univariate finite time series as presented in this section and in section 3.3 also give solutions for the partial realization problem for SISO-systems, i.e., systems with a single input and a single output. We comment on this in section 3.3.5. □

Notation. The "*incomplete Hankel array*" for $w \in \mathbb{R}^T$ is defined by

$$H_T(w) := \begin{bmatrix} w(1) & w(2) & \dots & w(T-1) & w(T) \\ w(2) & w(3) & \dots & w(T) & \\ \vdots & \vdots & & & \\ w(T-1) & w(T) & & & \\ w(T) & & & & \end{bmatrix} \quad \begin{matrix} r(1) \\ r(2) \\ \vdots \\ r(T-1) \\ r(T) \end{matrix} .$$

By $r(i)$ we have denoted the i–th row of $H_T(w)$.

A matrix M with elements $m_{ij}, i, j \in \mathbb{N}$, is called an extension of $H_T(w)$ if $m_{ij} = w(i+j-1)$, $i \in [1,T]$, $j \in [1,T-i+1]$. It is called a Hankel extension if M is a Hankel matrix, i.e., there is an $f: \mathbb{N} \to \mathbb{R}$ such that $m_{ij} = f(i+j-1)$. The rank of M is defined as the dimension of the space spanned by the columns (or rows) of M. Now the (*Kalman*) *rank* of $H_T(w)$, denoted by $\mathrm{rank}(H_T(w))$, is defined as the minimal rank of extensions of $H_T(w)$. Clearly, $\mathrm{rank}(H_T(w))$ is well–defined and $\mathrm{rank}(H_T(w)) \leq T$ for all $w \in \mathbb{R}^T$. Finally we call $r(i)$ linearly (in)dependent on $r(1),\dots,r(i-1)$ if in the $i \times (T-i+1)$ matrix consisting of the first i rows and first $T-i+1$ columns of $H_T(w)$ the last row is linearly (in)dependent on the foregoing ones. If $r(i)=0$ it is called linearly dependent on the foregoing rows. □

Lemma 3–11 $\{\mathrm{rank}(H_T(w))=n\} \Leftrightarrow \{r(n)$ linearly independent from $r(1),\dots, r(n-1)$, and $r(n+1)$ linearly dependent on $r(1),\dots,r(n)\}$.

Proof. See the appendix.

Lemma 3–12 In the class of minimal rank extensions of $H_T(w)$ there is a Hankel extension.

Proof. See the appendix

The following proposition describes how $P_T^K w$ can be determined by using $H_T(w)$. This amounts to determining the first row in $H_T(w)$ which is linearly dependent on the foregoing ones.

Proposition 3–13 (*i*) $\{\mathcal{B} \in P_T^K w\} \Leftrightarrow \{w \in \mathcal{B}$ and $c(\mathcal{B}) = \mathrm{rank}(H_T(w))\}$;
(*ii*) if $\mathrm{rank}(H_T(w))=d$, then $P_T^K w = \{\mathcal{B}_T(R)$; there exist $a_i \in \mathbb{R}$, $i \in [1,d]$, such

that $r(d+1)=\Sigma_{i=1}^{d}a_i r(i)$ and $R=\sigma^d-\Sigma_{i=1}^{d}a_i\sigma^{i-1}\}$.

Proof. See the appendix.

The procedure P^K only takes simplicity into account, not corroboration. As a result there often is no evidence for assigned models. We now first illustrate P^K by giving some examples which play an important role in section 3.3 to construct procedures with better properties than P^K.

Notation. For $x\in\mathbb{R}$ define $\text{ENT}(x):=\min\{n\in\mathbb{Z};\ n\geq x\}$, and $\text{ent}(x):=\max\{n\in\mathbb{Z};\ n\leq x\}$. □

Example 1. Consider the system $\mathcal{B}=\mathbb{R}^T$, so there are no laws. Then from the incomplete Hankel array it is easily seen that generically in $w\in\mathcal{B}$ $\text{rank}(H_T(w))=\text{ENT}(T/2)$, hence generically in $w\in\mathcal{B}$ $\{\mathcal{B}\in P_T^K w\}\Rightarrow\{\dim(\mathcal{B})=\text{ENT}(T/2)\}$. So even though there are no laws at all, P^K still generically imposes them. In fact $P_T^K w=\mathbb{R}^T$ if and only if $w(t)=0$, $t\in[1,T-1]$, and $w(T)\neq 0$. □

In the next examples we consider time series of length $T=6$. The models are identified by means of proposition 3–13.

Example 2. $w=(1,1,2,3,5,8)$, $\text{rank}(H_6(w))=2$, $r(3)=r(1)+r(2)$, $P_6^K w=\mathcal{B}_6(\sigma^2-\sigma-1)$. □
Example 3. $w=(1,1,1,0,1,1)$, $P_6^K w=\mathcal{B}_6(\sigma^3+\sigma^2+\sigma-2)$. □
Example 4. $w=(1,2,0,1,1,2)$, $P_6^K w=\mathcal{B}_6(\sigma^3-\frac{7}{5}\sigma^2-\frac{3}{5}\sigma+\frac{1}{5})$. □
Example 5. $w=(1,2,0,4,4,4)$, $P_6^K w=\mathcal{B}_6(\sigma^3-\sigma-2)$. □
Example 6. $w=(1,0,2,0,1,1)$, $P_6^K w=\mathcal{B}_6(\sigma^3+\frac{1}{3}\sigma^2-\frac{1}{2}\sigma-\frac{2}{3})$. □

The next example illustrates that P^K need not always assign unique models.

Example 7. Take $T=3$ and consider the system $\mathcal{B}=\mathbb{R}^3$. Then for $w=(a,b,c)\in\mathcal{B}$ with $ac-b^2\neq 0$ there holds that $P_3^K w=\{\mathcal{B}_3(\sigma^2-\beta\sigma-\alpha);\ \alpha a+\beta b=c\}$. In general, if T is odd then generically in $w\in\mathbb{R}^T$ $\text{rank}(H_T(w))=(T+1)/2$ and $P_T^K w$ consists of uncountably many models, while if T is even then generically in $w\in\mathbb{R}^T$ $\text{rank}(H_T(w))=T/2$ and $P_T^K w$ consists of one model. □

The analogue $\tilde{P}^K$ of P^K for least complex modelling in $\tilde{\mathbb{B}}_T$ can be defined as follows.

Definition 3-14 The procedure $\tilde{P}^K$ assigns least complex unfalsified models which are translation invariant, i.e., $\tilde{P}^K_T:\mathbb{R}^T \to 2^{\tilde{\mathbb{B}}_T}$ is defined by $\{\mathcal{B}\in\tilde{P}^K_T w\}: \Leftrightarrow \{w\in\mathcal{B}\in\tilde{\mathbb{B}}_T$ and $c(\mathcal{B})=\min\{c(\mathcal{B}');\ w\in\mathcal{B}'\in\tilde{\mathbb{B}}_T\}\}$.

Remark. Without going into details, $\tilde{P}^K_T w$ can be determined by looking in $H_T(w)$ for the first row which is linearly dependent on the foregoing ones and which explicitly involves the first row, i.e., by looking for the smallest d such that there exist $(a_1,\dots,a_d)\in\mathbb{R}^d$ with $a_1\neq 0$ and $r(d+1)= \Sigma_{i=1}^d a_i r(i)$. This way of determining $\tilde{P}^K_T w$ is based upon proposition II.3–33(*ii*). □

3.2.2. Properties

Before we state the properties of P^K we define the concept of a selection rule. If a procedure lacks desirable properties this could be due to the fact that for some observations the identified model is not unique. Such a procedure possibly could be improved by a selection rule. By this we mean a rule which for every observation chooses a unique model from the class of models which is identified by the procedure.

Definition 3-15 A *selection rule* for a procedure $P=\{P_T;\ T\in\mathbb{N}\}$ is a collection $S=\{S_T;\ T\in\mathbb{N}\}$ of maps $S_T:\mathbb{R}^T\to\mathbb{B}_T$ such that for all $w\in\mathbb{R}^T$ $S_T w\in P_T w$.

To analyze the properties of P^K the following proposition is helpful.

Proposition 3-16 Let $0\neq R\in\mathbb{R}[s]$ have degree $\bar{d}(R)=d$. Then generically in $w\in\mathcal{B}_T(R)$ $\operatorname{rank}(H_T(w))=\min\{d,\mathrm{ENT}(T/2)\}$. For $R=0$ generically in $w\in\mathbb{R}^T$ $\operatorname{rank}(H_T(w))=\mathrm{ENT}(T/2)$.

Proof. See the appendix.

Theorem 3–17 (*i*) P^K is exact. If $T\geq 3$, then

(*ii*) P^K is not monotone, not shift invariant, not linear;

(*iii*) P^K is not truthful;

(*iv*) P^K is not even weakly prudential, as $\text{im}(P_T^K)=\mathbb{B}_T$, $\mathbb{B}^w_{P_T^K}=\{\mathcal{B}\in\mathbb{B}_T;\ c(\mathcal{B})\leq \text{ENT}(T/2)\}$, $\mathbb{B}^s_{P_T^K}=\mathbb{B}^w_{P_T^K}$ if T is even, $\mathbb{B}^s_{P_T^K}=\{\mathcal{B}\in\mathbb{B}_T;\ c(\mathcal{B})\leq \text{ent}(T/2)\}$ if T is odd;

(*v*) there exists no selection rule for P^K which has at least one of the desired properties mentioned in (*ii*)–(*iv*).

Proof. See the appendix.

Remark. It is a matter of easy verification to show that for $T=2$ P^K is monotone, shift invariant, not linear, not truthful, and not weakly prudential. □

Remark. The proof of theorem 3–17 contained in the appendix showing that P^K lacks desirable properties concentrates on the case $\mathcal{B}=\mathbb{R}^T$. It can be shown that P^K only is monotone and shift invariant on models $\mathcal{B}$ for which $c(\mathcal{B})\in\{0,1\}$ and that P_T^K only is truthful for models $\mathcal{B}\in\mathbb{B}^s_{P_T^K}$. So $\mathbb{R}^T$ is not the only model for which P^K has undesirable performance. □

In the same way as P^K also $\tilde{P}^K$ only takes simplicity into account, not corroboration. As a result $\tilde{P}^K$ lacks desirable properties. It can be shown that $\tilde{P}^K$ is not (bilaterally) monotone, not linear, not truthful, and not even weakly prudential.

Remark. In section 3.3.4 we define an alternative procedure $\tilde{P}^*$ for modelling by means of translation invariant models. In our exposition in section 3.3 we concentrate on modelling by means of shift invariant models. First we describe a procedure P^0 which takes corroboration into account and which has better properties than P^K. Next we refine P^0 to a procedure P^* which has all the properties defined in section 3.1 on every system $\mathcal{B}\in\mathbb{B}_T$, $T\in\mathbb{N}$, while P_T^* coincides with P_T^K on those systems for which P^K is a satisfactory procedure, i.e., on $\mathcal{B}\in\mathbb{B}_T$ with $c(\mathcal{B})\leq\text{ENT}(T/2)-1$. We refer to theorem 3–28 and proposition 3–30. □

3.3. A procedure with optimal properties

3.3.1. Remarkability

The main reason why the partial realization procedure P^K lacks desirable properties is the fact that it pays attention only to simplicity, not to corroboration. In section 3.3.3 we describe a procedure P^* which has all the desirable properties defined in section 3.1. This procedure P^* consists of a slight refinement of a procedure P^0 which we describe in this section. P^0 consists of a refinement of P^K by taking corroboration into account, i.e., by only accepting "remarkable" laws. P^* in addition takes into account a concept of compatibility, defined in section 3.3.2.

The corroboration principle amounts to accepting laws which are satisfied by the data only provided there is some *evidence* for them. If for an observation an unfalsified law is of a type which also generically will be unfalsified for an observation on a phenomenon which obeys no law at all, i.e., generically on $\mathbb{R}^T$, then one has reason not to accept this law. We illustrate this idea by reconsidering example 4 of section 3.2.1.

Example 4. T=6, w=(1,2,0,1,1,2). The incomplete Hankel array is given by

$$H_T(w)=\begin{bmatrix} 1 & 2 & 0 & 1 & 1 & 2 \\ 2 & 0 & 1 & 1 & 2 & \\ 0 & 1 & 1 & 2 & & \\ 1 & 1 & 2 & & & \\ 1 & 2 & & & & \\ 2 & & & & & \end{bmatrix} \begin{matrix} r(1) \\ r(2) \\ r(3) \\ r(4) \\ r(5) \\ r(6) \end{matrix} .$$

Here $r(4)=\frac{7}{5}r(3)+\frac{3}{5}r(2)-\frac{1}{5}r(1)$, $\text{rank}(H_T(w))=3$, $P_T^K w=\mathcal{B}_T(\sigma^3-\frac{7}{5}\sigma^2-\frac{3}{5}\sigma+\frac{1}{5})$. Note that it is not at all remarkable that $\text{rank}(H_T(w))=3$, as this generically holds true on $\mathbb{R}^T$ for T=6. So one has reason not to accept this third order law, as it is of a type which generically holds true on $\mathbb{R}^6$. It is not remarkable that w satisfies a law of this type. □

Nonetheless, in $H_T(w)$ there is a remarkable dependence, as $r(5)$ is linearly dependent on $r(1)$ and this generically does not hold true on $\mathbb{R}^T$ for T=6. One has reason to accept this remarkable law, i.e., to accept the law $w(t+4)=w(t)$, $t\in\{1,2\}$, and to assign to w the model $\mathcal{B}_6(\sigma^4-1)$.

Notation. In order to define remarkability, let $w\in\mathbb{R}^T$ and let $r(1),\ldots,r(T)$

denote the rows of $H_T(w)$. Suppose that $r(d+1)$ is linearly dependent on $r(i_1+1),\ldots,r(i_{c-1}+1)$, where $0\leq i_1<i_2<\ldots<i_{c-1}<d$, say $r(d+1)=\Sigma_{k=1}^{c-1}a_k r(i_k+1)$. Let $R:=\sigma^d-\Sigma_{k=1}^{c-1}a_k\sigma^{i_k}$, then $w\in\mathcal{B}(R)$. The crucial question is in which case it is remarkable to find a law such as R. We define this to be remarkable if and only if the number of elements in $r(d+1)$ is strictly larger than the number of explaining rows, i.e., if and only if $T-d>c-1$, i.e., $c+d\leq T$. In case $r(d+1)=0$ we define $c:=1$ and we define σ^d to be remarkable for $d\leq T-1$. □

Remark. The class of remarkable laws consists exactly of the class of laws which is generically not satisfied on $\mathbb{R}^T$. This follows from theorem 3–20 in the sequel and corresponds to an intuitive notion of remarkability. □

We now define a procedure P^0 which assigns to data the most simple unfalsified models for which it is remarkable that they are unfalsified.

Definition 3–18 *(i)* For $R=\Sigma_{k=0}^{n}a_k s^k\in\mathbb{R}[s]$, $\bar{d}(R):=\max\{k;\ a_k\neq 0\}$, $\bar{d}(0):=-\infty$, and $c(R):=\#\{k;a_k\neq 0\}$;

(ii) R is called *remarkable* (for T) if $R\neq 0$ and $c(R)+\bar{d}(R)\leq T$;

(iii) $\mathbb{B}_T^*:=\{\mathcal{B}_T(R);\ c(R)+\bar{d}(R)\leq T\}$.

We call $\mathcal{B}_T(R)$ remarkable if R is remarkable for T. So $\mathbb{B}_T^*$ consists of the class of remarkable models together with $\mathbb{R}^T$.

Definition 3–19 The procedure P^0 assigns to data the *least complex unfalsified remarkable* models if these exist, else it assigns $\mathbb{R}^T$. So $P_T^0:\mathbb{R}^T\to 2^{\mathbb{B}_T^*}$ is defined by $\{\mathcal{B}\in P_T^0 w\}:\Leftrightarrow\{w\in\mathcal{B}\in\mathbb{B}_T^*$ and $c(\mathcal{B})=\min\{c(\mathcal{B}');\ w\in\mathcal{B}'\in\mathbb{B}_T^*\}\}$.

Interpretation. Comparing the definitions of P_T^0 and P_T^K, we see that P_T^K is refined to least complex unfalsified modelling in $\mathbb{B}_T^*$ instead of $\mathbb{B}_T$, i.e., assigned laws have to be remarkable. Hence corroboration is taken into account. Note that if w satisfies no remarkable law then $P_T^0 w=\mathbb{R}^T$, so if no remarkable law is satisfied then no law is accepted. □

Remark. To determine $P_T^0 w$ one can investigate $H_T(w)$ and determine the first row, say $d+1$, which is linearly dependent on $c-1$ foregoing ones such that $c+d\leq T$. If no such row exists, then $P_T^0 w=\mathbb{R}^T$. □

We illustrate P^0 by reconsidering some of the examples of section 3.2.1.

Example 2. $w=(1,1,2,3,5,8)$, $P_6^0 w=\mathcal{B}_6(\sigma^2-\sigma-1)$. □
Example 3. $w=(1,1,1,0,1,1)$, $P_6^0 w=\{\mathcal{B}_6(\sigma^4-1),\ \mathcal{B}_6(\sigma^4-\sigma)\}$. □
Example 4. $w=(1,2,0,1,1,2)$, $P_6^0 w=\mathcal{B}_6(\sigma^4-1)$. □
Example 5. $w=(1,2,0,4,4,4)$, $P_6^0 w=\mathcal{B}_6(\sigma^3-\sigma-2)$. □
Example 6. $w=(1,0,2,0,1,1)$, $P_6^0 w=\mathbb{R}^6$. □

Remark. Examples 3 and 5 play a role in refining P^0 to the procedure P^* defined in section 3.3.3. This has to do with the compatibility of remarkable laws. This concept is defined in the next section. □

An important question is whether laws which are identified by P_T^0 also generically are true laws, given that the data stem from a system in $\mathbb{B}_T$, i.e., the question of truthfulness of P^0. An essential property of P^0 is the following.

Theorem 3-20 Let $\mathcal{B}_0\in\mathbb{B}_T$. Then generically in $w\in\mathcal{B}_0$, $P_T^0 w=\{\mathcal{B}\in\mathbb{B}_T^*;\ \mathcal{B}_0\subset\mathcal{B},$ $c(\mathcal{B})$ minimal$\}$.

Proof. See the appendix.

So generically on $\mathcal{B}_0$ P_T^0 assigns the least complex remarkable models containing $\mathcal{B}_0$ and, in particular, generically the identified laws are true laws. This gives a strong motivation for our definition of remarkability.

Corollary 3-21 If $\mathcal{B}_0\in\mathbb{B}_T^*$, then generically in $w\in\mathcal{B}_0$ $P_T^0 w=\mathcal{B}_0$.

Proof. See the appendix.

Theorem 3-22 *(i)* P^0 is exact;
(ii) P^0 is truthful;

(*iii*) P^0 is strongly prudential, since $\text{im}(P_T^0)=\mathbb{B}^w_{P_T^0}=\mathbb{B}^s_{P_T^0}=\mathbb{B}_T^*$, for all $T\in\mathbb{N}$;

(*iv*) P^0 is not monotone, not shift invariant, and not linear; the inclusion conditions for monotonicity, shift invariance and linearity are satisfied if the action of P_T^0 is restricted to models in the set $\mathbb{B}_T^*$.

Proof. See the appendix.

Remark. We do not go into details of exact characterization of those models for which P^0 is monotone, shift invariant, or linear. This is because (*iv*) implies that P^0 lacks some desirable properties. In section 3.3.3 we slightly modify P^0 to get a procedure P^* with desirable properties everywhere. □

3.3.2. Compatibility

The first idea to refine P^K, i.e., to accept only remarkable laws, resulted in a procedure P^0 which has many desirable properties but which also lacks some of them. In this section we formulate a second idea, i.e., compatibility, in order to refine the procedure P^0. The resulting procedure P^* is defined in the next section. There it is shown that P^* has all the properties which were introduced in section 3.1.

We first give three examples to indicate in which direction P^0 could be refined. We use a concept of compatibility, defined as follows.

Notation. By Λ we denote an arbitrary index set. □

Definition 3-23 Let $R_\lambda\in\mathbb{R}[s]$, $\lambda\in\Lambda$, and $w\in\mathbb{R}^T$. Then $\{R_\lambda,\ \lambda\in\Lambda\}$ and w are called *compatible* if there exists a $\mathcal{B}\in\mathbb{B}_T$ such that $w\in\mathcal{B}\subset\bigcap\{\mathcal{B}_T(R_\lambda);\ \lambda\in\Lambda\}$.

Interpretation. A class of laws and an observation are called compatible if there exists a linear shift invariant system for which all the laws are valid and which is unfalsified by the observation. □

Example 3 (cf. sections 3.2.1 and 3.3.1). $T=6$, $w=(1,1,1,0,1,1)$, $P_6^0w=\{\mathcal{B}_6(\sigma^4-1),\ \mathcal{B}_6(\sigma^4-\sigma)\}$. As the model class is $\mathbb{B}_6$, the phenomenon is assumed to be linear and shift invariant. Maximizing simplicity under the restriction of corroboration leads to P_6^0w and identification of two laws. However, $\{\sigma^4-1, \sigma^4-\sigma\}$ and w are not compatible. Note that $\mathcal{B}_6(\sigma^4-1)\cap\mathcal{B}_6(\sigma^4-\sigma)\notin\mathbb{B}_6$ while due to proposition II.3–36(*ii*) the largest model in $\mathbb{B}_6$ contained in $\mathcal{B}_6(\sigma^4-1)\cap\mathcal{B}_6(\sigma^4-\sigma)$ is given by $\mathcal{B}_6(\sigma-1)$, and $w\notin\mathcal{B}_6(\sigma-1)$. So given that the phenomenon belongs to $\mathbb{B}_6$ and given the data w, the phenomenon cannot satisfy both laws σ^4-1 and $\sigma^4-\sigma$, and at least one of these two laws has to be false. It seems reasonable to reject at least one of them, even to reject both. (In case $w(7)$ could become available it could be sensible to store $\mathcal{B}_6(\sigma^4-1)$ and $\mathcal{B}_6(\sigma^4-\sigma)$ and to decide on the basis of $w(7)$.) □

Example 5 (cf. sections 3.2.1 and 3.3.1). $T=6$, $w=(1,2,0,4,4,4)$, $P_6^0w=\mathcal{B}_6(\sigma^3-\sigma-2)$. Note that in this case there is another unfalsified remarkable law, as $w\in\mathcal{B}_6(\sigma^4-\sigma^3)$, which is not accepted by P^0. Now $\{\sigma^3-\sigma-2, \sigma^4-\sigma^3\}$ and w are not compatible, as $\mathcal{B}_6(\sigma^3-\sigma-2)\cap\mathcal{B}_6(\sigma^4-\sigma^3)=\{\alpha(1,2,0,4,4,4);\ \alpha\in\mathbb{R}\}\notin\mathbb{B}_6$ while the largest model in $\mathbb{B}_6$ contained in this intersection is $\mathcal{B}_6(1)=\{0\}$, as $\mathrm{GCD}(\sigma^3-\sigma-2, \sigma^4-\sigma^3)=1$, and $w\neq0$. So given that the phenomenon belongs to $\mathbb{B}_6$ and given the data w, at least one of the remarkable laws $\sigma^3-\sigma-2$ and $\sigma^4-\sigma^3$ was observed just by bad luck, because for this phenomenon not both laws can be valid. It seems reasonable to accept no law at all. □

Taking compatibility into account need not always lead to rejection of unfalsified remarkable laws. It also can give reason to accept stronger laws, as the next example illustrates.

Example 8 (cf. the example given in the proof of theorem 3–22(*iv*) in the appendix). Assume that the data stem from a phenomenon with law $R=\sigma^{10}+\sigma^9+2\sigma^8+\sigma^7+\sigma^6+2\sigma^5+\sigma^4+\sigma^3+2\sigma^2+\sigma+1$ and that $T=20$, e.g., $w=(0,0,0,0,0,0,0,0,0,1,-1,-1,2,0,-4,4,4,4,-12,4,20)$. Note $\mathrm{rank}(H_{20}(w))=10$ and R is the only law of degree 10 which is unfalsified by w, so $P_{20}^Kw=\mathcal{B}_{20}(R)$. However, $c(R)+\bar{d}(R)=21$ so R is not remarkable for $T=20$. Let $R_1:=(\sigma-1)R$, $R_2:=(\sigma-\frac{1}{2})R$, then $c(R_1)+\bar{d}(R_1)=19$, $c(R_2)+\bar{d}(R_2)=20$, and $\bar{d}(R_1)=\bar{d}(R_2)=11$. So $P_{20}^0w=\{\mathcal{B}_{20}(R_1), \mathcal{B}_{20}(R_2)\}$ and this also holds true generically on $\mathcal{B}_{20}(R)$, see theorem 3–20.

Now note that $\{R_1,R_2\}$ and w are compatible, as $w\in\mathcal{B}_{20}(R)=\mathcal{B}_{20}(R_1)\cap\mathcal{B}_{20}(R_2)$. Although R itself is not remarkable, R_1 and R_2 are remarkable and moreover they are compatible with w. So it seems reasonable to assign $\mathcal{B}_{20}(R)$ to w, not as it would be remarkable in itself, but because one has evidence for it from the remarkable and compatible laws R_1 and R_2. □

The foregoing examples indicate the way P^0 could be refined. In the next section we define the resulting procedure P^* and analyse its properties.

3.3.3. Procedure and properties

Notation. The class of remarkable laws for $w \in \mathbb{R}^T$ is defined by $L(w) := \{R \neq 0;\ \bar{d}(R) \leq T-1 \text{ and } w \in \mathcal{B}_T(R) \in \mathbb{B}_T^*\} = \{R;\ w \in \mathcal{B}_T(R) \in \mathbb{B}_T^*,\ \mathcal{B}_T(R) \neq \mathbb{R}^T\}$. Moreover, for $w \in \mathbb{R}^T$ let $R(w) := \mathrm{GCD}\{R;\ R \in L(w)\}$, i.e., the least stringent law implied by all the remarkable laws for w. Define $\mathrm{GCD}\{\varnothing\} :\doteq 0$. □

Remark. Because of the restriction $\bar{d}(R) \leq T-1$, every law in $L(w)$ indeed is remarkable. In the sequel we throughout assume that in $\mathcal{B}_T(R)$ $\bar{d}(R) \leq T-1$, without always explicitly mentioning this. □

Remark. If for $\mathcal{B} \in \mathbb{B}_T$ both the laws R_1 and R_2 are valid, i.e., if $\mathcal{B} \subset \mathcal{B}_T(R_1) \cap \mathcal{B}_T(R_2)$, then $\mathcal{B} \subset \mathcal{B}(G)$ where G is the greatest common divisor of R_1 and R_2 in $\mathbb{R}[s]$, cf. proposition II.3–36(*ii*). □

For given $w \in \mathbb{R}^T$ there are two possible situations for the class of remarkable laws, i.e., they are *either compatible or not*.

(*i*) First suppose that the class of all remarkable laws for w, i.e., $L(w)$, and w are compatible. Then there exists $\mathcal{B} \in \mathbb{B}_T$ such that $w \in \mathcal{B} \subset \bigcap\{\mathcal{B}_T(R);\ R \in L(w)\}$. Note that according to proposition II.3–36(*ii*) $\mathcal{B}_T(R(w))$ is the largest model in $\mathbb{B}_T$ which satisfies this condition, i.e., $\mathcal{B}_T(R(w))$ is the largest model unfalsified by w for which all remarkable laws are valid, as $\mathcal{B}_T(R(w)) \subset \mathcal{B}_T(R)$ for all $R \in L(w)$. Models of less complexity either are falsified by the data or are not supported by means of unfalsified remarkable laws. So we would like to accept all remarkable laws and to assign $\mathcal{B}_T(R(w))$ to w.

(*ii*) Next suppose that $L(w)$ and w are not compatible, i.e., there is no $\mathcal{B} \in \mathbb{B}_T$ which is unfalsified by w and for which all remarkable laws for w are valid. So given that the phenomenon belongs to $\mathbb{B}_T$, at least one of the remarkable laws has to be false. As we have no information concerning which law is false it seems reasonable to accept no law at all.

Remark. Note that (*i*) is equivalent to $w \in \mathcal{B}_T(R(w))$ and (*ii*) to $w \notin \mathcal{B}_T(R(w))$. □

The foregoing motivates the following procedure P^* for exact modelling of a finite time series.

> **Definition 3-24** If the class of all remarkable laws is *compatible* with the data then the procedure P^* accepts exactly these laws, else it accepts no law. So $P_T^* : \mathbb{R}^T \to \mathbb{B}_T$ is defined by

$$P_T^* w := \begin{cases} \mathcal{B}_T(R(w)) & \text{if } w \in \mathcal{B}_T(R(w)),\ R(w) := \mathrm{GCD}\{R;\ R \in L(w)\}; \\ \\ \mathbb{R}^T & \text{if } w \notin \mathcal{B}_T(R(w)). \end{cases}$$

Interpretation. In case the class of all remarkable laws is compatible with the data, P^* assigns the largest model for which these laws are valid, i.e., it accepts all these remarkable laws and no other ones, as for other laws there is no evidence. □

Remark. Note that P^* always assigns a unique model to data. □

To illustrate P^* we briefly return to some examples, cf. sections 3.2.1, 3.3.1 and 3.3.2.

Example 2. $T=6$, $w=(1,1,2,3,5,8)$, $R := \sigma^2 - \sigma - 1$, $L(w) = \{R, \sigma R, (\sigma-1)R, (\sigma+1)R\}$, $R(w) = R$, $w \in \mathcal{B}_6(R)$; so $P_6^* w = \mathcal{B}_6(R)$. □

Example 3. $T=6$, $w=(1,1,1,0,1,1)$, $L(w) = \{\sigma^4 - 1, \sigma^4 - \sigma\}$, $R(w) = \sigma - 1$, $w \notin \mathcal{B}_6(\sigma-1)$; hence $P_6^* w = \mathbb{R}^6$. □

Example 4. $T=6$, $w=(1,2,0,1,1,2)$, $L(w) = \{\sigma^4 - 1\}$, $w \in \mathcal{B}_6(R(w))$; so $P_6^* w = \mathcal{B}_6(\sigma^4 - 1)$. □

Example 5. $T=6$, $w=(1,2,0,4,4,4)$, $L(w) = \{\sigma^3 - \sigma - 2,\ \sigma^4 - \sigma^3\}$, $R(w) = 1$, $w \notin \mathcal{B}(1) = \{0\}$; hence $P_6^* w = \mathbb{R}^6$. □

Example 8. Take w, R, R_1, R_2 as stated in section 3.3.2 for this example. Then $L(w) = \{R_1, \sigma R_1, R_2\}$, $R(w) = R$, $w \in \mathcal{B}_{20}(R)$; hence $P_{20}^* w = \mathcal{B}_{20}(R)$. □

The following proposition states that *generically all remarkable laws are true laws*. That is an important motivation for our definition of remarkability.

Proposition 3-25 For every $\mathcal{B}_0 \in \mathbb{B}_T$, generically in $w \in \mathcal{B}_0$ $\mathcal{B}_0 \subset \mathcal{B}_T(R(w))$.

Proof. See the appendix.

Proofs of the following corollaries also are contained in the appendix.

Corollary 3-26 Let $\mathcal{B}_0 \in \mathbb{B}_T$. Then generically in $w \in \mathcal{B}_0$ there holds $\{R \in L(w)\} \Rightarrow \{\mathcal{B}_0 \subset \mathcal{B}_T(R)$ and $P_T^* w \subset \mathcal{B}_T(R)\}$.

Corollary 3-27 Let $\mathcal{B}_0 \in \mathbb{B}_T$. Then generically in $w \in \mathcal{B}_0$ $P_T^* w = \mathcal{B}_T(R)$, where $R := \mathrm{GCD}\{\tilde{R} \neq 0;\ \bar{d}(\tilde{R}) \leq T-1$ and $\mathcal{B}_0 \subset \mathcal{B}_T(\tilde{R}) \in \mathbb{B}_T^*\}$.

Interpretation. Corollary 3-26 states that generically all remarkable laws hold true for the phenomenon and also that true remarkable laws are generically identified by P^*. This indicates a connection between remarkability and corroboration. Corollary 3-27 describes the generic way in which P^* models data. □

The properties of P^* are stated in our main theorem on exact modelling of a finite time series.

Theorem 3-28 (*i*) P^* is exact;
(*ii*) P^* is monotone, shift invariant, and linear;
(*iii*) P^* is truthful;
(*iv*) P^* is strongly prudential as $\mathrm{im}(P_T^*) = \mathbb{B}^w_{P_T^*} = \mathbb{B}^s_{P_T^*} = \{\mathcal{B}_T(R);\ R = \mathrm{GCD}\{R_\lambda;\ \lambda \in \Lambda\}$ for some $\{R_\lambda;\ \lambda \in \Lambda\}$, with for all $\lambda \in \Lambda$ $\bar{d}(R_\lambda) \leq T-1$, $\mathcal{B}_T(R_\lambda) \in \mathbb{B}_T^*\}$.

Proof. See the appendix.

From theorem 3-20 and corollary 3-27 we immediately get the following.

Proposition 3-29 Let $\mathcal{B}_0 \in \mathbb{B}_T$. Then generically in $w \in \mathcal{B}_0$, $\{\mathcal{B} \in P_T^0 w\} \Rightarrow \{P_T^* w \subset \mathcal{B}\}$.

The following proposition describes the relationship between P^K, P^0 and P^*. It also indicates the sense in which P^* is a procedure with desirable properties everywhere and which coincides with the partial realization procedure on those systems for which P^K is a reasonable procedure. Moreover it describes the sense in which the differences between P_T^K, P_T^0, and P_T^* disappear if $T \to \infty$. Proof of the proposition is immediate from proposition 3–16(*iv*) and theorems 3–22(*iii*) and 3–28(*iv*).

Proposition 3-30 Let $\mathbb{B}_T^K \subset \mathbb{B}_T^*$ be defined by $\mathbb{B}_T^K := \{\mathcal{B} \in \mathbb{B}_T;\ c(\mathcal{B}) \leq \mathrm{ENT}(T/2) - 1\}$.

(*i*) $\mathbb{B}_T^K = \mathbb{B}^s_{P_T^K} \subset \mathbb{B}^s_{P_T^0} \subset \mathbb{B}^s_{P_T^*}$ for T odd, and $\mathbb{B}_T^K = \mathbb{B}^s_{P_T^K} \setminus \{\mathcal{B} \in \mathbb{B}_T;\ c(\mathcal{B}) = T/2\} \subset \mathbb{B}^s_{P_T^0} \subset \mathbb{B}^s_{P_T^*}$ for T even;

(*ii*) for $\mathcal{B}_0 \in \mathbb{B}_T^K$, generically in $w \in \mathcal{B}_0$, $P_T^K w = P_T^0 w = P_T^* w$; for $\mathcal{B}_0 \in \mathbb{B}_T^*$, generically in $w \in \mathcal{B}_0$, $P_T^0 w = P_T^* w$;

(*iii*) for given $R \in \mathbb{R}[s]$ $P_T^K w = P_T^0 w = P_T^* w = \mathcal{B}_T(R)$ generically on $\mathcal{B}_T(R)$ for T sufficiently large, i.e., $T \geq 2\bar{d}(R) + 2$.

Remark. The problem of exact modelling of an infinite time series in $\mathbb{R}^{\mathbb{N}}$ by means of model $\mathcal{B}_\infty(R) := \{w \in \mathbb{R}^{\mathbb{N}};\ [R(\sigma)w](t) = 0 \text{ for all } t \in \mathbb{N}\}$ could be studied in an analogous way. It turns out that in this case the procedures P_∞^u, P_∞^K, P_∞^0 and P_∞^* are all equivalent. In particular all unfalsified laws are remarkable and compatible, and the undominated unfalsified model corresponds to accepting all these remarkable laws. □

3.3.4. Translation invariant models

In this section we describe a procedure $\tilde{P}^*$ for exact modelling of a finite time series by means of a linear translation invariant model. So in this case the model class is $\tilde{\mathbb{B}}_T$ as defined in section II.3.4.1. We define $\tilde{P}^*$ in a way analogous to the definition of P^* and with the same motivation.

Notation. Let $\tilde{\mathbb{B}}_T^*:=\{\mathcal{B}_T(R)\in\tilde{\mathbb{B}}_T;\ c(R)+\bar{d}(R)\leq T\}$, and for $w\in\mathbb{R}^T$ let $\tilde{L}(w):=\{R\neq 0;\ \bar{d}(R)\leq T-1$ and $w\in\mathcal{B}_T(R)\in\tilde{\mathbb{B}}_T^*\}$, $\tilde{R}(w):=\mathrm{GCD}\{R;\ R\in\tilde{L}(w)\}$. As $q=1$ it follows from proposition II.3–33(*ii*) that $\mathcal{B}_T(R)\in\tilde{\mathbb{B}}_T$ if and only if $R=0$, $\bar{d}(R)\geq T$, or $R=\Sigma_{k=0}^{T-1}a_k\sigma^k$ with $a_0\neq 0$. From this we conclude that $\mathcal{B}_T(\tilde{R}(w))\in\tilde{\mathbb{B}}_T$. □

Definition 3–31 The procedure $\tilde{P}^*$ for exact modelling by means of translation invariant models is defined by $\tilde{P}_T^*:\mathbb{R}^T\to\tilde{\mathbb{B}}_T$ with

$$\tilde{P}_T^*w:=\begin{cases}\mathcal{B}_T(\tilde{R}(w)) & \text{if } w\in\mathcal{B}_T(\tilde{R}(w));\\ \\ \mathbb{R}^T & \text{if } w\notin\mathcal{B}_T(\tilde{R}(w)).\end{cases}$$

Interpretation. The procedure $\tilde{P}^*$ accepts all translation invariant remarkable laws in case these are compatible with the data, else it accepts no law at all. □

Proposition 3–32 Let $\mathcal{B}_0\in\mathbb{B}_T$. Then generically in $w\in\mathcal{B}_0$, $\tilde{P}_T^*w=\mathcal{B}_T(R)$ where $R:=\mathrm{GCD}\{\tilde{R}\neq 0;\ \bar{d}(\tilde{R})\leq T-1$ and $\mathcal{B}_0\subset\mathcal{B}_T(\tilde{R})\in\tilde{\mathbb{B}}_T^*\}$.

Proof. See the appendix.

Remark. The properties of $\tilde{P}^*$ are stated in the next theorem, which is proved by using proposition 3–32 in a way which is completely analogous to the proof of theorem 3–28 by means of corollary 3–27. □

Theorem 3–33 $\tilde{P}^*$ is exact, bilaterally monotone, linear, truthful, and strongly prudential with $\mathrm{im}(\tilde{P}_T^*)=\mathbb{B}^{w}_{\tilde{P}_T^*}=\mathbb{B}^{s}_{\tilde{P}_T^*}=\{\mathcal{B}_T(R);\ R=\mathrm{GCD}\{R_\lambda;\ \lambda\in\Lambda\}$ for some $\{R_\lambda;\ \lambda\in\Lambda\}$, with for all $\lambda\in\Lambda$ $\bar{d}(R_\lambda)\leq T-1$ and $\mathcal{B}_T(R_\lambda)\in\tilde{\mathbb{B}}_T^*\}$.

Proof. See the appendix.

Remark. In section V.5.2.2 we desribe a less refined exact modelling procedure for the case of a multivariable time series. This procedure has the properties stated in theorem 3–33, cf. theorem V.5–5. □

We illustrate $\tilde{P}^*$ by some examples, cf. sections 3.2.1 and 3.3.3.

Example 2. $T=6$, $w=(1,1,2,3,5,8)$, $R:=\sigma^2-\sigma-1$, $\tilde{L}(w)=\{R,(\sigma-1)R,(\sigma+1)R\}$, $\tilde{R}(w)=R$, $w\in\mathcal{B}_6(R)$; so $\tilde{P}_6^*w=\mathcal{B}_6(R)$. □

Example 5. $T=6$, $w=(1,2,0,4,4,4)$; $\tilde{L}(w)=\{\sigma^3-\sigma-2\}$, $w\in\mathcal{B}_6(\tilde{R}(w))$; so $\tilde{P}_6^*w=\mathcal{B}_6(\sigma^3-\sigma-2)$. □

Example 9. $T=6$, $w=(1,2,2,3,2,2)$; then $P_6^Kw=\mathcal{B}_6(\sigma^3-\frac{4}{3}\sigma^2-\frac{4}{3}\sigma+\frac{7}{3})$, $P_6^0w=P_6^*w=\mathcal{B}_6(\sigma^4-\sigma)$, $\tilde{P}_6^*w=\mathbb{R}^6$. □

Remark. Note that in example 5 the prior assumption that the phenomenon is translation invariant leads to a model, $\mathcal{B}_6(\sigma^3-\sigma-2)$, which is of less complexity than in case we only assume the phenomenon to be shift invariant, which leads to the model $P_6^*w=\mathbb{R}^6$, see section 3.3.3. This, however, is a situation which generically does not occur, as for all $w\in\mathbb{R}^T$ $\tilde{L}(w)\subset L(w)$, so $\mathcal{B}_T(\tilde{R}(w))\supset\mathcal{B}_T(R(w))$. As generically $P_T^*w=\mathcal{B}_T(R(w))$, cf. proposition 3–25, and as also generically $\tilde{P}_T^*w=\mathcal{B}_T(\tilde{R}(w))$, there holds that generically on every $\mathcal{B}_0\in\mathbb{B}_T$ $\tilde{P}_T^*w\supset P_T^*w$. The next proposition states a stronger result. □

Proposition 3–34 (*i*) If $\mathcal{B}_0\in\tilde{\mathbb{B}}_T$, then generically on $\mathcal{B}_0$ $\tilde{P}_T^*w=P_T^*w$;
(*ii*) if $\mathcal{B}_0\in\mathbb{B}_T$, $\mathcal{B}_0\notin\tilde{\mathbb{B}}_T$, then generically on $\mathcal{B}_0$ $\tilde{P}_T^*w=\mathbb{R}^T$.

Proof. See the appendix.

Interpretation. If a phenomenon is not translation invariant, then generically $\tilde{P}^*$ identifies no laws. □

3.3.5. The partial realization problem for SISO-systems

The partial realization procedure P^K defined in section 3.2.1 solves the minimal partial realization problem as described in section 1.4 for SISO-systems, i.e., systems with a single input and a single output. We refer to Kalman [34].

The procedure P^* gives an alternative solution for this partial realization problem.

Notation. Let $g=(g_t;\ t\in[0,T])\in\mathbb{R}^{T+1}$ denote a partial impulse response and let $\tilde{g}:=g|_{[1,T]}$. Using the notation of sections 1.3 and 1.4 we recall that g is said to have a realization of dimension n if there exist $g^e\in\mathbb{R}^{\mathbb{Z}_+}$ and $(A,B,C)\in\mathbb{R}^{n\times n}\times\mathbb{R}^{n\times 1}\times\mathbb{R}^{1\times n}$ such that $g^e|_{[0,T]}=g$ and $S(g^e)=S(A,B,C,g_0)$. □

Lemma 3-35 The partial impulse response $g\in\mathbb{R}^{T+1}$ has a realization of dimension n if and only if there exists $R\in\mathbb{R}[s]$ with $\bar{d}(R)=n$ and $\tilde{g}\in\mathcal{B}_T(R)$.

Proof. See the appendix.

Remark. In the proof of lemma 3-35 we also give a construction of a realization of dimension n corresponding to $R\in\mathbb{R}[s]$ with $\bar{d}(R)=n$ and $\tilde{g}\in\mathcal{B}_T(R)$. □

For given partial impulse response $g\in\mathbb{R}^{T+1}$ we could use the procedure P^* to identify a model $\mathcal{B}_T(R^*):=P_T^*\tilde{g}$ of complexity $n^*:=c(\mathcal{B}(R^*))=\bar{d}(R^*)$. This complexity is minimal under the restriction that identified laws for $\tilde{g}$ should be corroborated by the data in the sense described in sections 3.3.1 and 3.3.2. The realization corresponding to R^* is not necessarily minimal, i.e., n^* need not be the smallest possible dimension of realizations of g. However, it is the least complex exact representation of the partial impulse response which is corroborated. There is no evidence for less complex realizations.

Remark. As the procedure P^* is monotone it generically identifies models which become less complex in case the impulse response is known over a larger time interval, this in contrast with the procedure P^K. Note that the procedure P^* prevents identification of a law which is of a type which also generically is exactly satisfied for a partial impulse response of an infinite dimensional system. Identified laws should be remarkable. □

4. Conclusion

In this chapter we considered exact modelling. We especially investigated the problem of modelling a finite time series in one variable by means of an autoregressive equation. The partial realization problem for SISO-systems forms a special case. We formulated desirable properties of modelling procedures and specified general principles of simplicity and corroboration. We described two procedures P^* and $\tilde{P}^*$ with many desirable properties.

Remark. An interesting question is whether there is a sense in which e.g. P^* is optimal. For example, one could restrict attention to procedures P which are monotone, shift invariant, linear, truthful, and prudential, and investigate if there exists a procedure in this class with a maximal set $\mathbb{B}_P^s$ of strongly corroborable models. In a sense such a procedure has maximal discriminatory power on $\mathbb{B}_T$. □

Remark. If $q>1$ then again we can define a partial realization procedure. For constructing procedures which take corroboration into account an important question is which laws are remarkable, and connected with this is the question which variables are free and which are not. For $q=1$, the variable is declared to be free if no remarkable law holds true for it. Procedures could be defined which take remarkability and compatibility into account. Analysis of these procedures could go along the same line as presented before. In section V.5.2.2 we describe a less refined exact modelling procedure for the case of multivariable time series. □

The case of approximate modelling is of more practical interest. It raises the question of defining appropriate model utilities. By increasing the complexity of a model one generally will be able to increase the fit. This leads to the interesting question of which increase in fit is large enough to make an increase in complexity acceptable. A crucial topic here is the definition of appropriate measures of complexity and fit. These measures should satisfy two requirements. They should have a sound interpretation and they also should be implementable by means of numerical algorithms.

In chapter IV we consider model approximation. In this case a maximal tolerated complexity is given. The aim is to approximate a given, complex model by one of tolerable complexity. The reduced model should be an optimal approximation of the complex model. Here the quality of approximation is expressed by means of a distance measure for models.

In chapter V we give procedures and algorithms for approximate modelling of time series. These procedures are based on model utilities which express the desires of low complexity and of good fit.

CHAPTER IV

MODEL APPROXIMATION

1. Introduction

The problem of model approximation can be described as follows. Let $\mathbb{M}$ be a given class of models, c a measure of *complexity* of models in $\mathbb{M}$ and d a measure of *distance* between models in $\mathbb{M}$. Given $M \in \mathbb{M}$, the problem of model approximation consists of finding models $\hat{M}$ of low complexity which have small distance from M. In the literature one often considers the special case where an upper bound on the complexity is given. The problem of model approximation then amounts to finding models $\hat{M} \in \mathbb{M}$ such that the distance $d(M,\hat{M})$ is minimal under the restriction that the complexity $c(\hat{M})$ of the approximate model does not surpass the complexity bound.

Remark. Hence model approximation is a special case of the general modelling problem described in section II.2.1. The data consists of a model and the misfit is expressed by a measure of distance between models. □

Remark. In practical applications, especially in control, it is sometimes more natural to specify an upper bound on the distance and to minimize complexity under this requirement, i.e., to follow the approach of section II.2.3 instead of that of section II.2.2. In case the complexity space is discrete, which is often the case, this problem can be solved by reducing the complexity until further reduction would lead to models with too large distance from the given model, i.e., by applying the procedure of modelling under a complexity constraint for various upper bounds on the complexity. We hence can restrict attention to this last problem. □

We will consider model approximation for dynamical systems. Hence we have to define a measure of complexity and a measure of distance for dynamical systems.

The chapter is organized as follows.

In section 2 we define a measure of complexity for dynamical systems. The complexity of a system expresses how many trajectories a system allows. We give some characterizations of this complexity. We formulate some possible orderings on the complexity space, which are used partly in this chapter and partly in the next one.

In section 3 we consider a special class of l_2–systems, i.e., those systems which consist of the l_2–trajectories contained in a linear, time invariant, complete system. So if $\mathbb{B}$ denotes the class of linear, time invariant, complete systems as defined in section II.3.1.2, then the model class is $\mathbb{B}_2 := \{H \subset l_2;\ \exists \mathcal{B} \in \mathbb{B} \text{ such that } H = \mathcal{B} \cap l_2\}$. We derive some representation results for systems in $\mathbb{B}_2$.

Section 4 constitutes the main part of this chapter. Using scattering theory we conclude that systems in $\mathbb{B}_2$ have special representations which we call *scattering representations* and which closely resemble forward and backward innovation representations of stochastic processes as described, e.g., in Lindquist and Pavon [46]. Moreover we give an explicit construction of the scattering representations. This construction consists of taking arbitrary forward and backward state space realizations of $H \in \mathbb{B}_2$, determining the positive definite solutions of corresponding Riccati equations, and transforming the parameters of the state space realizations by means of these solutions.

The problem of model approximation is discussed in section 5. We briefly comment on some well-known approximation (or: *reduction*) procedures. We describe a new model reduction problem, using appealing notions of complexity and distance. The complexity of $H \in \mathbb{B}_2$ measures how many trajectories H allows. The distance between H_1 and H_2 in $\mathbb{B}_2$ is defined as the aperture or "gap" between H_1 and H_2. We show that the resulting model reduction problem can explicitly be stated in terms of the parameters of the scattering representations. Finally we give a heuristic model reduction method by "balancing" the state space of scattering representations. This method is illustrated by means of two simple numerical simulations.

2. Complexity of dynamical systems

2.1. Definition and characterization

As before, let $\mathbb{B}$ denote the class of linear, time invariant, complete systems in $(\mathbb{R}^q)^{\mathbb{Z}}$. Roughly stated, we consider a system to be more complex if more time series are compatible with the system, i.e., if the system imposes less restrictions on the behaviour. A simple system is one with a few degrees of freedom. This is in accordance with the simplicity principle stated in section II.2.1. In particular, if $\mathcal{B}_1,\mathcal{B}_2\in\mathbb{B}$ and $\mathcal{B}_1\subset\mathcal{B}_2$, $\mathcal{B}_1\neq\mathcal{B}_2$, then we call $\mathcal{B}_1$ less complex than $\mathcal{B}_2$. More general, we call $\mathcal{B}_1$ less complex than $\mathcal{B}_2$ if it allows "less" time series.

Notation. For $\mathcal{B}\in\mathbb{B}$ let $\mathcal{B}_t:=\mathcal{B}|_{[0,t]}$ denote the space of time series of length $t+1$ which are compatible with the system $\mathcal{B}$, $t\in\mathbb{Z}_+:=\{0,1,2,...\}$. □

The complexity of a system is measured by the magnitude of the set of time series compatible with the system. It is defined as a sequence of numbers $(c_t(\mathcal{B});\ t\in\mathbb{Z}_+)$, where $c_t(\mathcal{B})$ measures the magnitude of $\mathcal{B}_t$.

Definition 2-1 The *complexity* of dynamical systems is defined by $c:\mathbb{B}\rightarrow(\mathbb{R}_+)^{\mathbb{Z}_+}$, $c(\mathcal{B}):=(c_t(\mathcal{B});\ t\in\mathbb{Z}_+)$, where $c_t(\mathcal{B}):=\frac{1}{t+1}\cdot\dim(\mathcal{B}_t)$.

In the next proposition we give some characteristics of this complexity. The statements of the proposition are contained in Willems [73, theorems 6, 8 and 25].

Notation. For $\mathcal{B}\in\mathbb{B}$ let $e^*(\mathcal{B})=(e_t^*(\mathcal{B});\ t\in\mathbb{Z}_+)$ denote the equation structure of any tightest equation representation of $\mathcal{B}$ as defined in section II.3.2.4. Further let $m(\mathcal{B})$ and $n(\mathcal{B})$ denote the number of driving variables and the number of state variables respectively in a minimal realization of $\mathcal{B}$ as defined in section II.3.3.

We will sometimes drop the argument $\mathcal{B}$ if this does not lead to confusion. □

Proposition 2-2 (*i*) $c_t(\mathcal{B})=q-\frac{1}{t+1}\cdot\Sigma_{k=0}^{t}(t+1-k)e_k^*(\mathcal{B})$;
(*ii*) $c_\infty(\mathcal{B}):=\lim_{t\to\infty} c_t(\mathcal{B})=m(\mathcal{B})$; $c'_\infty(\mathcal{B}):=\lim_{t\to\infty} t\{c_t(\mathcal{B})-c_\infty(\mathcal{B})\}=n(\mathcal{B})$;
(*iii*) $m(\mathcal{B})=q-\Sigma_{t=0}^{\infty}e_t^*(\mathcal{B})$; $n(\mathcal{B})=\Sigma_{t=0}^{\infty}te_t^*(\mathcal{B})$.

Interpretation. There is a bijective relationship between the complexity of a system and its tightest equation structure. Moreover, the limit behaviour of the complexity depends on the number of driving variables, or equivalently, the number of inputs or unrestricted variables, and on the number of state variables. □

Remark. This characterization of the complexity enables us to construct numerical algorithms for identification procedures involving this complexity. For formulating explicitly the objective of simplicity it is necessary to define an ordering of complexities. In the next section we consider some possible orderings which play a role in the sequel. □

2.2. Orderings

A *natural* ordering of complexities is the partial ordering $\overset{(p)}{\geq}$ defined by $\{c^{(1)}\overset{(p)}{\geq}c^{(2)}\}:\Leftrightarrow\{c_t^{(1)}\geq c_t^{(2)}$ for all $t\in\mathbb{Z}_+\}$. If $\mathcal{B}_1,\mathcal{B}_2\in\mathbb{B}$, then according to proposition 2–2 $\mathcal{B}_1$ is more complex than $\mathcal{B}_2$ in this ordering if and only if for all $t\in\mathbb{Z}_+$ $\Sigma_{k=0}^{t}(t+1-k)e_k^*(\mathcal{B}_1)\leq\Sigma_{k=0}^{t}(t+1-k)e_k^*(\mathcal{B}_2)$. So systems are complex in this sense if their behaviour is restricted by few laws which moreover are of high order. A simple system is one which satisfies many laws of small order.

Remark. Note that the complexity is related to the system considered as a set of trajectories and not to the number of parameters needed to represent the system. □

For the construction of modelling procedures it is highly desirable to have a total ordering of complexities. We now define two refinements of $\overset{(p)}{\geq}$ which we will use respectively in this chapter and in the next one.

Definition 2-3 Let $\mathcal{B}_1,\mathcal{B}_2\in\mathbb{B}$. Then $\{c(\mathcal{B}_1)\succeq c(\mathcal{B}_2)\}:\Leftrightarrow\{c_\infty(\mathcal{B}_1)>c_\infty(\mathcal{B}_2)$, or $c_\infty(\mathcal{B}_1)=c_\infty(\mathcal{B}_2)$ and $c'_\infty(\mathcal{B}_1)>c'_\infty(\mathcal{B}_2)$, or $c_\infty(\mathcal{B}_1)=c_\infty(\mathcal{B}_2)$ and $c'_\infty(\mathcal{B}_1)=c'_\infty(\mathcal{B}_2)\}$.

Interpretation. We recall from section II.3.3 that $q-c_\infty(\mathcal{B})$ is the number of output variables in $\mathcal{B}$. Given the state and the (free) input variables, the output variables are uniquely determined by the laws of $\mathcal{B}$. A simple system is one which leaves *few* variables *unrestricted*, i.e., for which c_∞ is small, and which has *small memory*, i.e., for which c'_∞ is small. Stated otherwise, a simple model is one for which the total number of laws $\Sigma_{t=0}^{\infty}e_t^*$ is large and for which $\Sigma_{t=0}^{\infty}te_t^*$ is small. This amounts to preference of many equations of short lag. □

Remark. The ordering $\succeq$ is a refinement of $\overset{(p)}{\geq}$ in the sense that $\{c(\mathcal{B}_1)\overset{(p)}{\geq}c(\mathcal{B}_2)\} \Rightarrow \{c(\mathcal{B}_1)\succeq c(\mathcal{B}_2)\}$, which is evident from proposition 2–2(*ii*). Moreover, $\succeq$ is a total ordering. We will use this ordering in this chapter. □

In the approximate modelling procedures of chapter V we use utility functions involving the complexity. These utility functions are based on a (total) *lexicographic* ordering of complexities which is another refinement of the natural ordering, and which is defined as follows.

Definition 2–4 Let $c^{(1)},c^{(2)}\in(\mathbb{R}_+)^{\mathbb{Z}_+}$. Then $\{c^{(1)}\geq c^{(2)}\} :\Leftrightarrow \{c^{(1)}=c^{(2)}$, or there is a $t_0\in\mathbb{Z}_+$ such that $c_{t_0}^{(1)}>c_{t_0}^{(2)}$ and $c_t^{(1)}=c_t^{(2)}$ for all $t<t_0\}$.

The following is an immediate consequence of proposition 2–2(*i*).

Corollary 2–5 Let $\mathcal{B}_1,\mathcal{B}_2\in\mathbb{B}$. Then $\{c(\mathcal{B}_1)\geq c(\mathcal{B}_2)\} \Leftrightarrow \{e^*(\mathcal{B}_1)=e^*(\mathcal{B}_2)$, or there is a $t_0\in\mathbb{Z}_+$ such that $e_{t_0}^*(\mathcal{B}_1)<e_{t_0}^*(\mathcal{B}_2)$ and $e_t^*(\mathcal{B}_1)=e_t^*(\mathcal{B}_2)$ for all $t<t_0\}$.

Interpretation. In assessing the complexity of a system by means of the lexicographic ordering the number of short lag equations is decisive. Note that in the natural ordering $\overset{(p)}{\geq}$ a simple system is one which satisfies many equations of small order. Moreover, in the ordering $\succeq$ a simple system is one for which the number of outputs $\Sigma_{t=0}^{\infty}e_t^*$ is large and for which the number of states $\Sigma_{t=0}^{\infty}te_t^*$ is small, which also corresponds to many equations of small order. This is also reflected by the lexicographic ordering. □

Remark. This clearly is a total ordering which is a refinement of the natural ordering. This lexicographic ordering allows relatively simple recursive algorithms, as presented in section V.4. □

Remark. The reverse lexicographic ordering, defined by $\{c^{(1)}(\underset{\geq}{r})c^{(2)}\}:\Leftrightarrow$ $\{c^{(1)}=c^{(2)}$, or there is a $t_0\in\mathbb{Z}_+$ such that $c^{(1)}_{t_0}>c^{(2)}_{t_0}$ and $c^{(1)}_t\geq c^{(2)}_t$ for all $t>t_0\}$, is also appealing. It is directly connected with m and n, as for this ordering $\{m_1>m_2\} \Rightarrow \{c^{(1)}(\underset{>}{r})c^{(2)}\}$ and $\{m_1=m_2,\ n_1>n_2\} \Rightarrow \{c^{(1)}(\underset{>}{r})c^{(2)}\}$. This does not hold true for the lexicographic ordering. However, the construction of algorithms for identification procedures based on the reverse lexicographic ordering seems to be very difficult. This forms the main motivation for considering the lexicographic ordering. □

The lexicographic ordering can easily be expressed in terms of the canonical forms of sections II.3.2.5 and II.3.2.6. The next result is used in the algorithms for deterministic time series analysis of chapter V. It is an immediate consequence of corollary 2–5 and definitions II.3–11 and II.3–15.

Corollary 2–6 Let $\mathcal{B}_i\in\mathbb{B}$, $\mathcal{B}_i=\mathcal{B}(R^{(i)}_d)=\mathcal{B}(R^{(i)}_p)$ with $R^{(i)}_d$ in (CDF) and $R^{(i)}_p$ in (CPF), i=1,2. Let $e^{(i)}_d$ and $e^{(i)}_p$ denote the equation structure of $R^{(i)}_d$ and $R^{(i)}_p$ respectively, i=1,2. Then $\{c(\mathcal{B}_1)\geq c(\mathcal{B}_2)\} \Leftrightarrow \{e^{(1)}_p=e^{(1)}_d\leq e^{(2)}_d=e^{(2)}_p$ in lexicographic ordering$\}$.

3. A class of l_2-systems

3.1. Definition

Notation. Let $\|\cdot\|:(\mathbb{R}^d)^{\mathbb{Z}} \to [0,\infty]$ be defined by $\|z\|:=\{\sum_{t\in\mathbb{Z}}z(t)^Tz(t)\}^{1/2}$ and let $l_2^d:=l_2(\mathbb{Z},\mathbb{R}^d):=\{z\in(\mathbb{R}^d)^{\mathbb{Z}};\ \|z\|<\infty\}$ denote the set of square summable sequences in $\mathbb{R}^d$. It is well-known that l_2^d is a Hilbert space with inner product $<z_1,z_2>:=$ $\sum_{t\in\mathbb{Z}}z_1(t)^Tz_2(t)$. Further for $z\in(\mathbb{R}^d)^{\mathbb{Z}}$ let $z^{--}:=z|_{(-\infty,-1]}$, $z^{-}:=z|_{(-\infty,0]}$, $z^{+}:=$ $z|_{[0,\infty)}$, and $z^{++}:=z|_{[1,\infty)}$. □

In the sequel of this chapter we consider the model class $\mathbb{B}_2$, which is defined as follows.

Definition 3-1 $\mathbb{B}_2 := \mathbb{B} \cap l_2^q := \{H \subset l_2^q;\ \exists \mathcal{B} \in \mathbb{B} \text{ such that } H = \mathcal{B} \cap l_2^q\}$.

So a system $H \in \mathbb{B}_2$ consists of the l_2-trajectories of a linear, time invariant, complete system in $(\mathbb{R}^q)^{\mathbb{Z}}$.

Remark. Let $\mathbb{C}_2$ denote the class of linear, shift invariant, closed subspaces of l_2^q, i.e., $\mathbb{C}_2 := \{C \subset l_2^q;\ C$ linear, $\sigma C = C$, C closed in the l_2^q-topology$\}$. As l_2^q-convergence implies pointwise convergence, it follows from proposition II.3–3 that $\mathbb{B}_2 \subset \mathbb{C}_2$. However, $\mathbb{C}_2 \neq \mathbb{B}_2$, i.e., not every, linear, shift invariant, closed subspace of l_2^q belongs to $\mathbb{B}_2$, as the following example shows.

Let $g := (g_k;\ k \in \mathbb{N})$ be a given sequence in $\mathbb{R}$ such that $\Sigma_{k=1}^{\infty} |g_k| < \infty$, and let $C := \{(u,y) \in l_2^2;\ y(t) = \Sigma_{k=1}^{\infty} g_k u(t-k),\ t \in \mathbb{Z}\}$. It easily follows that $C \in \mathbb{C}_2$. Consider $(\delta, y_\delta) \in C$ defined by $\delta(0) := 1$, $\delta(t) := 0$ for $t \neq 0$, $y_\delta^{++} := g$ and $y_\delta^{-} := 0$. Now suppose $C \in \mathbb{B}_2$, then it follows from theorem II.3–5 that there is a polynomial $0 \neq r \in \mathbb{R}[s]$ such that $[r(\sigma) y_\delta](t) = 0$ for t sufficiently large or, equivalently, that $\text{rank}(H_g) < \infty$ where $H_g \in \mathbb{R}^{\mathbb{N} \times \mathbb{N}}$ is defined by $(H_g)_{i,j} := g_{i+j-1}$, $i,j \in \mathbb{N}$. It is a well-known result from realization theory, see e.g. Silverman [63, theorem 11], that this is equivalent to existence of $n \in \mathbb{N}$, $c,b \in \mathbb{R}^n$, and $A \in \mathbb{R}^{n \times n}$, such that $g_k = c^T A^{k-1} b$, $k \in \mathbb{N}$. If $(g_k;\ k \in \mathbb{N})$ does not satisfy this last condition then we conclude $C \notin \mathbb{B}_2$. An example is given by $g_k := k^{-2}$, $k \in \mathbb{N}$. □

Remark. A topological characterization of $\mathbb{B}_2$ is the following. Let V be a linear, shift invariant subspace of l_2^q and let V' be its closure in $(\mathbb{R}^q)^{\mathbb{Z}}$ with respect to the topology of pointwise convergence. Then $\{V \in \mathbb{B}_2\} \Leftrightarrow \{V = V' \cap l_2^q\}$. Indeed, if $V \in \mathbb{B}_2$, say $V = \mathcal{B} \cap l_2^q$ with $\mathcal{B} \in \mathbb{B}$, then according to proposition II.3–3 $V' \subset \mathcal{B}$, hence $V \subset V' \cap l_2^q \subset \mathcal{B} \cap l_2^q = V$, so $V = V' \cap l_2^q$. On the other hand, if $V = V' \cap l_2^q$ then $V \in \mathbb{B}_2$, as V' is closed and it is easily shown that V' is linear and time invariant, hence $V' \in \mathbb{B}$. □

3.2. Representation

Next we give representation results for systems in $\mathbb{B}_2$ analogous to the results for $\mathbb{B}$ stated in theorem II.3–21 and propositions II.3–22 and II.3–25. We use these results to derive special scattering representations in section 4, which play a role in model approximation in section 5.

Proposition 3-2 Any $H\in\mathbb{B}_2$ has an l_2-realization $\mathcal{B}_s^2(A,B,C,D):=\{(v,x,w)\in l_2^m\times l_2^n\times l_2^q;\ \begin{bmatrix}\sigma x\\ w\end{bmatrix}=\begin{bmatrix}A & B\\ C & D\end{bmatrix}\begin{bmatrix}x\\ v\end{bmatrix}\}$, i.e., $H=\{w;\ \exists(v,x)$ such that $(v,x,w)\in\mathcal{B}_s^2\}$. Analogously it has l_2-input/state/output realizations $\mathcal{B}_{i/s/o}^2$ and backward l_2-realizations ${}^R\mathcal{B}_s^2$ and ${}^R\mathcal{B}_{i/s/o}^2$.

Proof. See the appendix.

Remark. For the notation we refer to definitions II.3–24 and II.3–26. □

Notation. We call an l_2-realization minimal if both m and n are individually as small as possible. We call $(A,B)\in\mathbb{R}^{n\times n}\times\mathbb{R}^{n\times m}$ *controllable* if $R:=[B\ \ AB\ \ \ldots\ \ A^{n-1}B]\in\mathbb{R}^{n\times mn}$ is surjective. □

Proposition 3-3 In $\mathbb{B}_2$ minimality is equivalent to the conditions that (A,B,C,D) is perfectly observable, (A,B) controllable, and D injective.

Proof. See the appendix.

Notation. For $\mathcal{B}\in\mathbb{B}$ let $\mathcal{B}_0:=\{w\in\mathcal{B};\ \exists t_0,t_1\in\mathbb{Z}$ such that $w(t)=0$ for $t<t_0$ and for $t>t_1\}$. Let $\mathcal{B}_0'$ denote the closure of $\mathcal{B}_0$ in $(\mathbb{R}^q)^{\mathbb{Z}}$ with respect to the topology of pointwise convergence. $\mathcal{B}$ is called *controllable* if $\mathcal{B}=\mathcal{B}_0'$. Let $\mathbb{B}_c$ denote the class of controllable systems in $\mathbb{B}$, i.e., $\mathbb{B}_c:=\{\mathcal{B}\in\mathbb{B};\ \mathcal{B}$ controllable$\}$. □

Remark. It can be shown that $\mathcal{B}\in\mathbb{B}$ is controllable if and only if in a minimal realization, as defined in definition II.3–20, the pair (A,B) is controllable. We refer to Willems [74, section 4.8.2]. □

Corollary 3-4 (*i*) $\mathbb{B}_2=\mathbb{B}_c\cap l_2^q$;

(*ii*) if $H=\mathcal{B}\cap l_2^q$, $\mathcal{B}\in\mathbb{B}_c$, then $H|_{[t_0,t_1]}=\mathcal{B}|_{[t_0,t_1]}$ for all $-\infty<t_0\le t_1<+\infty$;

(*iii*) if $H\in\mathbb{B}_2$ has a minimal realization $\mathcal{B}_s^2(A,B,C,D)$, then all minimal realizations are given by $\mathcal{B}_s^2(S(A+BF)S^{-1},\ SBR,\ (C+DF)S^{-1},\ DR)$ for some arbitrary F and invertible S,R, all of appropriate dimensions.

Proof. See the appendix.

Interpretation. Taking $\mathbb{B}_2$ as model class amounts to modelling exactly the *finite time behaviour of controllable systems* and imposing a (norm) restriction on the behaviour at infinity. □

Remark. It can be shown that $\mathcal{B}\in\mathbb{B}_c$ if and only if for every $w_1^-\in\mathcal{B}|_{(-\infty,-1]}$ and $w_2^+\in\mathcal{B}_{[0,\infty)}$ there exist $t\in\mathbb{Z}_+$ and $w\in\mathcal{B}$ such that $w|_{(-\infty,-1]}=w_1^-$ and $w|_{[t,\infty)}=w_2^+$, i.e., if and only if every past in $\mathcal{B}$ can be driven into any future in $\mathcal{B}$. We refer to Willems [74, section 4.3.1]. Corollary 3–4(i) implies that for all $H\in\mathbb{B}_2$ the future can be controlled in this way. □

Finally we consider the question when $\mathcal{B}\in\mathbb{B}_2$ is the graph of a causal or anticausal map.

Notation. Let $\mathbb{C}_0:=\{z\in\mathbb{C};\ |z|=1\}$, $\mathbb{C}_+:=\{z\in\mathbb{C};\ |z|<1\}$, and $\mathbb{C}_-:=\{z\in\mathbb{C};\ |z|>1\}$. Let $\sigma(A)$ denote the spectrum of a square matrix A. Let $F:l_2^m\to l_2^p$ be a linear map. We call F *causal* if for all $t\in\mathbb{Z}$ $\{u|_{(-\infty,t]}=0\}\Rightarrow\{(Fu)|_{(-\infty,t]}=0\}$, *anticausal* if for all $t\in\mathbb{Z}$ $\{u|_{[t,\infty)}=0\}\Rightarrow\{(Fu)|_{[t,\infty)}=0\}$. The graph of F is defined as $\mathrm{gr}(F):=\{(u,y)\in l_2^m\times l_2^p;\ y=Fu\}$. □

Proposition 3–5 Let $H\in\mathbb{B}_2$ have a minimal l_2–input/state/output realization $\mathcal{B}^2_{i/s/o}(\tilde{A},\tilde{B},\tilde{C},\tilde{D})$, i.e., for a permutation matrix Π there holds $\Pi H=\{\begin{pmatrix}u\\y\end{pmatrix}\in l_2^q;\ \exists x\in l_2^n$ such that $\begin{pmatrix}\sigma x\\y\end{pmatrix}=\begin{pmatrix}\tilde{A}&\tilde{B}\\\tilde{C}&\tilde{D}\end{pmatrix}\begin{pmatrix}x\\u\end{pmatrix}\}$. Then

(i) $\{\exists F:l_2^m\to l_2^{q-m}:u\to y$ with $\Pi H=\mathrm{gr}(F)\}\Leftrightarrow\{\sigma(\tilde{A})\cap\mathbb{C}_0=\varnothing\}$;

(ii) $\{F$ is causal$\}\Leftrightarrow\{\sigma(\tilde{A})\subset\mathbb{C}_+\}$;

(iii) $\{F$ is anticausal$\}\Leftrightarrow\{\sigma(\tilde{A})\subset\mathbb{C}_-\}$.

Proof. See the appendix.

Remark. A similar result does not hold true on $\mathbb{B}$. Consider, e.g., $\mathcal{B}:=\{(u,y)\in(\mathbb{R}^q)^{\mathbb{Z}};\ (\sigma-\frac{1}{2})y=(\sigma-2)u\}$. It is easily shown that $\mathcal{B}$ has minimal realizations $\mathcal{B}_{i/s/o}(\frac{1}{2},-\frac{3}{2},1,1)$, for $\Pi=I$, and $\mathcal{B}_{i/s/o}(2,-\frac{3}{2},-1,1)$, for $\Pi=\begin{bmatrix}0&1\\1&0\end{bmatrix}$. However, in $\mathcal{B}$ y is not a (causal) function of u and u is not an (anticausal) function of y, as $(u_i,y_i)\in\mathcal{B}$, $i=1,2$, where $u_1:=0$, $y_1(t):=(\frac{1}{2})^t$, $t\in\mathbb{Z}$, and $u_2(t):=2^t$, $t\in\mathbb{Z}$, $y_2:=0$. □

Remark. Instead of graph representations we consider in the next section representations of $H \in \mathbb{B}_2$ as the image of a map. By means of scattering theory we show that every $H \in \mathbb{B}_2$ can be represented as the image of a causal isometric map and also of an anticausal isometric map. □

4. Scattering representations of l_2-systems

4.1. Scattering theory for l_2-systems

In this section we use scattering theory to conclude that systems in $\mathbb{B}_2$ can be represented as the image of special isometric maps. In the next sections we give an explicit construction of these isometries. They are used in section 5 for model approximation.

Notation. Let $\mathbb{Z}_+ := \{0,1,2,...\}$, $\mathbb{Z}_- := \{...,-3,-2,-1\}$, and for a Hilbert space X let $l_2(\mathbb{Z},X) := \{x \in X^{\mathbb{Z}};\ \sum_{t \in \mathbb{Z}} \|x(t)\|^2 < \infty\}$, $l_2(\mathbb{Z}_+,X) := \{x \in l_2(\mathbb{Z},X);\ x(t)=0 \text{ for } t \in \mathbb{Z}_-\}$, $l_2(\mathbb{Z}_-,X) := \{x \in l_2(\mathbb{Z},X);\ x(t)=0 \text{ for } t \in \mathbb{Z}_+\}$. By σ^* we denote right shift, i.e., for $x \in X^{\mathbb{Z}}$ $\sigma^* x$ is defined by $(\sigma^* x)(t) := x(t-1)$. Note that σ^* is the adjoint of σ. We denote the closure of a set S by $\mathrm{cl}(S)$. □

We now state a result proved by Lax and Phillips [45, theorems II.1.1 and II.1.2]. Let H be an arbitrary Hilbert space and $V: H \to H$ a unitary operator. Suppose there exist closed subspaces D_+ and D_- of H which satisfy the following properties.

$$(i)_+ \quad VD_+ \subset D_+ \quad (ii)_+ \quad \bigcap_{k \in \mathbb{Z}} V^k D_+ = \{0\} \quad (iii)_+ \quad \mathrm{cl}\Big(\bigcup_{k \in \mathbb{Z}} V^k D_+\Big) = H;$$

$$(i)_- \quad VD_- \supset D_- \quad (ii)_- \quad \bigcap_{k \in \mathbb{Z}} V^k D_- = \{0\} \quad (iii)_- \quad \mathrm{cl}\Big(\bigcup_{k \in \mathbb{Z}} V^k D_-\Big) = H.$$

Scattering theorem

Suppose $(i)_\pm$, $(ii)_\pm$ and $(iii)_\pm$ hold. Then there exist Hilbert spaces N_+, N_- with inner product $\langle \cdot,\cdot \rangle_+$, $\langle \cdot,\cdot \rangle_-$, and maps $L_+ : l_2(\mathbb{Z},N_+) \to H$, $L_- : l_2(\mathbb{Z},N_-) \to H$, such that

(i) L_+ and L_- are *isometries* onto H;

(ii) $L_+\sigma^*=VL_+$, $L_-\sigma^*=VL_-$;

(iii) $L_+l_2(\mathbb{Z}_+,N_+)=D_+$, $L_-l_2(\mathbb{Z}_-,N_-)=D_-$.

Moreover, N_+ and N_- are unitarily equivalent, and $(L_+,N_+,<\cdot,\cdot>_+)$ and $(L_-,N_-,<\cdot,\cdot>_-)$ are unique up to isomorphisms of N_+ and N_-.

Now consider a system $H\in\mathbb{B}_2$. As H is closed in the l_2^q–topology, see section 3.1, it follows that H is a Hilbert space. Moreover, σ^* clearly is a unitary operator on H. Let $H_+:=\{w\in H;\ w^{--}=0\}$ and $H_-:=\{w\in H;\ w^+=0\}$, then H_+ and H_- are closed subspaces of H.

Proposition 4-1 $(H,D_+,V):=(H,H_+,\sigma^*)$ satisfies $(i)_+$, $(ii)_+$, $(iii)_+$, and $(H,D_-,V):=(H,H_-,\sigma^*)$ satisfies $(i)_-$, $(ii)_-$, $(iii)_-$.

Proof. See the appendix.

The scattering theorem and proposition 4-1 imply the following.

Corollary 4-2 Let $H\in\mathbb{B}_2$. Then there exist Hilbert spaces $N_\pm$ with inner product $<\cdot,\cdot>_\pm$ and maps $L_\pm:l_2(\mathbb{Z},N_\pm)\to l_2^q$ such that

(i) L_+ and L_- are *isometries* and $\mathrm{im}(L_+)=\mathrm{im}(L_-)=H$;

(ii) $L_+\sigma=\sigma L_+$, $L_-\sigma=\sigma L_-$;

(iii) $L_+l_2(\mathbb{Z}_+,N_+)=H_+$, i.e., L_+ is *causal*;
$L_-l_2(\mathbb{Z}_-,N_-)=H_-$, i.e., L_- is *anticausal*.

Moreover, $(L_\pm,N_\pm,<\cdot,\cdot>_\pm)$ are unique up to isomorphisms of $N_\pm$.

Interpretation. A system $H\in\mathbb{B}_2$ can be represented as the image of a causal, or anticausal, time invariant isometric operator. □

Definition 4-3 We call L_+ the *forward scattering representation* of H and L_- the *backward scattering representation*.

Remark. This is a slight abuse of the term scattering, as generally $L_+^{-1}L_-$ is called the scattering operator, see Lax and Phillips [45, section II.2]. □

Remark. In sections 4.3 and 4.4 we derive explicit characterizations of L_+ and L_-. It turns out that for $H\in\mathbb{B}_2$ the operators L_+ and L_- can be described in terms of realizations $\mathcal{B}_s^2$ and ${}^R\mathcal{B}_s^2$ of H. Among those subspaces of l_2^q which admit scattering representations with respect to σ, $\mathbb{B}_2$ exactly consists of those subspaces for which these representations have a finite dimensional state space realization, cf. theorem 4–9 in section 4.3. □

4.2. Finite dimensional isometries

In sections 4.3 and 4.4 we explicitly describe the scattering representations L_+ and L_- of a given system $H\in\mathbb{B}_2$ in terms of minimal realizations of H. As a preliminary step we investigate in this section the question when there exists an isometric map $v\to w$ for a minimal realization $\mathcal{B}_s^2(A,B,C,D):=\{(v,x,w)\in l_2^m\times l_2^n\times l_2^q;\ \begin{pmatrix}\sigma x\\ w\end{pmatrix}=\begin{pmatrix}A & B\\ C & D\end{pmatrix}\begin{pmatrix}x\\ v\end{pmatrix}\}$.

Proposition 4–4 Let $H\in\mathbb{B}_2$ have a minimal realization $\mathcal{B}_s^2(A,B,C,D):=\{(v,x,w)\in l_2^m\times l_2^n\times l_2^q;\ \begin{pmatrix}\sigma x\\ w\end{pmatrix}=\begin{pmatrix}A & B\\ C & D\end{pmatrix}\begin{pmatrix}x\\ v\end{pmatrix}\}$, then

(*i*) if $\sigma(A)\cap\mathbb{C}_0=\emptyset$ then there exists an operator $L(A,B,C,D)$: $l_2^m\to l_2^q$: $v\to w$ such that $H=\mathrm{im}(L)$;

(*ii*) if $\sigma(A)\subset\mathbb{C}_+$ then $L(A,B,C,D)$ is causal, if $\sigma(A)\subset\mathbb{C}_-$ then it is anticausal;

(*iii*) H has realizations with $\sigma(A)\subset\mathbb{C}_+$ and also ones with $\sigma(A)\subset\mathbb{C}_-$.

Proof. See the appendix.

Definition 4–5 For $\sigma(A)\cap\mathbb{C}_0=\emptyset$ we call $L(A,B,C,D)$ the *driving operator* corresponding to $\mathcal{B}_s^2(A,B,C,D)$.

Remark. We call $L(A,B,C,D)$ *finite dimensional* as the map $v\to w$ can be realized by means of a finite dimensional state variable, i.e., by means of $\mathcal{B}_s^2(A,B,C,D)$. □

Interpretation. Proposition 4–4 states that systems in $\mathbb{B}_2$ can be represented as the image of *causal*, or *anticausal*, *time invariant finite dimensional* operators. According to corollary 4–2 such systems also can be represented as the image of causal, or anticausal, time invariant *isometric* operators. In the next sections we show that these operators in fact also are finite dimensional. □

We now investigate when a driving operator is isometric.

Notation. Let $K=K^T\in\mathbb{R}^{n\times n}$ be nonsingular and for $x,y\in\mathbb{R}^n$ let $<x,y>_K:=y^TKx$. Define $\text{in}(K):=\max\{\dim(V);\ V$ linear subspace of $\mathbb{R}^n$, $<x,x>_K<0$ for all $0\neq x\in V\}$. Then $(\mathbb{R}^n,<\cdot,\cdot>_K)$ is called a *Pontryagin space* of *index* $\text{in}(K)$. We call a linear map $M:(\mathbb{R}^{d_1},<\cdot,\cdot>_{K_1})\to(\mathbb{R}^{d_2},<\cdot,\cdot>_{K_2})$ a (K_1,K_2) Pontryagin isometry if for all $a\in\mathbb{R}^{d_1}$ $<Ma,Ma>_{K_2}=<a,a>_{K_1}$. □

Proposition 4-6 Let $\mathcal{B}_s^2:=\mathcal{B}_s^2(A,B,C,D):=\{(v,x,w)\in l_2^m\times l_2^n\times l_2^q;\ \begin{bmatrix}\sigma x\\ w\end{bmatrix}=\begin{bmatrix}A & B\\ C & D\end{bmatrix}\begin{bmatrix}x\\ v\end{bmatrix}\}$, where (A,B,C,D) is perfectly observable, (A,B) is controllable, and D is injective, then

(*i*) $(\|w\|^2=\|v\|^2$ for all $(v,x,w)\in\mathcal{B}_s^2\}\Leftrightarrow\{\exists K=K^T\in\mathbb{R}^{n\times n}$ nonsingular such that $\begin{bmatrix}A & B\\ C & D\end{bmatrix}$ is a $(\begin{bmatrix}K & 0\\ 0 & I_m\end{bmatrix},\begin{bmatrix}K & 0\\ 0 & I_q\end{bmatrix})$ Pontryagin isometry$\}$;

(*ii*) if one of the two equivalent conditions of (*i*) is satisfied, then

(*ii*–1) $\sigma(A)\cap\mathbb{C}_0=\emptyset$ and hence the driving operator $L(A,B,C,D):v\to w$ exists and moreover is an isometry;

(*ii*–2) $\text{in}(K)$ equals the number of eigenvalues of A in $\mathbb{C}_-$;

(*ii*–3) $\{L$ causal$\}\Leftrightarrow\{\text{in}(K)=0\}$; $\{L$ anticausal$\}\Leftrightarrow\{\text{in}(K)=n\}$.

Proof. See the appendix.

Remark. If for $H\in\mathbb{B}_2$ we can construct a realization $\mathcal{B}_s^2(A_+,B_+,C_+,D_+)$ which satisfies condition (*i*) of proposition 4–6 with $K>0$, then we conclude that the corresponding driving operator is causal, time invariant and isometric. In the next section we show that such a realization exists and that it easily can be constructed from any minimal realization of H. Moreover we show that $L(A_+,B_+,C_+,D_+)=L_+$, up to an isomorphism on $\mathbb{R}^m$. In section 4.4 we consider the construction of L_-. □

4.3. Forward scattering representation

In this section we prove that for $H \in \mathbb{B}_2$ there exists a minimal realization $\mathcal{B}_s^2(A_+,B_+,C_+,D_+)$ such that the corresponding driving operator $L(A_+,B_+,C_+,D_+)$ is the forward scattering representation L_+ of H. We moreover give an algorithm to compute (A_+,B_+,C_+,D_+), starting from an arbitrary minimal realization $\mathcal{B}_s^2(A,B,C,D)$ of H.

Remark. According to corollary 4–2 and proposition 4–6 $L_+=L(A_+,B_+,C_+,D_+)$ if $\begin{bmatrix} A_+ & B_+ \\ C_+ & D_+ \end{bmatrix}$ is a $(\begin{bmatrix} K_+ & 0 \\ 0 & I_m \end{bmatrix}, \begin{bmatrix} K_+ & 0 \\ 0 & I_q \end{bmatrix})$ Pontryagin isometry for some $K_+>0$. Let $\mathcal{B}_s^2(A,B,C,D)$ be an arbitrary minimal realization of H. According to corollary 3–4(*iii*) the class of all minimal realizations is then given by $\mathcal{B}_s^2(S(A+BF)S^{-1}$, SBR, $(C+DF)S^{-1}$, $DR)$, where S and R are invertible. The next proposition describes in which case the corresponding driving operator is an isometry. □

Proposition 4–7 $\begin{bmatrix} S(A+BF)S^{-1} & SBR \\ (C+DF)S^{-1} & DR \end{bmatrix}$ is a $(\begin{bmatrix} \bar{K} & 0 \\ 0 & I_m \end{bmatrix}, \begin{bmatrix} \bar{K} & 0 \\ 0 & I_q \end{bmatrix})$ Pontryagin isometry if and only if with $K:=S^T\bar{K}S$ there holds

(ARE) $K=A^TKA-(B^TKA+D^TC)^T(B^TKB+D^TD)^{-1}(B^TKA+D^TC)+C^TC$

(R) $RR^T=(B^TKB+D^TD)^{-1}$

(F) $F=-(B^TKB+D^TD)^{-1}(B^TKA+D^TC)$.

Proof. See the appendix.

Remark. (ARE) is called the algebraic Riccati equation. □

Lemma 4–8 Let $H \in \mathbb{B}_2$ have minimal realization $\mathcal{B}_s^2(A,B,C,D)$, then among the solutions of (ARE) there exists a unique solution $K_+=K_+^T>0$. Moreover, $F_+:=-(B^TK_+B+D^TD)^{-1}(B^TK_+A+D^TC)$ is the unique feedback of the form (F) such that $\sigma(A+BF_+)\subset\mathbb{C}_+$.

Proof. See the appendix.

Remark. (ARE) does not always have a solution $K_-=K_-^T<0$, see example 2 in section 5.3. □

Notation. Let $H \in \mathbb{B}_2$ have minimal realization $\mathcal{B}_s^2(A,B,C,D)$. Let K_+ be the positive definite solution of the corresponding (ARE). Let F_+ be defined by (F) for K_+ and let R_+ be a solution of (R) for K_+. We define $(A_+,B_+,C_+,D_+):=(A+BF_+,BR_+,C+DF_+,DR_+)$. □

Theorem 4-9 The forward scattering representation of H is given by $L_+=L(A_+,B_+,C_+,D_+)$.

Proof. See the appendix.

Remark. Hence the scattering representation L_+ of H is *finite dimensional* and it corresponds to a minimal realization $\mathcal{B}_s^2(A_+,B_+,C_+,D_+)$ of H in the sense that $\mathrm{gr}(L_+)=\{(v,w)\in l_2^m \times l_2^q;\ \exists x \in l_2^n$ such that $(v,x,w)\in \mathcal{B}_s^2(A_+,B_+,C_+,D_+)\}$. □

Remark. The representation of L_+ by (A_+,B_+,C_+,D_+) is *unique* up to an orthogonal transformation on $\mathbb{R}^m$, due to (R), and up to a nonsingular transformation S on $\mathbb{R}^n$. This follows from the uniqueness of K_+. □

Remark. For $H \in \mathbb{B}_2$ let $c_t(H):=\frac{1}{t+1}\cdot \dim(H|_{[0,t]})$, $t \in \mathbb{Z}_+$. From corollary 3-4 and proposition 2-2 it follows that $c_\infty(H):=\lim_{t\to\infty} c_t(H)$ and $c'_\infty(H):=\lim_{t\to\infty} t\{c_t(H)-c_\infty(H)\}$ exist. This complexity is reflected in L_+, as $\dim(N_+)=c_\infty(H)$ and L_+ can be realized by means of a state variable of dimension $c'_\infty(H)$. □

The foregoing can be summarized in an algorithm for determining L_+.

Notation. For $H \in \mathbb{B}_2$ let $\mathcal{B}^*$ denote the closure of H in $(\mathbb{R}^q)^{\mathbb{Z}}$ with respect to the topology of pointwise convergence. The next lemma summarizes some results which are proved in the appendix, see the proofs of proposition 3-3 and corollary 3-4. □

Lemma 4-10 $\mathcal{B}^*$ is controllable and $H=\mathcal{B}^* \cap l_2^q$. Moreover, $\mathcal{B}_s^2(A,B,C,D)$ is a minimal realization of H if and only if $\mathcal{B}_s(A,B,C,D)$ is a minimal realization of $\mathcal{B}^*$. Hence $(m(\mathcal{B}^*),n(\mathcal{B}^*))=(m(H),n(H))$.

Algorithm for L_+

1. Let $H \in \mathbb{B}_2$ be given. Determine a minimal realization $\mathcal{B}_s^2(A,B,C,D)$ of H, e.g., by determining a minimal realization of $\mathcal{B}^*$ by an algorithm in Willems [73, theorem 17].
2. Determine the positive definite solution K_+ of (ARE), define F_+ by (F) for K_+ and determine R_+ as a solution of (R) for K_+. Define $(A_+,B_+,C_+,D_+):=(A+BF_+,BR_+,C+DF_+,DR_+)$.
3. Define $L_+:l_2^m \to l_2^q$ by $\mathrm{gr}(L_+):=\{(v_+,w);\ \exists x_+ \in l_2^n$ with $\begin{pmatrix}\sigma x_+\\ w\end{pmatrix}=\begin{pmatrix}A_+ & B_+\\ C_+ & D_+\end{pmatrix}\begin{pmatrix}x_+\\ v_+\end{pmatrix}\}$.

Remark. It is shown in Willems [73, theorem 17], that a minimal realization of $\mathcal{B}^*$ can be constructed from $\mathcal{B}^*|_{[t_0,t_1]}$ for some finite $t_0,t_1 \in \mathbb{Z}_+$. As $\mathcal{B}^*$ is controllable it follows from corollary 3–4(*ii*) that step 1 of the algorithm for L_+ can be performed using $H|_{[t_0,t_1]}$. In case H is given in the form $H=\mathcal{B}(R)\cap l_2^q$ with $\mathcal{B}(R)$ controllable, a minimal realization is more easily constructed by using an algorithm in Willems [73, corollary 18]. □

4.4. Backward scattering representation

One way to construct the backward scattering representation L_- of a system $H \in \mathbb{B}_2$ is to consider the time reversed system $\tilde{H}:=\mathcal{R}H$, where $\mathcal{R}$ is the time reverse operator as defined in section II.3.3.

Proposition 4–11 If $\tilde{L}_+$ is the forward scattering representation of $\tilde{H}$, then $L_-=\mathcal{R}\tilde{L}_+\mathcal{R}$.

Proof. See the appendix.

Remark. Hence L_- can be constructed by using the algorithm of section 4.3 to determine $\tilde{L}_+$ on the basis of an arbitrary minimal realization $\mathcal{B}_s^2(\tilde{A},\tilde{B},\tilde{C},\tilde{D})$ of $\tilde{H}$. Let $\mathrm{gr}(\tilde{L}_+)=\{(\tilde{v},\tilde{w}) \in l_2^m \times l_2^q;\ \exists \tilde{x} \in l_2^n$ such that $(\tilde{v},\tilde{x},\tilde{w}) \in \mathcal{B}_s^2(A_-,B_-,C_-,D_-)\}$, then $\mathrm{gr}(L_-)=\mathcal{R}\mathrm{gr}(\tilde{L}_+)$. □

Remark. It is easily verified that $\mathcal{R}\mathcal{B}_s^2(\tilde{A},\tilde{B},\tilde{C},\tilde{D})={}^R\mathcal{B}_s^2(\tilde{A},\tilde{B},\tilde{C},\tilde{D})$. We hence conclude from the foregoing remark that we can construct L_-, starting from an arbitrary minimal *backward* realization ${}^R\mathcal{B}_s^2(\tilde{A},\tilde{B},\tilde{C},\tilde{D})$ of H. Let $\tilde{K}_+>0$ be the positive definite solution of the corresponding (ARE) and determine $\tilde{R}_+$ and $\tilde{F}_+$ from $\tilde{K}_+$ as solutions of the corresponding (R) and (F) respectively. Let $A_-:=\tilde{A}+\tilde{B}\tilde{F}_+$, $B_-:=\tilde{B}\tilde{R}_+$, $C_-:=\tilde{C}+\tilde{D}\tilde{F}_+$, and $D_-:=\tilde{D}\tilde{R}_+$. Then $\mathrm{gr}(L_-)=\{(v_-,w)\in l_2^m\times l_2^q;\ \exists x_-\in l_2^n$ such that $(v_-,x_-,w)\in {}^R\mathcal{B}_s^2(A_-,B_-,C_-,D_-)\}$. □

Remark. For some systems $H\in\mathbb{B}_2$ we need no backward realizations and can instead obtain L_- from a minimal forward realization of H. Indeed, suppose $\mathcal{B}_s^2(A,B,C,D)$ is a minimal forward realization of H with the property that (ARE) has a solution $K_-=K_-^T<0$ with $B^TK_-B+D^TD>0$. Determine R_- and F_- from K_- as solutions of (R) and (F) respectively, and define $\tilde{A}_-:=A+BF_-$, $\tilde{B}_-:=BR_-$, $\tilde{C}_-:=C+DF_-$, $\tilde{D}_-:=DR_-$. According to propositions 4–6 and 4–7 $L:=L(\tilde{A}_-,\tilde{B}_-,\tilde{C}_-,\tilde{D}_-)$ then is an anticausal time invariant isometry with $\mathrm{im}(L)=H$. To prove that $L=L_-$ it suffices to show that $H_-=Ll_2(\mathbb{Z}_-,\mathbb{R}^m)$. As L is anticausal $Ll_2(\mathbb{Z}_-,\mathbb{R}^m)\subset H_-$, while for $w\in H_-$ and $(v,x,w)\in\mathcal{B}_s^2(\tilde{A}_-,\tilde{B}_-,\tilde{C}_-,\tilde{D}_-)$ it follows from $w^+=0$ and perfect observability that $x^+=0$, and due to the injectivity of $\tilde{D}_-$ and $\tilde{D}_-v=w-\tilde{C}_-x$ that also $v^+=0$. □

We summarize the foregoing in the form of an algorithm for the construction of the backward scattering operator L_- for a given system $H\in\mathbb{B}_2$.

Algorithm for L_-

1. If in a minimal realization $\mathcal{B}_s^2(A,B,C,D)$ of H the corresponding (ARE) has a solution $K_-<0$ with $B^TK_-B+D^TD>0$, then define F_- by (F) for K_- and determine R_- as a solution of (R) for K_-. Let $(\tilde{A}_-,\tilde{B}_-,\tilde{C}_-,\tilde{D}_-):=(A+BF_-,BR_-,C+DF_-,DR_-)$ and $(A_-,B_-,C_-,D_-):=(\tilde{A}_-^{-1},-\tilde{A}_-^{-1}\tilde{B}_-,\ \tilde{C}_-\tilde{A}_-^{-1},\ \tilde{D}_--\tilde{C}_-\tilde{A}_-^{-1}\tilde{B}_-)$.
2. If a solution $K_-<0$ with $B^TK_-B+D^TD>0$ does not exist, then determine a minimal backward realization ${}^R\mathcal{B}_s^2(\tilde{A},\tilde{B},\tilde{C},\tilde{D})$ of H, e.g., by applying step 1 of the algorithm for L_+ for $\tilde{H}:=\mathcal{R}H$. Determine the positive definite solution $\tilde{K}_+$ of the corresponding (ARE), define $\tilde{F}_+$ by (F) for $\tilde{K}_+$ and determine $\tilde{R}_+$ as a solution of (R) for $\tilde{K}_+$. Define $(A_-,B_-,C_-,D_-):=(\tilde{A}+\tilde{B}\tilde{F}_+,\ \tilde{B}\tilde{R}_+,\ \tilde{C}+\tilde{D}\tilde{F}_+,\ \tilde{D}\tilde{R}_+)$.
3. Define $L_-:l_2^m\to l_2^q$ by $\mathrm{gr}(L_-):=\{(v_-,w);\ \exists x_-\in l_2^n$ with $\begin{pmatrix}\sigma^{-1}x_-\\ w\end{pmatrix}=\begin{pmatrix}A_- & B_-\\ C_- & D_-\end{pmatrix}\begin{pmatrix}x_-\\ v_-\end{pmatrix}\}$.

Remark. Hence L_- is a (finite dimensional) backward driving operator corresponding to a minimal backward realization of H. □

Remark. For step 1 we also refer to the concluding remark in section II.3.3.

Note that $\tilde{A}_-$ is invertible as $\sigma(\tilde{A}_-)\subset\mathbb{C}_-$, see proposition 4–6(*ii*). □

Remark. A direct calculation shows that for given $H\in\mathbb{B}_2$ and fixed coordinates in $\mathbb{R}^n$ the solution sets $\mathcal{K}$ and $\mathcal{K}_R$ of (ARE) for a minimal realization $\mathcal{B}_s^2$ and ${}^R\mathcal{B}_s^2$ of H respectively do not depend upon the particular realizations chosen. There exist systems $H\in\mathbb{B}_2$ for which $\mathcal{K}$ has *no* solution $K_-<0$ with $B^TK_-B+D^TD>0$. We give a simple example in section 5.3. If such a solution exists then $-K_-$ is the unique positive definite element of $\mathcal{K}_R$. An interesting question is the relationship between $\mathcal{K}$ and $\mathcal{K}_R$. □

Remark. Note that under minimality conditions the continuous time algebraic Riccati equation always has a negative definite solution, see e.g. Willems [70]. Moreover, any realization in continuous time has both a forward and a backward interpretation. □

Remark. Let $w\in H$ and let $(v_+,x_+,w)\in\mathcal{B}_s^2(A_+,B_+,C_+,D_+)$, $(v_-,x_-,w)\in{}^R\mathcal{B}_s^2(A_-,B_-,C_-,D_-)$. Then $\sigma x_+=x_-$, up to an isomorphism on $\mathbb{R}^n$, see proposition II.3–27. We use this result in section 5.3 for a balancing method of model approximation. □

5. Model approximation for l_2-systems

5.1. Introduction

We consider the model approximation problem as described in section 1 for the model class $\mathbb{B}_2$ of section 3. Hence we have to define a measure of complexity of models in $\mathbb{B}_2$ and a measure of distance between models in $\mathbb{B}_2$. We define the complexity in accordance with definition 2–1 with the ordering $\succeq$ of definition 2–3.

Remark. So for $H\in\mathbb{B}_2$ the complexity is $c(H):=(c_t(H);\ t\in\mathbb{Z}_+)$, where $c_t(H):=\frac{1}{t+1}\cdot\dim(H|_{[0,t]})$. Note that corollary 3–4(*ii*) implies that $c(H)=c(\mathcal{B})$ for any controllable system $\mathcal{B}\in\mathbb{B}$ for which $H=\mathcal{B}\cap l_2^q$. Let $m(H):=c_\infty(H)$ and $n(H):=c'_\infty(H)$ be as defined in section 2.1. From proposition 2–2 and lemma 4–10 we conclude that

$m(H)$ is the number of driving variables and $n(H)$ the number of state variables in any minimal realization of H. According to the ordering of definition 2–3, a system H_1 is more complex than a system H_2 if $m(H_1)>m(H_2)$ or if $m(H_1)=m(H_2)$ and $n(H_1)>n(H_2)$. We then write $(m(H_1),n(H_1)) \succ (m(H_2),n(H_2))$. □

The model approximation problem in $\mathbb{B}_2$ is as follows, cf. section 1. Let $H\in\mathbb{B}_2$ be given along with $(\bar{m},\bar{n})\prec(m(H),n(H))$. The aim is to find a model $\hat{H}\in\mathbb{B}_2$ with $(m(\hat{H}),n(\hat{H}))\preceq(\bar{m},\bar{n})$ such that $\hat{H}$ is as close as possible to H.

Remark. Unfortunately it seems difficult to construct explicit numerical procedures if we require $\bar{m}<m(H)$. Hence we suppose that $\bar{m}=m(H)$ and $\bar{n}<n(H)$. The model approximation problem then consists of reducing the dimension of the state space. This coincides with the problem formulation which is considered until now in the literature on approximation of dynamical systems. □

In section 5.2 we consider a distance measure which is quite natural if a system is considered as a set of trajectories. We obtain a very explicit description of the corresponding optimal model approximation problem by means of the scattering representations of section 4. However, we were not able to solve the resulting approximation problem. In section 5.3 we present a heuristic and simple approximation method based on scattering representations. This method is illustrated in section 5.4 by means of two simple numerical simulations.

We conclude this introduction by briefly discussing two well-known solutions to the problem of reducing the dimension of the state space for systems in $\mathbb{B}_2$, i.e., Hankel norm approximation as thoroughly discussed, e.g., in Glover [17] and reduction by balancing as presented, e.g., by Moore [54] and Pernebo and Silverman [58].

Remark. These methods do not consider a system as a set, but as an operator which transforms inputs into outputs. Let H_1 and H_2 be two input/output systems with m inputs and p outputs. The corresponding input/output maps $L_i: u \to y$ are called l_2–stable if $L_i l_2^m \subset l_2^p$, $i=1,2$. In this case an appealing distance measure would be the l_2–induced norm $d(H_1,H_2):=\|L_1-L_2\|_2$. However, the corresponding model reduction problem is still unsolved. □

A heuristic and simple method is ***reduction by balancing.*** A rough outline of this procedure is the following. Let $H \in \mathbb{B}_2$ have a minimal realization $\mathcal{B}^2_{i/s/o}(A,B,C,D)$ with $\sigma(A) \subset \mathbb{C}_+$ and with $\mathbb{R}^n$ as state space. For $\bar{n}<n$ the most "relevant" $\bar{n}$-dimensional subspace $\hat{X} \subset \mathbb{R}^n$ is determined by using norms measuring the importance of states in $\mathbb{R}^n$. These norms measure the control and observation energy of states. Let $(\hat{A},\hat{B},\hat{C})$ be the restriction of (A,B,C) to $\hat{X}$, then the approximate system $\hat{H}$ is defined by its realization $\mathcal{B}^2_{i/s/o}(\hat{A},\hat{B},\hat{C},D)$.

Remark. It is not a priori clear in which sense the reduced system is close to the original one. In Glover [17, theorem 9.6], an upper bound is given for the l_2-induced error $\|L-\hat{L}\|_2$ of the input/output maps L and $\hat{L}$ corresponding to H and $\hat{H}$ respectively. Approximate systems obtained by balancing often have an input/output behaviour which is surprisingly close to the original system. □

Hankel norm approximation is a method of model approximation with a more natural interpretation than balancing. Moreover, optimal solutions can be calculated by means of algorithms which are only slightly more complicated than the algorithms for reduction by balancing, cf. Glover [17]. For $H \in \mathbb{B}_2$ with a minimal realization $\mathcal{B}^2_{i/s/o}(A,B,C,D)$ with $\sigma(A) \subset \mathbb{C}_+$, the Hankel operator $\Gamma_H : l_2(\mathbb{N},\mathbb{R}^m) \to l_2(\mathbb{N},\mathbb{R}^p)$ is defined as follows. Let $u \in l_2^m$ with $u^+=0$ and let $(u,y) \in H$, then $\Gamma_H u^{--} := y^+$. So the Hankel operator describes the influence of past inputs on future outputs. The distance between H_1 and H_2 in $\mathbb{B}_2$ is defined as $\|\Gamma_{H_1} - \Gamma_{H_2}\|_2$.

Remark. The condition $\sigma(A) \subset \mathbb{C}_+$ is imposed to guarantee that Γ_H is bounded. □

Remark. In the next sections we consider the problem of model approximation in $\mathbb{B}_2$, where a system $H \in \mathbb{B}_2$ is considered as a subset of l_2^q or as the *image* of a (driving) operator, instead of as the *graph* of an input/output map. □

5.2. Gap

In this section we derive an explicit formulation of an optimal model approximation problem where the distance between systems in $\mathbb{B}_2$ is defined as the gap of the corresponding subspaces of l_2^q.

Notation. For $w,w' \in l_2^q$ let $d(w,w'):=\|w-w'\|$, for $H' \in \mathbb{B}_2$ and $w \in l_2^q$ let $d(w,H'):=\inf\{d(w,w');\ w' \in H'\}$, and for $H,H' \in \mathbb{B}_2$ let $\delta(H,H'):=\max\{d(w,H');\ w \in H,\ \|w\|=1\}$ denote the maximal distance from H' of elements in the unit ball of H. □

Definition 5-1 The *gap* between $H,H' \in \mathbb{B}_2$ is defined as $g(H,H'):=\max\{\delta(H,H'),\ \delta(H',H)\}$.

Remark. The gap g is a distance measure on $\mathbb{B}_2$, see Akhiezer and Glazman [2, section 34], or Kato [40, section IV.2.1]. The gap is sometimes called *aperture* or opening. Clearly $g(H,H') \leq 1$. □

Notation. For $H \in \mathbb{B}_2$ let P denote the orthogonal projection on H in l_2^q. For $L: l_2^q \to l_2^q$ let $\|L\|_2:=\sup\{\|Lw\|;\ \|w\|=1\}$ and let L^* denote the adjoint operator of L. For $H \in \mathbb{B}_2$ let L_+ and L_- denote the forward and backward scattering representation respectively of H. □

The following result is proved in Akhiezer and Glazman [2, section 34].

Lemma 5-2 $g(H,H')=\|P-P'\|_2$.

Corollary 5-3 $g(H,H')=\|L_+L_+^*-L_+'L_+'^*\|_2=\|L_-L_-^*-L_-'L_-'^*\|_2$.

Proof. See the appendix.

Remark. Hence the gap can be expressed in terms of scattering representations. The next theorem gives an explicit formula for the gap in terms of a finite number of parameters. □

Notation. Let $L(A_+,B_+,C_+,D_+)$ and $L(A_-,B_-,C_-,D_-)$ be the forward and backward scattering representation respectively of $H \in \mathbb{B}_2$, obtained e.g. by the algorithms of sections 4.3 and 4.4. Denote the corresponding transfer functions by $G_+(z):=D_+ + C_+(zI-A_+)^{-1}B_+$ and $G_-(z):=D_- + C_-(zI-A_-)^{-1}B_-$. For $G: \mathbb{C} \to \mathbb{C}^{q \times m}$ define G^* by $G^*(z):=[G(z)]^*$, where $[G(z)]^*$ is the (complex) adjoint matrix of $G(z) \in \mathbb{C}^{q \times m}$. Let $\|G\|_\infty := \sup\{\|G(z)\|;\ \|z\|=1\}$, where $\|G(z)\|:=\max\{\|G(z)x\|;\ x \in \mathbb{C}^m,\ \|x\|=1\}$ denotes the induced matrix norm of $G(z) \in \mathbb{C}^{q \times m}$. □

Theorem 5-4 $g(H,H')=\|G_+G_+^* - G_+'G_+'^*\|_\infty = \|G_-G_-^* - G_-'G_-'^*\|_\infty$.

Proof. See the appendix.

Proposition 5-5 If $m(H) \neq m(H')$ then $g(H,H')=1$.

Proof. See the appendix.

Remark. Proposition 5–5 implies that g is too rough a measure if we want to approximate $H \in \mathbb{B}_2$ by means of a system $\hat{H}$ with less driving variables. This obviously is quite a serious drawback of the gap metric. □

For given $H \in \mathbb{B}_2$ and $\bar{n} < n(H)$, the *optimal model approximation problem* with respect to the gap metric consists of determining $\hat{H} \in \mathbb{B}_2$ in such a way that $n(\hat{H}) \leq \bar{n}$ and $g(H,\hat{H}) = \inf\{g(H,H');\ H' \in \mathbb{B}_2,\ n(H') \leq \bar{n}\}$. This problem can be expressed in terms of a finite number of parameters by using theorem 5–4. The problem amounts to reducing the number of state variables under the condition that the set of trajectories of the approximate system is as close as possible to the set of trajectories which are compatible with the system H.

Remark. Hence by scattering theory we have obtained an explicit formula for the gap metric on $\mathbb{B}_2$ and an explicit description of an optimal model approximation problem, both in terms of the parameters of special realizations of systems in $\mathbb{B}_2$. □

Remark. The gap metric on $\mathbb{B}_2$ induces a metric on the class $\mathbb{B}_c$ of linear, time invariant, complete, controllable systems, as there is a bijection $\mathbb{B}_c \to \mathbb{B}_2$, cf. corollary 3–4(*ii*) and proposition II.3–3. Hence the optimal model approximation problem with respect to the gap metric also is relevant for reducing the complexity of controllable systems. □

Remark. It is an open question whether the optimal model approximation problem can be solved by an explicit algorithm. In the next section we present a heuristic balancing method of model approximation for which the approximate model can be obtained by a simple algorithm. □

5.3. Reduction by balancing

Let $H\in\mathbb{B}_2$ be given with complexity $(c_\infty(H),c'_\infty(H))=(m,n)$. Suppose that a maximal tolerated dimension $\bar{n}$ of the state space is specified and that $n>\bar{n}$. We will construct an approximate system $\hat{H}\in\mathbb{B}_2$ with $c_\infty(\hat{H})=m$ and $c'_\infty(\hat{H})\leq\bar{n}$ in such a way that the states which are "less relevant" are neglected. Using a *balancing* argument we will *approximate the state space* $\mathbb{R}^n$ of H by means of an $\bar{n}$-dimensional subspace $\hat{X}\subset\mathbb{R}^n$ which contains the most relevant states as measured by norms which we characterize in definition 5-6 and proposition 5-7.

Notation. Let H have forward and backward scattering representations $L_+:v_+\to w$ and $L_-:v_-\to w$ respectively, corresponding to minimal realizations $\mathcal{B}_s^+:=\mathcal{B}_s^2(A_+,B_+,C_+,D_+):=\{(v_+,x_+,w)\in l_2^m\times l_2^n\times l_2^q;\ \begin{pmatrix}\sigma x_+\\ w\end{pmatrix}=\begin{pmatrix}A_+ & B_+\\ C_+ & D_+\end{pmatrix}\begin{pmatrix}x_+\\ v_+\end{pmatrix}\}$ and $\mathcal{B}_s^-:={}^R\mathcal{B}_s^2(A_-,B_-,C_-,D_-):=\{(v_-,x_-,w)\in l_2^m\times l_2^n\times l_2^q;\ \begin{pmatrix}\sigma^{-1}x_-\\ w\end{pmatrix}=\begin{pmatrix}A_- & B_-\\ C_- & D_-\end{pmatrix}\begin{pmatrix}x_-\\ v_-\end{pmatrix}\}$ of H. Recall that for $z\in(\mathbb{R}^d)^{\mathbb{Z}}$ $z^{--}:=z|_{(-\infty,-1]}$, $z^-:=z|_{(-\infty,0]}$, $z^+:=z|_{[0,\infty)}$, and $z^{++}:=z|_{[1,\infty)}$. □

Definition 5-6 Let $x_0\in\mathbb{R}^n$, then

$\|x_0\|_+^2:=\|w^+\|^2-\|v_+^+\|^2$ for $(v_+,x_+,w)\in\mathcal{B}_s^+$, $x_+(0)=x_0$;

$\|x_0\|_-^2:=\|v_-^{++}\|^2-\|w^{++}\|^2$ for $(v_-,x_-,w)\in\mathcal{B}_s^-$, $x_-(0)=x_0$;

$\|x_0\|_{++}^2:=\inf\{\|v_+^{--}\|^2;\ (v_+,x_+,w)\in\mathcal{B}_s^+,\ x_+(0)=x_0\}$;

$\|x_0\|_{--}^2:=\inf\{\|v_-^{++}\|^2;\ (v_-,x_-,w)\in\mathcal{B}_s^-,\ x_-(0)=x_0\}$.

The following proposition proves that these expressions indeed are well-defined and specify norms on $\mathbb{R}^n$.

Notation. Let $K_+>0$ and $\tilde{K}_+>0$ be the positive definite solutions of (ARE) for any minimal forward respectively backward realization of H. Further define

$Q_+:=\Sigma_{t=0}^{\infty}A_+^tB_+B_+^T(A_+^T)^t$. □

Proposition 5-7 There holds
$\|x_0\|_+^2=x_0^TK_+x_0$;
$\|x_0\|_-^2=x_0^T\tilde{K}_+x_0$;
$\|x_0\|_{++}^2=\|x_0\|_{--}^2=\inf\{\|w\|^2;\ (v_\pm,x_\pm,w)\in\mathcal{B}_s^\pm,\ x_\pm(0)=x_0\}=x_0^TQ_+^{-1}x_0=x_0^T(K_++\tilde{K}_+)x_0$.

Proof. See the appendix.

Interpretation. These norms give an expression of the *relevance* of states in the state space $\mathbb{R}^n$ of H. If $\|x_0\|_+$ is large then any future external energy $\|w^+\|$ will be large compared with $\|v_+^+\|$, so x_0 is relatively relevant for future external behaviour. If $\|x_0\|_-$ is large then the driving inputs $\|v_-^{++}\|$ are large in norm compared to the external effect $\|w^{++}\|$, so x_0 is relatively unimportant for future behaviour. If $\|x_0\|_{++}=\|x_0\|_{--}$ is large then x_0 is relatively important as it corresponds to large total external energy $\|w\|$. □

Remark. We conclude that $\tilde{K}_+=Q_+^{-1}-K_+$. So $\tilde{K}_+$ can be calculated directly in terms of the parameters of the forward scattering representation of H. □

The method of model approximation by balancing determines the approximate model $\hat{H}$ by neglecting those states which are least relevant and determining an $\bar{n}$-dimensional subspace $\hat{X}\subset\mathbb{R}^n$ which contains the most relevant states. The next lemma indicates that the different measures of being relevant can be reconciled (or: balanced) by an appropriate choice of basis in the state space $\mathbb{R}^n$.

Lemma 5-8 There is a choice of basis such that $(K_+,\tilde{K}_+)=(\Lambda,\Lambda^{-1})$ with $\Lambda:=\mathrm{diag}(\lambda_1,\ldots,\lambda_n)$, $\lambda_1\geq\ldots\geq\lambda_n>0$, where λ_i^2, $i=1,\ldots,n$, are the eigenvalues of $K_+^{1/2}\tilde{K}_+^{-1}K_+^{1/2}$.

Proof. See the appendix.

Remark. A basis as in the lemma is called *balanced*. In balanced coordinates a state $x_0=e_i:=(0,\ldots,0,1,0,\ldots,0)$, with a 1 only in the i-th spot, is more relevant for lower indices i, as for large indices λ_i is small, hence x_0 is unimportant both in the sense of norm $\|\cdot\|_+$ and in the sense of norm $\|\cdot\|_-$. The balancing method of approximation described in the sequel also neglects the states which are least relevant with respect to $\|\cdot\|_{++}=\|\cdot\|_{--}$, see the third remark after the algorithm in this section. □

Remark. The numbers λ_i, $i=1,\dots,n$, do not depend upon a choice of basis in $\mathbb{R}^n$ and hence are *invariants* for the system H, as K_+ and $\tilde{K}_+$ do not depend on a choice of basis in $\mathbb{R}^m$ or on the choice of a feedback $F:\mathbb{R}^m\rightarrow\mathbb{R}^n$. In the sequel we derive that $\lambda_i=\sigma_i\cdot(1-\sigma_i^2)^{-1/2}$, where $1>\sigma_1\geq\dots\geq\sigma_n>0$ are so-called Hankel singular values corresponding to H, see the fourth remark after the algorithm in this section. □

This leads to the following balancing method of model approximation, which constructs $\hat{X}$ in such a way that the relevance of states in $\mathbb{R}^n$ is maximal on $\hat{X}$.

Algorithm for approximation by balancing

1. For $H\in\mathbb{B}_2$ determine forward and backward scattering representations $\mathcal{B}_s^2(A_+,B_+,C_+,D_+)$ and ${}^R\mathcal{B}_s^2(A_-,B_-,C_-,D_-)$ respectively, e.g., by the algorithms of sections 4.3 and 4.4. Determine the solutions $K_+>0$ and $\tilde{K}_+>0$ of the corresponding (ARE)'s.
2. Determine balanced coordinates in $\mathbb{R}^n$ such that $(K_+,\tilde{K}_+)=(\Lambda,\Lambda^{-1})$ with $\Lambda=\text{diag}(\lambda_1,\dots,\lambda_n)$, $\lambda_1\geq\dots\geq\lambda_n>0$, e.g., by the method in the proof of lemma 5-8. Denote the parameters of the forward scattering representation in balanced coordinates by (A_b,B_b,C_b,D_+).
3. In the balanced coordinates take $\hat{X}:=\text{span}\{e_1,\dots,e_{\bar{n}}\}$ and take corresponding partitions $A_b=\begin{pmatrix}\hat{A} & A_{12}\\ A_{21} & A_{22}\end{pmatrix}$, $B_b=\begin{pmatrix}\hat{B}\\ B_2\end{pmatrix}$, $C_b=(\hat{C},C_2)$, $\hat{D}:=D_+$.
4. Define the approximate system $\hat{H}$ by $\hat{H}:=\{\hat{w}\in l_2^q;\ \exists(v,\hat{x})\in l_2^m\times l_2^{\bar{n}}$ such that $\begin{pmatrix}\sigma\hat{x}\\ \hat{w}\end{pmatrix}=\begin{pmatrix}\hat{A} & \hat{B}\\ \hat{C} & \hat{D}\end{pmatrix}\begin{pmatrix}\hat{x}\\ v\end{pmatrix}\}$.

Remark. This algorithm can also be used for reducing the complexity of controllable systems $\mathcal{B}\in\mathbb{B}_c$, cf. corollary 3-4(*ii*). We give some examples in section 5.4. □

Remark. Instead of determining ${}^R\mathcal{B}_s^2(A_-,B_-,C_-,D_-)$ in step 1 we also can calculate $\tilde{K}_+$ directly from $\mathcal{B}_s^2(A_+,B_+,C_+,D_+)$, as $\tilde{K}_+=Q_+^{-1}-K_+$ where $Q_+:=\Sigma_{t=0}^{\infty}A_+^tB_+B_+^T(A_+^T)^t$. □

Remark. The approximate system $\hat{H}$ is the same one as would be obtained by choosing coordinates such that $(K_+,\tilde{K}_+)=(\Lambda^2,I)$. Hence the states in $\hat{X}$ *also* are the most relevant ones if measured by the norms $\|\cdot\|_{++}$ and $\|\cdot\|_{--}$. □

Remark. $\hat{H}$ also coincides with the approximate system obtained by classical balancing of the system $\mathcal{B}_s^+$ if v_+ is considered as input and w as output. We refer to Pernebo and Silverman [58]. In this case the coordinates are chosen such that $K_+=Q_+=\Sigma:=\text{diag}(\sigma_1,\dots,\sigma_n)$ with $\sigma_1\geq\dots\geq\sigma_n>0$, and in these coordinates the approximate state space is taken as $\bar{X}:=\text{span}\{e_1,\dots,e_{\bar{n}}\}$.

That $\bar{X}=\hat{X}$ and hence that the approximate systems coincide is seen as follows. In the classical coordinates $K_+^{1/2}\tilde{K}_+^{-1}K_+^{1/2}=\Sigma^2(I-\Sigma^2)^{-1}>0$, and as $f:x\to x^2(1-x^2)^{-1}$ is strictly increasing on $0\leq x<1$ it follows from the proof of lemma 5–8 that the coordinates mentioned in that lemma are obtained from the classical coordinates by the transformation $S:=(I-\Sigma^2)^{1/4}$. As S is diagonal it follows that $\bar{X}=\hat{X}$.

Finally note that the invariants λ_i, $i=1,\dots,n$, of H are given by $\lambda_i=\sigma_i\cdot(1-\sigma_i^2)^{-1/2}$, where σ_i, $i=1,\dots,n$, are the so-called Hankel singular values of the input/output system $\{(v,w);\ \exists x$ such that $(v,x,w)\in\mathcal{B}_s^+\}$, cf. Glover [17, sections 2.2 and 2.3]. □

Remark. Hence our method closely resembles classical balancing. In classical balancing however one considers stable input/output systems, while here we consider the l_2-behaviour of arbitrary systems in $\mathbb{B}$. Moreover the reduction is not based on an input/output representation, but on a scattering representation, as a system is considered as a set and not as an input/output operator. In the next section we illustrate the difference between reduction by classical balancing and reduction by scattering by describing two simple numerical simulations. □

Remark. Note that in this balancing method of model approximation the relevance of a state x_0 is judged by considering the *future* of trajectories compatible with x_0, see the definitions of $\|\cdot\|_+$ and $\|\cdot\|_-$. So, although the time direction is irrelevant in $\mathbb{B}_2$ and $\mathbb{B}$, the approximation method is asymmetric with respect to time. □

Remark. Let $w \in H$ and $(v,x,w) \in \mathcal{B}_s^+$ in balanced coordinates. Then in $(v,\hat{x},\hat{w}) \in \mathcal{B}_s^2(\hat{A},\hat{B},\hat{C},\hat{D})$ the state $\hat{x}$ unfortunately does not coincide with the projection of x on $\hat{X}$. This means that in the approximate system the approximate state does not faithfully describe the "relevant" dynamics of the state of the original system. The same holds true for classical balancing. □

Remark. If $\lambda_{\bar{n}} = \lambda_{\bar{n}+1}$ then $\hat{X}$ and hence $\hat{H}$ are in general not uniquely defined. □

In the next proposition we describe some properties of the approximate system.

Proposition 5-9 For $H \in \mathbb{B}_2$ let $\hat{H} \in \mathbb{B}_2$ and $\hat{A} \in \mathbb{R}^{\bar{n} \times \bar{n}}$ be computed by the algorithm for approximation by balancing. Then $\sigma(\hat{A}) \subset \mathbb{C}_+$ and $m(\hat{H}) = m$, while $n(\hat{H}) \leq \bar{n}$.

Proof. See the appendix.

Remark. It may happen that $n(\hat{H}) < \bar{n}$, cf. example 2 of this section. □

Remark. Let $\hat{L} := L(\hat{A},\hat{B},\hat{C},\hat{D})$ be the driving operator corresponding to the realization of $\hat{H}$ as obtained in step 4 of the algorithm. Then in general $\hat{L}$ is not the forward scattering representation of $\hat{H}$. However, $\hat{L}$ is causal and $\mathrm{im}(\hat{L}) = \hat{H}$. □

We conclude by giving two simple examples illustrating the algorithm. In the next section we give the results of two simulations.

Example 1. Let $H := \{w \in l_2^2;\ w = (w_1,w_2),\ \ (\sigma^2-2)w_2 - \sigma w_1 = 0\}$. Define $(u,y) := (w_1,w_2)$, then H consists of the l_2-behaviour of the input/output system $y(t) = 2y(t-2) + u(t-1)$. Note that this system is unstable. Taking $x(t) := \mathrm{col}(y(t),y(t-1))$ and $v(t) := u(t)$ we obtain a minimal forward realization $\mathcal{B}_s^2(A,B,C,D)$ with $A := \begin{pmatrix} 0 & 2 \\ 1 & 0 \end{pmatrix}$, $B := \begin{pmatrix} 1 \\ 0 \end{pmatrix}$, $C := \begin{pmatrix} 0 & 0 \\ 1 & 0 \end{pmatrix}$, $D := \begin{pmatrix} 1 \\ 0 \end{pmatrix}$. The corresponding (ARE) has solution set $\{\begin{bmatrix}1&2\\2&2\end{bmatrix}, \begin{bmatrix}1&-2\\-2&2\end{bmatrix}, K_+, K_-\}$, where $K_+:=\text{diag}(2+\sqrt{5}, 1+\sqrt{5})>0$ and $K_-:=\text{diag}(2-\sqrt{5}, 1-\sqrt{5})<0$, hence $B^TK_-B+D^TD=3-\sqrt{5}>0$ and $\tilde{K}_+=-K_-$. The last result generally holds true and also can be concluded by considering the minimal backward realization ${}^R\mathcal{B}_s^2(\begin{bmatrix}0&1\\1/2&0\end{bmatrix}, \begin{bmatrix}0\\-1/2\end{bmatrix}, \begin{bmatrix}0&0\\0&1\end{bmatrix}, \begin{bmatrix}1\\0\end{bmatrix})$ of H. Let $\alpha:=(3+\sqrt{5})^{-1}$, then (R) gives $R_+=\sqrt{\alpha}$, (F) gives $F_+=(0 \quad -2+2\alpha)$, and hence $A_+=\begin{bmatrix}0&2\alpha\\1&0\end{bmatrix}$, $B_+=\begin{bmatrix}\sqrt{\alpha}\\0\end{bmatrix}$, $C_+=\begin{bmatrix}0&-2+2\alpha\\1&0\end{bmatrix}$, $D_+=\begin{bmatrix}\sqrt{\alpha}\\0\end{bmatrix}$. Balanced coordinates are obtained by the transformation $S=\begin{bmatrix}1&0\\0&\sqrt{2}\end{bmatrix}$. Requiring a reduction of the state space to one of dimension $\bar{n}=1$, the approximate system is described by $\hat{A}=0$, $\hat{B}=\sqrt{\alpha}$, $\hat{C}=\begin{bmatrix}0\\1\end{bmatrix}$, and $\hat{D}=\begin{bmatrix}\sqrt{\alpha}\\0\end{bmatrix}$. Hence $\hat{H}=\{(w_1,w_2)\in l_2^2;\ \sigma w_2-w_1=0\}$ and $(m(\hat{H}),n(\hat{H}))=(1,1)$. □

Example 2. Let $H:=\{(w_1,w_2)\in l_2^2;\ \sigma^2w_2-w_1=0\}$. Taking $x(t):=\text{col}(w_1(t-2),w_1(t-1))$ and $v(t):=w_1(t)$ we obtain a minimal forward realization $\mathcal{B}_s^2(A,B,C,D)$ with $A:=\begin{bmatrix}0&1\\0&0\end{bmatrix}$, $B:=\begin{bmatrix}0\\1\end{bmatrix}$, $C:=\begin{bmatrix}0&0\\1&0\end{bmatrix}$ and $D:=\begin{bmatrix}1\\0\end{bmatrix}$. The corresponding (ARE) has a unique solution given by $K_+=I$. Taking $\tilde{x}(t):=(w_1(t-1),w_1(t))$ and $\tilde{v}(t):=w_2(t)$ we obtain a minimal backward realization ${}^R\mathcal{B}_s^2(\tilde{A},\tilde{B},\tilde{C},\tilde{D})$ with $\tilde{A}:=\begin{bmatrix}0&0\\1&0\end{bmatrix}$, $\tilde{B}:=\begin{bmatrix}1\\0\end{bmatrix}$, $\tilde{C}:=\begin{bmatrix}0&1\\0&0\end{bmatrix}$, $\tilde{D}:=\begin{bmatrix}0\\1\end{bmatrix}$. The corresponding (ARE) also has a unique solution given by $\tilde{K}_+=I$. From (R) and (F) we get $(A_+,B_+,C_+,D_+)=(A, \beta B, C, \beta D)$, where $\beta:=1/\sqrt{2}$. As $K_+=\tilde{K}_+=I$, any unitary transformation S gives balanced coordinates, as for $x \to Sx$ there holds $(K_+,\tilde{K}_+)=(I,I) \to (K_+',\tilde{K}_+')= ((S^{-1})^TK_+S^{-1},(S^{-1})^T\tilde{K}_+S^{-1})=((SS^T)^{-1},(SS^T)^{-1})$, cf. the proof of lemma 5-8, so $(\tilde{K}_+')^{-1}=K_+'=I$ for any S with $(SS^T)^2=I$. Now suppose that we want to approximate H by a system $\hat{H}$ for which the dimension of the state space $\hat{X}$ is at most one. By taking S to be an arbitrary unitary matrix we conclude that any one-dimensional subspace of $\mathbb{R}^2$ gives a solution for approximation by balancing. For $S=I$ we obtain $(\hat{A},\hat{B},\hat{C},\hat{D})=(0,0,\begin{bmatrix}0\\1\end{bmatrix},\begin{bmatrix}\beta\\0\end{bmatrix})$, corresponding to the system $\hat{H}=\{(w_1,w_2)\in l_2^2;\ w_2=0\}$ with $n(\hat{H})=0$. For $S=\beta\begin{bmatrix}1&1\\1&-1\end{bmatrix}$ we obtain $(SA_+S^{-1}, SB_+, C_+S^{-1})=(\frac{1}{2}\begin{bmatrix}1&-1\\1&-1\end{bmatrix}, \frac{1}{2}\begin{bmatrix}1\\-1\end{bmatrix}, \beta\begin{bmatrix}0&0\\1&1\end{bmatrix})$ and as approximate system $\tilde{H}=\{(w_1,w_2)\in l_2^2;\ \sigma w_2-\frac{1}{2}w_2-\frac{1}{2}w_1=0\}$ with $n(\tilde{H})=1$. □

Remark. The first example shows that step 1 of the algorithm for L_- in section 4.4 may be successful, the second example shows that sometimes it is not. Moreover, the second example shows that approximate models obtained by balancing under the requirement that the state space has dimension at most $\bar{n}$ may result in approximations $\hat{H}$ with $n(\hat{H})<\bar{n}$. □

5.4. Simulations

5.4.1. Introduction

In this section we illustrate the method of model approximation by balancing scattering representations, as described in section 5.3, by means of two simple numerical simulations. We compare the performance of the approximate models with that of the models obtained by the classical method of model approximation by balancing for input/output systems as described, e.g., in Pernebo and Silverman [58]. As is usual in literature on balancing we compare the models by considering the corresponding impulse responses.

Notation. We denote the scattering reduction method of section 5.3 by (SR), the classical balancing method by (BR). □

Remark. In the simulations we consider controllable systems with two variables, i.e., systems $\mathcal{B}\in\mathbb{B}_c$ with $\mathcal{B}\subset(\mathbb{R}^2)^{\mathbb{Z}}$. We recall that (SR), which is presented in section 5.3 for the model class $\mathbb{B}_2$, can also be used for approximating models in $\mathbb{B}_c$, cf. corollary 3–4(*ii*) and proposition II.3–3. □

Remark. Let $\mathcal{B}\in\mathbb{B}$, $\mathcal{B}\subset(\mathbb{R}^2)^{\mathbb{Z}}$, have AR–representation $\mathcal{B}=\mathcal{B}(r_1,r_2):=\{(w_1,w_2)\in(\mathbb{R}^2)^{\mathbb{Z}};\ r_1(\sigma)w_1=r_2(\sigma)w_2\}$, where $r_1,r_2\in\mathbb{R}[s]$. It follows from Willems [74, section 4.3.1], that $\mathcal{B}$ is controllable if and only if $r_1=r_2=0$ or $r_1\neq0\neq r_2$ and r_1 and r_2 have no common non-zero roots. In the latter case, i.e., $\mathcal{B}\in\mathbb{B}_c$ and $\mathcal{B}\neq(\mathbb{R}^2)^{\mathbb{Z}}$, (w_1,w_2) can be split into an *output* variable and a ("causal" or "non-anticipating") *input* variable, cf. definition II.3–24 and Willems [74, sections 3.1.2 and 3.1.5]. Let $r_i(s)=\sum_{k=0}^{d_i}r_k^{(i)}s^k$, $i=1,2$, with $r_{d_1}^{(1)}\neq0\neq r_{d_2}^{(2)}$, so $\bar{d}(r_i)=d_i$. If $d_1>d_2$ then w_2 is the input and w_1 is the output, if $d_1<d_2$ then w_1 is the input and w_2 is the output, and if $d_1=d_2$ then either variable can be chosen as input and the other variable as output. We refer to Willems [74, sections 4.5.1 and 4.5.2]. □

Notation. Let $\mathcal{B}=\mathcal{B}(r_1,r_2)\in\mathbb{B}_c$, $r_1\neq0\neq r_2$, with $\bar{d}(r_1)\geq\bar{d}(r_2)$, and let w_2 be taken as input variable and w_1 as output variable. Then the *impulse response* g of $\mathcal{B}$ is defined as the output on $\mathbb{Z}_+$ due to a unit pulse input at time $t=0$, i.e., if

$(w_1,\delta)\in\mathcal{B}$, where $\delta(0):=1$ and $\delta(t):=0$ for all $t\neq 0$ and where $w_1^{--}:=0$, then $g:=w_1^+$. It can be shown that w_1 exists and that g is uniquely defined for $\mathcal{B}$ for the given choice of input and output variables. We also refer to section III.1.3. □

Two main distinctions between (SR) and (BR) are the following. For (SR) the approximate system does not depend on the *choice of input and output variables*, while for (BR) the approximate system in general will depend on this choice. Moreover, for (BR) the input/output system should be *stable*, i.e., if $\mathcal{B}_{i/s/o}(\tilde{A},\tilde{B},\tilde{C},\tilde{D})$ is a minimal input/state/output realization of $\mathcal{B}$ then it is required that $\sigma(\tilde{A})\subset\mathbb{C}_+$, cf. Pernebo and Silverman [58]. For $\mathcal{B}=\mathcal{B}(r_1,r_2)$ with w_2 as input and w_1 as output this amounts to the condition that r_1 has all its roots in $\mathbb{C}_+$. This is not required for (SR).

In the next sections we describe two simulations and compare the impulse responses of the approximate systems with those of the original systems. The first simulation concerns the case of roots outside of the unit circle. The second one illustrates the effect of the choice of input and output variables.

5.4.2. Simulation 1

Let $\mathcal{B}_k\in\mathbb{B}_c$ be defined by $\mathcal{B}_k:=\{(w_1,w_2)\in(\mathbb{R}^2)^{\mathbb{Z}};\ p(\sigma)w_1=q_k(\sigma)w_2\}$, $k=1,2$, where $p(s):=(s-0.95-0.1i)(s-0.95+0.1i)(s-0.6)(s-0.2)(s-0.1)$, $q_1(s):=(s-0.99)q(s)$ and $q_2(s):=(s-1.01)q(s)$, where $q(s):=(s-0.7)(s-0.15)(s+0.15)(s-0.05)$. Both $\mathcal{B}_1$ and $\mathcal{B}_2$ have minimal realizations with dimension of the state space equal to five, and we consider reduction of the state space to dimension three or two.

As a measure of performance of the approximate systems we consider the impulse response if w_2 is taken as input variable and w_1 as output variable. The corresponding impulse response of $\mathcal{B}_k$ is denoted by $g(k)$, $k=1,2$. Note that the systems $\mathcal{B}_1$ and $\mathcal{B}_2$ are nearly similar from this point of view, see figure 7.

We perform three experiments.

In experiment 1 w_2 is taken as input variable and w_1 as output variable.

In experiment 2 w_1 is taken as input variable and w_2 as output variable in the reduction procedure, and for the reduced systems the impulse response with w_2 as input and w_1 as output is determined.

In experiment 3 we suppose that w_2 is observed with one time unit delay, i.e., the observed systems are $\mathcal{B}_k^o := \mathcal{B}(p, s\cdot q_k)$, $k=1,2$. In $\mathcal{B}_k^o$ w_1 has to be considered as the input variable and w_2 as the output variable, as $\bar{d}(s\cdot q_k) > \bar{d}(p)$, $k=1,2$. For $\mathcal{B}_k^o$ approximate systems $\mathcal{B}(\hat{p}^o, \hat{q}^o)$ are determined. Then $\mathcal{B}(s\cdot\hat{p}^o, \hat{q}^o)$ is taken as approximation of $\mathcal{B}_k$, and the impulse response of $\mathcal{B}(s\cdot\hat{p}^o, \hat{q}^o)$ with w_2 as input and w_1 as output is taken as an approximation of the corresponding impulse response of the original system $\mathcal{B}_k$.

The results of experiments 1, 2 and 3 are depicted in columns 1, 2 and 3 respectively in figure 7. A solid line corresponds to the impulse response of the original system, a dashed line to that of approximate systems.

For experiments 1 and 2 the impulse responses of the approximate systems of order 3 nearly coincide with those of the original system, both for (SR) and for (BR). The corresponding impulse responses are not depicted. The impulse responses of the approximate systems of order 2 obtained by (SR) are denoted by $gs(k)$ for experiment 1 and $rgs(k)$ for experiment 2, $k=1,2$, and those obtained by (BR) are denoted by $gb(k)$ for experiment 1, $k=1,2$, and by $rgb(1)$ for experiment 2. Note that $\mathcal{B}_2$ cannot be reduced by (BR) in experiment 2, as q_2 has a root outside of the unit circle.

For experiment 3 the impulse responses of the approximate systems of order 3 obtained by (SR) are denoted by $dgs(k)$, $k=1,2$, and for (BR) by $dgb(1)$. Also in this experiment $\mathcal{B}_2$ cannot be reduced by (BR).

Interpretation. The results summarized in figure 7 have the following interpretation. The choice of input and output is not relevant for (SR), see row 1, but this choice has effect for (BR), see row 2. In case of instability the results of (SR) are satisfactory, see row 3, while for (BR) reduction is not possible, cf. row 4.

The results of (SR) are slightly inferior to those of (BR) in case of stable systems with a prescribed selection of inputs and outputs, see column 1, and they are slightly superior in case this selection is not prescribed, see column 2. □

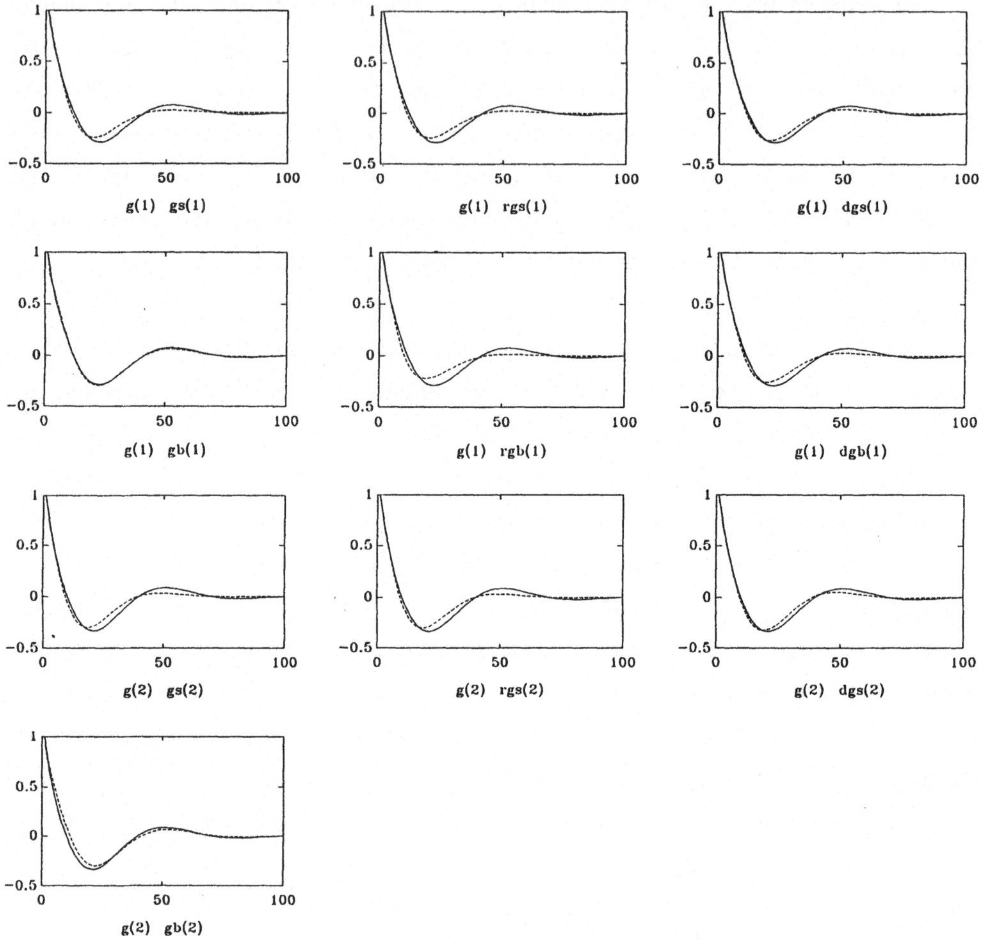

figure 7: simulation 1.

5.4.3. Simulation 2

Let $\mathcal{B}_k \in \mathbb{B}_c$ be defined by $\mathcal{B}_k := \{(w_1,w_2) \in (\mathbb{R}^2)^{\mathbb{Z}};\ r_k(\sigma)w_1 = r(\sigma)w_2\}$, $k=1,2$, where $r(s) := (s-0.95-0.3i)(s-0.95+0.3i)(s-0.90)$, $r_1(s) := (s-0.975-0.2i)(s-0.975+0.2i)(s-0.85)$ and $r_2(s) := (s-0.975-0.1i)(s-0.975+0.1i)(s-0.91)$.

We suppose that there is no prior reason to specify which variable is input and which one is output. We consider the objective of reducing the dimension of the state space from three to two in such a way, that the impulse response for w_1 as input and w_2 as output and the one for w_2 as input and w_1 as output are both reasonably approximated.

In figure 8 the results for $\mathcal{B}_1$ are depicted in column 1, those for $\mathcal{B}_2$ in column 2.

The impulse responses of $\mathcal{B}_k$ are denoted by $g12(k)$ if w_1 is taken as input and w_2 as output, and by $g21(k)$ if w_2 is taken as input and w_1 as output, $k=1,2$. These impulse responses are indicated in figure 8 by a solid line.

The impulse responses of the approximate systems are indicated by a dashed line.

For (SR) the result does not depend on the choice of input and output variables. The impulse responses of the corresponding reduced systems are denoted by $gs12(k)$ if w_1 is taken as input and w_2 as output, and by $gs21(k)$ if w_2 is taken as input and w_1 as output, $k=1,2$.

For (BR) the result depends on the choice of input and output variables. If for reduction in the original system w_1 is taken as input and w_2 as output, then the impulse responses are denoted by $gb1212(k)$ if w_1 is taken as input in the reduced system and w_2 as output, and by $gb2112(k)$ if w_2 is taken as input in the reduced system and w_1 as output, $k=1,2$. Similarly, if for reduction in the original system w_2 is taken as input and w_1 as output, then the impulse responses of the resulting approximate systems are denoted by $gb1221(k)$ and $gb2121(k)$ respectively, $k=1,2$.

Remark. In figure 8 $gb1212(k)$ nearly coincides with $g12(k)$, $k=1,2$, and $gb2121(k)$ nearly coincides with $g21(k)$, $k=1,2$. □

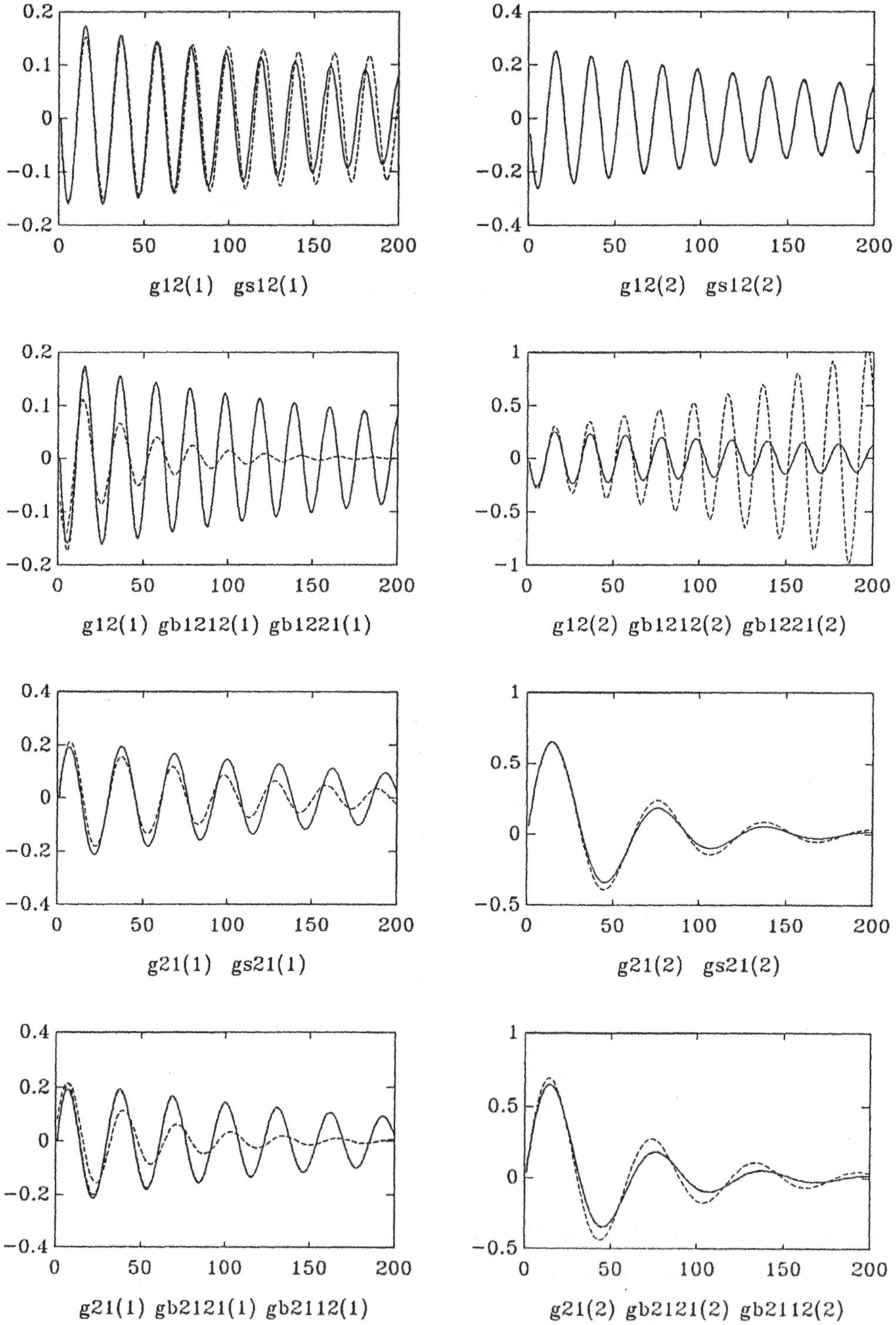

figure 8: simulation 2.

Interpretation. The results summarized in figure 8 have the following interpretation. Fixing w_2 as input and w_1 as output for reduction by (BR) leads to a bad approximation $gb1221(1)$ of $g12(1)$ and even to an unstable approximation $gb1221(2)$ of $g12(2)$, see row 2. Hence for (BR) w_1 should be taken as input and w_2 as output. This leads to very satisfactory approximations $gb1212(k)$ of $g12(k)$, k=1,2. The approximations $gs12(k)$ of 912(k) obtained by (SR) are slightly inferior, see row 1. However, the approximations $gs21(k)$ of $g21(k)$ are superior to $gb2112(k)$, see rows 3 and 4.

So, although (BR) gives very satisfactory results in case the input and output variables are fixed, it gives less satisfactory results if this is not the case. The scattering reduction method (SR) establishes a balance in the desire to approximate both the effect of w_1 on w_2 and the effect of w_2 on w_1. □

6. Conclusion

In this chapter we considered representation and approximation of models in the class $\mathbb{B}_2$ consisting of the l_2-behaviour of linear, time invariant, complete systems.

By means of scattering theory we derived special isometric representations of systems in $\mathbb{B}_2$. In these representations trajectories are generated by unstructured forward (or backward) driving variables. We presented algorithms to compute these scattering representations.

For the problem of approximation of systems in $\mathbb{B}_2$ we formulated complexity and distance measures. The complexity of a system indicates the magnitude of the system considered as a subspace. An optimal model approximation problem was formulated where the distance between systems in $\mathbb{B}_2$ is measured by the gap metric. This problem can be described explicitly in terms of scattering representations. We finally presented a heuristic balancing method of system approximation in case the objective is to reduce the dimension of the state space. We described an algorithm for approximation by balancing which is based on scattering representations. The algorithm is fairly simple and involves determination of minimal realizations of systems and solving corresponding algebraic Riccati equations. We illustrated the balancing method by means of numerical simulations.

Remark. Three main open problems are the following. First, it would be interesting to investigate the model approximation problem with the gap as distance measure. Second, it would be desirable to construct a distance measure for which the distance between systems with different numbers of driving variables is measured in an appropriate way. Third, it remains somewhat vague in which theoretical sense the approximate system obtained by balancing is close to the original system. □

In the next chapter we consider the problem of approximate modelling in case the data does not consist of a model but of a time series. We use the complexity measure as introduced in this chapter. We formulate measures of fit which are based on the purpose of description or prediction. We describe procedures and algorithms for determining models which are optimal with respect to a utility of models. This utility expresses the usefulness of a model for modelling a given time series. The usefulness depends on the complexity and fit of the model.

CHAPTER V

APPROXIMATE MODELLING

1. Introduction

In this chapter we present and investigate procedures and algorithms for approximate modelling of data. Given the objectives of modelling, the aim is to identify an optimal model for the data. The objectives of modelling are specified in terms of a utility function expressing principles of simplicity and corroboration, see section II.2.1.

In contrast with chapter III we no longer require that the identified models are unfalsified by the data. Instead we measure the degree of *corroboration* of a model by data by means of *misfit* measures which are related to the objectives of modelling. The problem then becomes to identify a model with optimal combination of complexity and misfit.

This method of model identification by approximate modelling of data is one of the leading topics in, e.g., statistics, econometrics, systems theory, and engineering. It includes problems as structure identification, estimation, and model validation. In contrast with the dominant approach to this identification problem we will not make prior assumptions concerning a presumed data generating mechanism in formulating identification procedures. In particular we make *no stochastic assumptions* for such a mechanism, for reasons discussed in sections I.1.2 and II.1.3.

We mainly consider deterministic time series analysis. The problem then consists of deterministic identification of dynamical systems by means of approximate modelling of an observed time series.

The chapter is organized as follows.

In section 2 we present algorithms for (deterministic) static approximate modelling of data. Both for the purpose of description and for that of prediction we define complexity and misfit measures. These measures are defined *directly* in terms of models considered as sets of possible outcomes, not indirectly in terms of parametrizations. We formulate procedures which minimize misfit under a complexity constraint and procedures which minimize complexity under a misfit constraint, cf. sections II.2.2 and II.2.3. We give simple algorithms to compute optimal models. These algorithms are based on singular value decomposition of the empirical covariance matrix of the data.

The results for static modelling in section 2 form the basis for descriptive and predictive procedures and algorithms for deterministic time series analysis described in sections 3 and 4. The complexity of a dynamical system is defined as in section IV.2.1. The descriptive or predictive misfit of a dynamical system with respect to an observed time series is defined in analogy with the static case. Hence these misfits are defined directly in terms of the system considered as a set of possible trajectories and not indirectly in terms of parametrizations. The misfits can be numerically expressed in terms of the canonical forms of sections II.3.2.5 and II.3.2.6. The resulting algorithms for deterministic time series analysis are fairly simple and basically consist of recursive application of the static algorithms.

The procedures have a clear optimality property as *data modelling procedures*. The identified model represents the data in a way which is optimal with respect to a utility reflecting the purpose of modelling. In section 5 we suppose that the data are generated by a phenomenon and we investigate whether the procedures also have an optimal performance for modelling dynamical phenomena, i.e., we check whether the procedures are *consistent*. A consistent procedure identifies an optimal model for the phenomenon in case the number of observations tends to infinity. We consider this question of consistency in case the phenomenon is a deterministic, linear, time invariant, complete system. For the predictive procedures we moreover consider the case of a phenomenon corresponding to an ergodic stationary stochastic process. In the latter case we especially define an optimal deterministic approximation of a stochastic process in terms of (deterministic) predictive relationships which are optimal for the specified objectives of modelling.

In section 6 finally we illustrate the procedures and algorithms for deterministic time series analysis by means of some examples. We present some illustrative numerical simulations.

Remark. The procedures and algorithms for descriptive modelling of an observed time series were developed in Willems [73, section 25]. □

2. Deterministic static modelling

2.1. Two descriptive identification procedures

2.1.1. Descriptive complexity and misfit

Suppose we want to describe a finite number of points in $\mathbb{R}^n$ by means of a linear subspace. So the data set $\mathcal{D}$ consists of the finite subsets of $\mathbb{R}^n$ and the model class $\mathbb{M}$ consists of the linear subspaces of $\mathbb{R}^n$. A model M has the interpretation that $x \in \mathbb{R}^n$ is declared to be possible if and only if $x \in M$.

As measure of complexity we take $c^D : \mathbb{M} \to \{0,1,\ldots,n\}$ defined as follows.

Definition 2-1 The *descriptive complexity* of a model $M \in \mathbb{M}$ is defined as its dimension, i.e., $c^D(M) := \dim(M)$.

Interpretation. A simple model is one which excludes much. □

Notation. Equip $\mathbb{R}^n$ with the Euclidean inner product, i.e., for $x, y \in \mathbb{R}^n$ let $<x,y> := \Sigma_{i=1}^n x_i y_i$. Define $\|x\| := \{\frac{1}{n} \Sigma_{i=1}^n x_i^2\}^{1/2}$. For a linear subspace $M \subset \mathbb{R}^n$ let $M^\perp := \{y \in \mathbb{R}^n;\ <x,y>=0 \text{ for all } x \in M\}$ denote the orthogonal complement of M in $\mathbb{R}^n$. □

To define a descriptive misfit we first consider models M of codimension 1, i.e., there is $0 \neq a \in \mathbb{R}^n$ with $M = (\text{span}\{a\})^\perp$. Such a model claims the law $<x,a>=0$ to hold true. A measure of the degree of corroboration of this law by the data

d is $\varepsilon_1^D(d,M):=e^D(d,a)$, which is defined as follows.

Definition 2-2 For data $d=\{\tilde{x}_1,...,\tilde{x}_N\}\in(\mathbb{R}^n)^N$ and $0\neq a\in\mathbb{R}^n$, the *descriptive misfit* of the law $<x,a>=0$ with respect to d is defined as $e^D(d,a):=\{\frac{1}{N}\Sigma_{i=1}^N<\tilde{x}_i,a>^2/\|a\|^2\}^{1/2}$.

Interpretation. The descriptive misfit measures the mean–square *equation error*. Note that $M=(\text{span}\{a\})^\perp$ is unfalsified by d if and only if $e^D(d,a)=0$. □

Remark. In particular cases the specification of the modelling problem might suggest inner products on $\mathbb{R}^n$ which differ from the Euclidean one, i.e., measuring the equation error by $<x,a>_Q:=x^TQa$ with $I\neq Q=Q^T>0$. A typical example is normalizing scaling of variables, i.e., to take $Q:=\text{diag}(\nu_1^{-1},...,\nu_n^{-1})$ where $\nu_j:=\{\frac{1}{N}\Sigma_{i=1}^N\tilde{x}_{ij}^2\}^{1/2}$ for $\tilde{x}_i=(\tilde{x}_{ij})\in\mathbb{R}^n$. The corresponding identification problem can be reformulated in terms of the Euclidean inner product by applying a prior data transformation $x \to Qx$. □

If the codimension of M is larger than 1, then $\varepsilon_1^D(d,M)$ is defined as the descriptive misfit of the *worst* law claimed by M, i.e., $\varepsilon_1^D(d,M):=\max\{\varepsilon_1^D(d,M');\ M\subset M',\ \text{codim}(M')=1\}$. Note that the model M claims that $\tilde{x}_i\in M$, so in particular also that for $M'\supset M$ $\tilde{x}_i\in M'$, $i=1,...,n$.

Definition 2-3 For $d\in(\mathbb{R}^n)^N$ and $M\in\mathbb{M}$, the *first descriptive misfit* is $\varepsilon_1^D(d,M):=\max\{\varepsilon^D(d,a);\ 0\neq a\in M^\perp\}$.

Remark. For $M=\mathbb{R}^n$ we define $\varepsilon_1^D(d,\mathbb{R}^n):=0$. □

Definition 2-4 The *descriptive misfit* is a map $\varepsilon^D:\mathcal{D}\times\mathbb{M}\to\mathbb{R}_+^n$, where for $d\in\mathcal{D}$ and $M\in\mathbb{M}$ $\varepsilon_1^D(d,M)$ is as defined in definition 2–3, $\varepsilon_k^D(d,M):=0$ for $k\in[n-c^D(M)+1,n]$, and for $k\in[2,n-c^D(M)]$ the k–th descriptive misfit $\varepsilon_k^D(d,M)$ is inductively defined as follows: if for $j<k$ $\varepsilon_j^D(d,M)=e^D(d,a_j)$ with $0\neq a_j\in M^\perp\cap(\text{span}\{a_1,...,a_{j-1}\})^\perp$, then $\varepsilon_k^D(d,M):=\max\{e^D(d,a);\ 0\neq a\in M^\perp\cap(\text{span}\{a_1,...,a_{k-1}\})^\perp\}$.

Interpretation. The model M corresponds to claiming that all the laws in $M^{\perp}$ hold true. This is a space of dimension $n-c^D(M)$. The descriptive misfit measures the equation errors of a suitably chosen basis for the claimed laws. The first descriptive misfit is the misfit of the worst law claimed by M. The second descriptive misfit is the misfit of the worst-but-one law, as it measures the worst quality of the laws claimed by M which are orthogonal to the worst law a_1 in M. The other misfit numbers have a similar interpretation. □

Remark. In section 2.1.3 it is shown that ε^D is a well-defined map, i.e., even if the a_j's are not unique the numbers $\varepsilon_k^D(d,M)$, $k\in[2,n-c^D(M)]$, are well-defined, independent of the choice of a_j. See proposition 2-11. □

Remark. Note that for $M=\mathbb{R}^n$ there holds $\varepsilon_k^D(d,\mathbb{R}^n):=0$ for $k\in[1,n]$. □

Remark. Note that complexity and misfit are defined on the level of models, considered as sets, and not on the parameter level. This especially avoids problems in case of non-unique parametrization. □

In section 2.1.3 we give explicit algorithms for the descriptive procedures $P^D_{\varepsilon_{tol}}$, corresponding to minimizing complexity for a given tolerated misfit, and $P^D_{c_{tol}}$, corresponding to minimizing misfit for a given tolerated complexity. We use the following orderings of complexity and misfit. On the complexity space $\{0,1,...,n\}$ we take the natural ordering.

Definition 2-5 The misfit space $\mathbb{R}^n_+$ is ordered *lexicographically*, i.e., $\{(\varepsilon_1,...,\varepsilon_n)\geq(\tilde{\varepsilon}_1,...,\tilde{\varepsilon}_n)\}$ $:\Leftrightarrow$ $\{\varepsilon_k=\tilde{\varepsilon}_k$ for all $k\in[1,n]$, or there exists a k such that $\varepsilon_k>\tilde{\varepsilon}_k$ and $\varepsilon_i=\tilde{\varepsilon}_i$ for all $i<k\}$, where $\mathbb{R}_+$ is ordered in the natural way.

2.1.2. Some results on singular values

The algorithms for deterministic approximate modelling which we give in sections 2.1.3 and 2.2.3 for static data and in section 4 for time series analysis are all based on some basic results in numerical linear algebra concerning the singular value decomposition. In this section we define this decomposition and summarize results relevant for the sequel.

Definition 2-6 A *singular value decomposition* (SVD) of a matrix $A\in\mathbb{R}^{n_1\times n_2}$ is a decomposition $A=U\Sigma V^T$, where $U\in\mathbb{R}^{n_1\times n_1}$ and $V\in\mathbb{R}^{n_2\times n_2}$ are orthogonal, i.e., $UU^T=U^TU=I_{n_1}$ and $VV^T=V^TV=I_{n_2}$, and where Σ is diagonal, i.e., $\Sigma=\begin{pmatrix}\Sigma' & 0\\ 0 & 0\end{pmatrix}\in\mathbb{R}^{n_1\times n_2}$ where $\Sigma'=\text{diag}(\sigma_1,\ldots,\sigma_r)\in\mathbb{R}^{r\times r}$, with $\sigma_1\geq\ldots\geq\sigma_r>0$.

For the next result we refer to Stewart [67, section 6.6, theorem 6.1].

Lemma 2-7 Every matrix A has a (SVD).

Remark. The numbers σ_k, $k=1,\ldots,r$, are uniquely defined and consist of the square roots of the non-zero eigenvalues of A^TA, as $A^TAV=V\Sigma^T\Sigma$. The numbers σ_k, $k=1,\ldots,r$, are called the (non-zero) *singular values* of A. Note that $r=\text{rank}(A)$. For $k>r$ we define $\sigma_k:=0$. Finally note that U and V need not be unique. □

Notation. For $V=(v_1,\ldots,v_{n_2})$ with $v_i\in\mathbb{R}^{n_2}$, $i=1,\ldots,n_2$, let $L_k^*:=\text{span}\{v_{n_2-k+1},\ldots,v_{n_2}\}$. □

Remark. If $\sigma_{n_2-k}=\sigma_{n_2-k+1}$ then L_k^* is not uniquely defined by A. However, if $\sigma_{n_2-k}>\sigma_{n_2-k+1}$ then L_k^* is uniquely defined as the sum of the eigenspaces of A^TA corresponding to eigenvalues $\lambda\leq\sigma_{n_2-k+1}^2$. □

The next proposition expresses the so-called minimax property of singular values.

Proposition 2-8 Let $\sigma_1\geq\ldots\geq\sigma_r>0$ be the singular values of $A\in\mathbb{R}^{n_1\times n_2}$. Then $\sigma_{k+1}=\min\{\max_{0\neq x\in L}\frac{\|Ax\|}{\|x\|};\ L\subset\mathbb{R}^{n_2},\ \dim(L)\geq n_2-k\}=\max_{0\neq x\in L_{n_2-k}^*}\frac{\|Ax\|}{\|x\|}$.

Proof. See the appendix.

Notation. Let $A \in \mathbb{R}^{n_1 \times n_2}$ be given. For a linear subspace $L \subset \mathbb{R}^{n_2}$ with $\dim(L)=d$ define $\varepsilon(L):=\varepsilon_A(L) \in \mathbb{R}_+^{n_2}$ as follows. Let $\varepsilon_1(L):=\max_{0 \neq x \in L} \frac{\|Ax\|}{\|x\|}$, for $k \in [d+1,n_2]$ let $\varepsilon_k(L):=0$, and for $k \in [2,d]$ let $\varepsilon_k(L)$ be inductively defined as follows: if for $j<k$ $\varepsilon_j(L)=\frac{\|Ax_j\|}{\|x_j\|}$ with $0 \neq x_j \in L \cap (\text{span}\{x_1,\dots,x_{j-1}\})^{\perp}$, then $\varepsilon_k(L):=\max\{\frac{\|Ax\|}{\|x\|};\ 0 \neq x \in L \cap (\text{span}\{x_1,\dots,x_{k-1}\})^{\perp}\}$. □

Lemma 2-9 For any matrix $A \in \mathbb{R}^{n_1 \times n_2}$ and linear space $L \subset \mathbb{R}^{n_2}$ $\varepsilon_A(L)$ is well-defined.

Proof. See the appendix.

The following proposition forms the basis of the algorithms in sections 2.1.3 and 2.2.3.

Proposition 2-10 Let $A \in \mathbb{R}^{n_1 \times n_2}$ be given and let L be an arbitrary linear subspace of $\mathbb{R}^{n_2}$ with $\dim(L) \geq k$. With respect to the lexicographic ordering on $\mathbb{R}_+^{n_2}$ there holds

(*i*) $\varepsilon_A(L) \geq \varepsilon_A(L_k^*)=(\sigma_{n_2-k+1},\dots,\sigma_r,0,\dots,0)$;

(*ii*) if $\sigma_{n_2-k} > \sigma_{n_2-k+1}$ then $\{\varepsilon_A(L)=\varepsilon_A(L_k^*)\} \Leftrightarrow \{L=L_k^*\}$;

(*iii*) if $\sigma_{n_2-k}=\sigma_{n_2-k+1}>0$ then $\{\varepsilon_A(L)=\varepsilon_A(L_k^*)\} \Leftrightarrow \{L=L'+L''$ where $L':=\text{span}\{v_j;\ j$ such that $\sigma_j<\sigma_{n_2-k}\}$ and where L'' is any subspace of dimension $k-\dim(L')$ of $\text{span}\{v_j;\ j$ such that $\sigma_j=\sigma_{n_2-k}\}$ $\}$.

Proof. See the appendix.

2.1.3. Algorithms for descriptive modelling

Using the results on singular values of section 2.1.2 we give simple algorithms for descriptive modelling of static data. These algorithms form the basis of the algorithms for descriptive modelling of time series as described in section 4.2.

Notation. For data $d=\{\tilde{x}_1,\dots,\tilde{x}_N\} \in (\mathbb{R}^n)^N$ let the *empirical covariance matrix* be defined as $S:=\frac{1}{N}\sum_{i=1}^{N}\tilde{x}_i\tilde{x}_i^T$. □

Proposition 2–11 The descriptive misfit of section 2.1.1 is a well-defined mapping $\varepsilon^D:\mathcal{D}\times\mathbb{M}\to\mathbb{R}_+^n$. For data $d\in(\mathbb{R}^n)^N$ and model $M\in\mathbb{M}$ $\varepsilon^D(d,M)=\varepsilon_S(M^\perp)$ where S is the empirical covariance matrix of the data.

Proof. See the appendix.

Remark. This result enables us to use proposition 2–10 to determine optimal descriptive models. □

Next we give algorithms for the descriptive procedures $P^D_{c_{tol}}$, corresponding to minimizing lexicographically the descriptive misfit under the condition that $\dim(M)\leq c_{tol}$, and $P^D_{\varepsilon_{tol}}$, corresponding to minimizing complexity under the condition that $\varepsilon^D(d,M)<\varepsilon_{tol}$.

Remark. More precise, for given data $d\in(\mathbb{R}^n)^N$ the procedures $P^D_{c_{tol}}$ and $P^D_{\varepsilon_{tol}}$ identify the following models, cf. sections II.2.2 and II.2.3. Let $c_{tol}\in\{0,1,\ldots,n\}$ be given. Then $M\in P^D_{c_{tol}}(d)$ if and only if *(i)* $c^D(M)\leq c_{tol}$; *(ii)* $\varepsilon^D(d,M)$ is lexicographically minimal among models of complexity at most c_{tol}, i.e., $\{c^D(M')\leq c_{tol}\}\Rightarrow\{\varepsilon^D(d,M')\geq\varepsilon^D(d,M)\}$; *(iii)* among models with minimal achievable descriptive misfit the complexity is minimal, i.e., $\{c^D(M')\leq c_{tol}, \varepsilon^D(d,M')=\varepsilon^D(d,M)\}\Rightarrow\{c^D(M)\leq c^D(M')\}$.

On the other hand, suppose that $\varepsilon_{tol}\in\mathbb{R}^n$ is given. We suppose that $\varepsilon_{tol}=\varepsilon_1^{tol}(1,\ldots,1)$, $\varepsilon_1^{tol}\in\mathbb{R}$. In this case the requirement that $\varepsilon^D(d,M)<\varepsilon_{tol}$ is equivalent to $\varepsilon_1^D(d,M)<\varepsilon_1^{tol}$, as clearly $\varepsilon_j^D(d,M)\leq\varepsilon_k^D(d,M)$ for $j\geq k$. Then $M\in P^D_{\varepsilon_{tol}}(d)$ if and only if *(i)* $\varepsilon_1^D(d,M)<\varepsilon_1^{tol}$; *(ii)* $c^D(M)$ is minimal among models with worst fit smaller than ε_1^{tol}, i.e., $\{\varepsilon_1^D(d,M')<\varepsilon_1^{tol}\}\Rightarrow\{c^D(M')\geq c^D(M)\}$; *(iii)* among models with minimal achievable complexity the misfit is minimal, i.e., $\{\varepsilon_1^D(d,M')<\varepsilon_1^{tol}, c^D(M')=c^D(M)\}\Rightarrow\{\varepsilon^D(d,M)\leq\varepsilon^D(d,M')\}$. □

Notation. For data $d=\{\tilde{x}_1,\ldots,\tilde{x}_N\}\in(\mathbb{R}^n)^N$ let $S:=\frac{1}{N}\Sigma_{i=1}^N\tilde{x}_i\tilde{x}_i^T$ have (SVD) $S=U\Sigma U^T$, where $\Sigma=\mathrm{diag}(\sigma_1,\ldots,\sigma_n)$ with $\sigma_1\geq\ldots\geq\sigma_n\geq0$. If $\mathrm{rank}(S)=r$ then $\sigma_r>0=\sigma_{r+1}=\ldots=\sigma_n$. Let $U=(u_1,\ldots,u_n)$ where $u_j\in\mathbb{R}^n$ denotes the j–th column of U. Define $M_k^*:=\mathrm{span}\{u_1,\ldots,u_k\}$ and $M(\sigma):=\mathrm{span}\{u_j;\ j \text{ such that } \sigma_j=\sigma\}$. □

Proposition 2-12 For given data $d=\{\tilde{x}_1,\ldots,\tilde{x}_N\}\in(\mathbb{R}^n)^N$ and tolerated complexity c_{tol}, $P^D_{c_{tol}}(d)$ is given by

(*i*) $P^D_{c_{tol}}(d)=\{0\}$ if $c_{tol}=0$;

(*ii*) $P^D_{c_{tol}}(d)=\text{span}\{\tilde{x}_1,\ldots,\tilde{x}_N\}$ if $c_{tol}\geq r$;

(*iii*) $P^D_{c_{tol}}(d)=M^*_{c_{tol}}$ if $0<c_{tol}<r$ and $\sigma_{c_{tol}}>\sigma_{c_{tol}+1}$;

(*iv*) if $0<c_{tol}<r$ and $\sigma_1\geq\ldots\geq\sigma_{c_1}>\sigma_{c_1+1}=\ldots=\sigma_{c_{tol}}=\sigma_{c_{tol}+1}\geq\sigma_{c_{tol}+2}\geq\ldots\geq\sigma_n$ then $P^D_{c_{tol}}(d)=\{M^*_{c_1}+L;\ L\subset M(\sigma_{c_{tol}}),\ \dim(L)=c_{tol}-c_1\}$.

Proof. See the appendix.

Proposition 2-13 Let data $d=\{\tilde{x}_1,\ldots,\tilde{x}_N\}\in(\mathbb{R}^n)^N$ be given. Assume moreover that a maximal tolerated misfit level is given with $\varepsilon_{tol}=\varepsilon_1^{tol}.(1,\ldots,1)$, $\varepsilon_1^{tol}>0$, so the misfit restriction concerns only the worst law claimed by a model. Then

(*i*) $P^D_{\varepsilon_{tol}}(d)=\{0\}$ if $\varepsilon_1^{tol}>\sigma_1$;

(*ii*) $P^D_{\varepsilon_{tol}}(d)=\text{span}\{\tilde{x}_1,\ldots,\tilde{x}_N\}$ if $\varepsilon_1^{tol}\leq\sigma_r$;

(*iii*) if $\sigma_r<\varepsilon_1^{tol}\leq\sigma_1$, then $P^D_{\varepsilon_{tol}}(d)=M^*_k$ where k is such that $\sigma_k\geq\varepsilon_1^{tol}>\sigma_{k+1}$.

Proof. See the appendix.

Remark. The algorithms hence amount to singular value decomposition of the empirical covariance matrix. □

Remark. The procedure $\bar{\bar{P}}^D_{\varepsilon_{tol}}$ corresponding to the requirement $\varepsilon_1^D(d,M)\leq\varepsilon_1^{tol}$ with $\varepsilon_1^{tol}\geq0$ can be described in a way analogous to proposition 2-13. So the identified model is $\{0\}$ if $\varepsilon_1^{tol}\geq\sigma_1$, $\text{span}\{\tilde{x}_1,\ldots,\tilde{x}_N\}$ if $\varepsilon_1^{tol}<\sigma_r$, and M^*_k if $\sigma_r\leq\varepsilon_1^{tol}<\sigma_1$ and $\sigma_k>\varepsilon_1^{tol}\geq\sigma_{k+1}$. We also refer to Willems [73, theorem 24]. □

Remark. Note that no model can be accepted by $P^D_{\varepsilon_{tol}}$ in case $\varepsilon_1^{tol}\leq0$ and that no model can be accepted by $\bar{\bar{P}}^D_{\varepsilon_{tol}}$ in case $\varepsilon_1^{tol}<0$. □

Remark. There is a close relationship between these procedures and *total least squares*. As a simple example consider the case $c_{tol}=n-1$. For $0\neq a\in\mathbb{R}^n$ let $M(a):=(\text{span}\{a\})^{\perp}:=\{x\in\mathbb{R}^n;\ <x,a>=0\}$ and let π_a denote the orthogonal projection operator onto $M(a)$. For given data $d=\{\tilde{x}_1,\ldots,\tilde{x}_N\}\in(\mathbb{R}^n)^N$, in total least squares

a is determined such that $\delta(d,a):=\frac{1}{N}\Sigma_{i=1}^{N}\|\tilde{x}_i-\pi_a\tilde{x}_i\|^2$ is minimal, cf. Golub and Van Loan [19]. See figure 9 for the case $n=2$. As $I-\pi_a=\|a\|^{-2}.aa^T$ it follows that $\delta(d,a)=\{a^T(\frac{1}{N}\Sigma_{i=1}^{N}\tilde{x}_i\tilde{x}_i^T)a\}/\|a\|^2=\{\varepsilon_1^D(d,M(a))\}^2$. So in this case of $c_{tol}=n-1$ the procedure $P_{c_{tol}}^D$ corresponds to total least squares. Analogous results can be obtained for $c_{tol}<n-1$ and for $P_{\varepsilon_{tol}}^D$, cf. Willems [73, section 21]. □

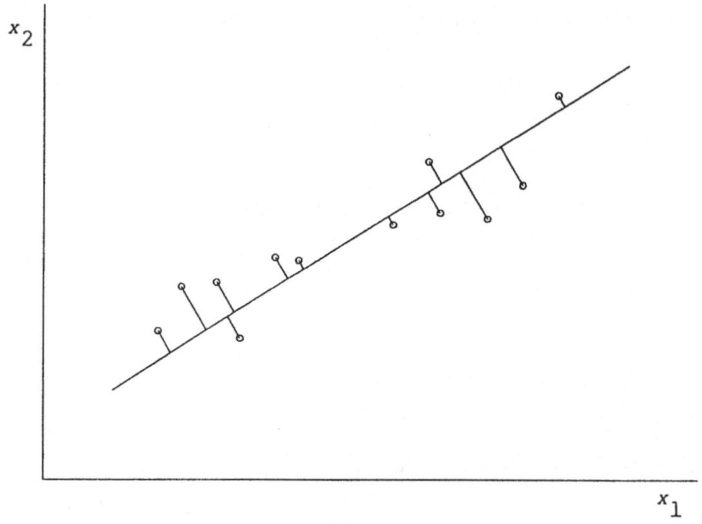

figure 9

Remark. The procedure $P_{c_{tol}}^D$ resembles parameter fitting procedures for identification. On the other hand the procedure $P_{\varepsilon_{tol}}^D$ determines the model structure from the data. Especially the number of independent relations is not fixed a priori but is determined by the data and a pragmatic requirement of fit. □

2.2. Two predictive identification procedures

2.2.1. Predictive complexity and misfit

Suppose we want to predict (or estimate) n_2 variables $y\in\mathbb{R}^{n_2}$ on the basis of n_1 other variables $x\in\mathbb{R}^{n_1}$ by means of linear subspace of $\mathbb{R}^{n_1+n_2}$.

Let N observations $(\tilde{x}_i,\tilde{y}_i)$, $\tilde{x}_i\in\mathbb{R}^{n_1}$, $\tilde{y}_i\in\mathbb{R}^{n_2}$, $i=1,\ldots,N$, be available, so the data set is $\mathcal{D}=(\mathbb{R}^{n_1+n_2})^N$.

As model class $\mathbb{M}$ we take the collection of linear subspaces of $\mathbb{R}^{n_1+n_2}$ for which the projection on the first n_1 coordinates is surjective, i.e., $M \in \mathbb{M}$ if M is a linear subspace of $\mathbb{R}^{n_1+n_2}$ and $\{x \in \mathbb{R}^{n_1};\ \exists y \in \mathbb{R}^{n_2} \text{ with } (x,y) \in M\} = \mathbb{R}^{n_1}$.

Interpretation. A model M predicts that, for given x, y will belong to the set $M(x) := \{y \in \mathbb{R}^{n_2};\ (x,y) \in M\}$. Stated otherwise, for observed x the model M predicts that the (unobserved) y associated with x will be such that $<a_1,x> + <a_2,y> = 0$ for all $(a_1,a_2) \in M^\perp$, $a_1 \in \mathbb{R}^{n_1}$, $a_2 \in \mathbb{R}^{n_2}$. The surjectivity condition is equivalent to requiring that prediction is possible for every $x \in \mathbb{R}^{n_1}$. □

It is easily seen that $M(x) = y + M(0)$ for any $x \in \mathbb{R}^{n_1}$, $y \in M(x)$. So for given model $M \in \mathbb{M}$ the dimension of the (affine) predicted set is independent of the observation x. We define the predictive complexity $c^P : \mathbb{M} \to \{0,1,\dots n_2\}$ as follows.

Definition 2-14 The *predictive complexity* of a model $M \in \mathbb{M}$ is defined as the dimension of the affine predicted set, i.e., $c^P(M) := \dim(M(0))$.

Interpretation. A simple model corresponds to predictions with few degrees of freedom. □

To define a predictive misfit we again first consider models of codimension 1. Let $0 \neq a = (a_1,a_2) \in \mathbb{R}^{n_1} \times \mathbb{R}^{n_2}$ and $M = (\text{span}\{a\})^\perp$. Note that $M \in \mathbb{M}$ implies that $a_2 \neq 0$. The model M predicts that, for given x, y will satisfy $<a_2,y> = -<a_1,x>$. For data d the relative mean prediction error of this model is $\varepsilon_1^P(d,M) := e^P(d,a)$, which is defined as follows.

Definition 2-15 For data $d = \{(\tilde{x}_i,\tilde{y}_i);\ i=1,\dots,N\} \in (\mathbb{R}^{n_1} \times \mathbb{R}^{n_2})^N$ and for $a = (a_1,a_2) \in \mathbb{R}^{n_1} \times \mathbb{R}^{n_2}$ with $a_2 \neq 0$, the *relative mean prediction error* is defined by $e^P(d,a) := [\{\frac{1}{N}\Sigma_{i=1}^N(<a_1,\tilde{x}_i> + <a_2,\tilde{y}_i>)^2\} \,/\, \{\frac{1}{N}\Sigma_{i=1}^N <a_2,\tilde{y}_i>^2\}]^{1/2}$.

Interpretation. The relative mean prediction error measures the mean-square prediction error relative to the magnitude of the predicted signal. □

If the codimension of M is larger than 1, then $\varepsilon^P(d,M)$ is defined in analogy with the misfit in section 2.1.1, i.e., $\varepsilon_1^P(d,M)$ measures the predictive misfit of the worst prediction made by M, $\varepsilon_2^P(d,M)$ the misfit of the prediction worst-but-one, and so on.

Definition 2-16 For $d\in(\mathbb{R}^{n_1}\times\mathbb{R}^{n_2})^N$ and $M\in\mathbb{M}$, the *first predictive misfit* is $\varepsilon_1^P(d,M):=\max\{e^P(d,a);\ 0\neq a\in M^\perp\}$.

Remark. If $M(0)=\mathbb{R}^{n_2}$, or equivalently if $M=\mathbb{R}^{n_1+n_2}$, then $\varepsilon_1^P(d,M):=0$. □

Notation. For $M\in\mathbb{M}$ let $M_2^\perp:=\{a_2\in\mathbb{R}^{n_2};\ \exists a_1\in\mathbb{R}^{n_1}$ such that $(a_1,a_2)\in M^\perp\}$, so $M_2^\perp$ consists of the set of predicted functionals on y. It is easily seen that $M_2^\perp=\{M(0)\}^\perp$, hence $\dim(M_2^\perp)=n_2-c^P(M)$. Further let $S_{yy}:=\frac{1}{N}\Sigma_{i=1}^N\tilde{y}_i\tilde{y}_i^T$ and for $\alpha,\beta\in\mathbb{R}^{n_2}$ let $\{\alpha\perp_{(y)}\beta\}:\Leftrightarrow\{\alpha^TS_{yy}\beta=0\}$. □

Definition 2-17 The *predictive misfit* $\varepsilon^P(d,M)\in\mathbb{R}_+^{n_2}$ is defined by $\varepsilon_1^P(d,M)$ as in definition 2-16, $\varepsilon_k^P(d,M):=0$ for $k\in[n_2-c^P(M)+1,n_2]$, and for $k\in[2,n_2-c^P(M)]$ the k-th predictive misfit $\varepsilon_k^P(d,M)$ is inductively defined as follows: if for $j<k$ $\varepsilon_j^P(d,M)=e^P(d,a^{(j)})$ with $0\neq a^{(j)}\in M^\perp$, $a_2^{(j)}\perp_{(y)}\text{span}\{a_2^{(1)},\dots,a_2^{(j-1)}\}$, then $\varepsilon_k^P(d,M):=\max\{e^P(d,a);\ 0\neq a\in M^\perp,\ a_2\perp_{(y)}\text{span}\{a_2^{(1)},\dots,a_2^{(k-1)}\}\}$.

Interpretation. The predictive misfit measures the prediction errors of a suitably chosen basis for the space of all predictive relationships in $M^\perp$ which are imposed by the model M. The first predictive misfit expresses the error of the worst prediction in M. The other misfits measure the errors of predictions *orthogonal* to the worst one. The orthogonality is taken with respect to $\perp_{(y)}$, i.e., data dependent, as in this way the misfit is *independent of a choice of coordinates in* $\mathbb{R}^{n_2}$. Indeed, let $B\in\mathbb{R}^{n_2\times n_2}$ be nonsingular and let $y':=By$. Then the model $M':=\{(x,y');\ (x,B^{-1}y')\in M\}$ in new coordinates is equivalent to M in old coordinates, as $M'(x)=BM(x)$ and hence $y\in M(x)$ if and only if $y'\in M'(x)$, i.e., for any x the models M and M' lead to the same predictions. Let $d':=\{(\tilde{x}_i,\tilde{y}_i');\ i\in[1,N]\}$ and for $a=(a_1,a_2)$ let $a':=(a_1,(B^T)^{-1}a_2)$, then $e^P(d',a')=e^P(d,a)$. It is reasonable to require that $\varepsilon^P(d',M')=\varepsilon^P(d,M)$. As $a_2'=(B^T)^{-1}a_2$ and $S_{y'y'}=BS_{yy}B^T$ this is achieved by taking the orthogonality with respect to S_{yy} and $S_{y'y'}$ respectively. □

Remark. In section 2.2.2 we show that ε^P is well-defined provided that $N \geq \max\{n_1, n_2\}$ and provided that the data are generic in the sense that $\text{span}\{\tilde{x}_1, \ldots, \tilde{x}_N\} = \mathbb{R}^{n_1}$ and $\text{span}\{\tilde{y}_1, \ldots, \tilde{y}_N\} = \mathbb{R}^{n_2}$, see proposition 2-18. □

Remark. Note that for $M = \mathbb{R}^{n_1+n_2}$ there holds $\varepsilon_k^P(d, M) := 0$ for $k \in [1, n_2]$. □

Remark. Note that again complexity and misfit are defined on the level of models, not on the parameter level. □

We order the complexity and misfit spaces as in section 2.1.1, i.e., naturally and lexicographically respectively, cf. definition 2-5. In section 2.2.3 we give explicit algorithms for the procedures $P^D_{\varepsilon_{tol}}$ corresponding to minimizing complexity for a given tolerated misfit, and $P^D_{c_{tol}}$ corresponding to minimizing predictive misfit for a given tolerated complexity. These algorithms are based on results on canonical correlation which we describe in the next section.

2.2.2. Some results on canonical correlation

In this section we show that the predictive misfit of definition 2-17 is well-defined for generic data in case the number of observations is sufficiently large. Moreover we show that models of minimal predictive misfit can be expressed in terms of canonical variables.

Let the data be $d = \{(\tilde{x}_i, \tilde{y}_i);\ i = 1, \ldots, N\} \in (\mathbb{R}^{n_1+n_2})^N$. Throughout we assume that $N \geq \max\{n_1, n_2\}$ and that the data are *generic* in the sense that $\text{span}\{\tilde{x}_1, \ldots, \tilde{x}_N\} = \mathbb{R}^{n_1}$ and $\text{span}\{\tilde{y}_1, \ldots, \tilde{y}_N\} \in \mathbb{R}^{n_2}$

Remark. Recall from section III.3.1.1 that a set $\pi \subset \mathbb{R}^n$ is called generic if there is a polynomial $0 \neq p\colon \mathbb{R}^n \to \mathbb{R}$ such that $\pi \supset (\mathbb{R}^n \backslash p^{-1}(0))$. Let $L_1 := (\tilde{x}_1, \ldots, \tilde{x}_N) \in \mathbb{R}^{n_1 \times N}$ and $L_2 := (\tilde{y}_1, \ldots, \tilde{y}_N) \in \mathbb{R}^{n_2 \times N}$, then the imposed condition is equivalent to $p(d) := \det(L_1 L_1^T).\det(L_2 L_2^T) \neq 0$. As $p \not\equiv 0$ for $N \geq \max\{n_1, n_2\}$ it follows that in this case the condition is satisfied generically in $(\mathbb{R}^{n_1+n_2})^N$. □

Notation. Let the empirical covariance matrix be $\begin{pmatrix} S_{xx} & S_{xy} \\ S_{yx} & S_{yy} \end{pmatrix} := \frac{1}{N}\Sigma_{i=1}^{N} \begin{pmatrix} \tilde{x}_i \\ \tilde{y}_i \end{pmatrix} \begin{pmatrix} \tilde{x}_i \\ \tilde{y}_i \end{pmatrix}^T \in \mathbb{R}^{(n_1+n_2)\times(n_1+n_2)}$. For generic data S_{xx} and S_{yy} are invertible. Let $S_{xx}^{-1/2}$ and $S_{yy}^{-1/2}$ be the positive square roots of S_{xx}^{-1} and S_{yy}^{-1} respectively. Let $S_{xx}^{-1/2}S_{xy}S_{yy}^{-1/2}$ have (SVD) $U\Sigma V^T$, where $UU^T=U^TU=I_{n_1}$, $VV^T=V^TV=I_{n_2}$, and $\Sigma=\begin{pmatrix} \Sigma' & 0 \\ 0 & 0 \end{pmatrix}$, $\Sigma'=\text{diag}(\sigma_1,\ldots,\sigma_r)$, $\sigma_1\geq\ldots\geq\sigma_r>0$. Here $r=\text{rank}(S_{xy})$. As $\begin{pmatrix} S_{xx} & S_{xy} \\ S_{yx} & S_{yy} \end{pmatrix}\geq 0$ it follows that $\sigma_1\leq 1$. □

Proposition 2-18 The predictive misfit $\varepsilon^P(d,M)$ is well-defined for generic data and for all $M\in\mathbb{M}$.

Proof. See the appendix.

Remark. It can be shown that $\varepsilon^P(d,M)$ also is well-defined if we only assume that $N\geq n_2$ and $\text{span}\{\tilde{y}_1,\ldots,\tilde{y}_N\}=\mathbb{R}^{n_2}$. □

Notation. For a given linear subspace $L_2\subset\mathbb{R}^{n_2}$ let $\mathbb{M}(L_2):=\{M\in\mathbb{M};\ M_2^{\perp}=L_2\}$ denote the class of models in $\mathbb{M}$ for which L_2 is the set of predicted functionals on y. Define $M^*(L_2)\in\mathbb{M}(L_2)$ as the model corresponding to least-squares prediction of $<a_2,y>$ for functionals $a_2\in L_2$, i.e., $M^*(L_2):=\{(x,y);\ <a_2,y>=<S_{xx}^{-1}S_{xy}a_2,x> \text{ for all } a_2\in L_2\}$. □

Lemma 2-19 For every $M\in\mathbb{M}(L_2)$ and for generic data, $\varepsilon^P(d,M)\geq\varepsilon^P(d,M^*(L_2))$ with equality if and only if $M=M^*(L_2)$. Moreover, $\varepsilon^P(d,M^*(L_2))= \varepsilon_{(I-\Sigma^T\Sigma)^{1/2}}(V^TS_{yy}^{1/2}L_2)$ with the last expression as defined in section 2.1.2.

Proof. See the appendix.

Interpretation. To minimize the predictive misfit it suffices to choose the space $M_2^{\perp}$ of predicted functionals on y in an optimal way and to take $M^*(M_2^{\perp})$ as prediction model. The choice of $M_2^{\perp}$ can be made by using proposition 2-10. This forms the basis of the algorithms in section 2.2.3. □

Notation. Denote the columns of $S_{xx}^{-1/2}U$ by $a_1^{(i)}\in\mathbb{R}^{n_1}$, $i=1,\ldots,n_1$, and those of $S_{yy}^{-1/2}V$ by $a_2^{(i)}$, $i=1,\ldots,n_2$. For $k=1,\ldots,r$ define $M_k^*:=\{(x,y)\in\mathbb{R}^{n_1}\times\mathbb{R}^{n_2};\ <a_2^{(i)},y>=\sigma_i<a_1^{(i)},x>,\ i=1,\ldots,k\}$, so $c^P(M_k^*)=n_2-k$. □

Corollary 2-20 For generic data and for any $M\in\mathbb{M}$ with $c^P(M)\leq n_2-k$ there holds that $\varepsilon^P(d,M)\geq\varepsilon^P(d,M_k^*)=((1-\sigma_k^2)^{1/2}, (1-\sigma_{k-1}^2)^{1/2},\ldots,(1-\sigma_1^2)^{1/2}, 0,\ldots,0)$.

Proof. See the appendix.

Remark. The last result establishes the relation with *canonical correlation* analysis. In statistics canonical correlation analysis usually is introduced in the following way. Let $x:=\mathrm{col}(x_1,\ldots,x_{n_1})$ and $y:=\mathrm{col}(y_1,\ldots,y_{n_2})$ be two random vectors with zero mean and $\mathrm{cov}(x,y):=E\begin{pmatrix}x\\y\end{pmatrix}\begin{pmatrix}x\\y\end{pmatrix}^T=\begin{pmatrix}S_{xx} & S_{xy}\\ S_{yx} & S_{yy}\end{pmatrix}>0$. The first canonical correlation coefficient ρ_1 is the maximal correlation between linear combinations of the components of x and y, i.e., if for $0\neq\alpha\in\mathbb{R}^{n_1}$, $0\neq\beta\in\mathbb{R}^{n_2}$ $\rho(\alpha,\beta):=E(\alpha^Tx\cdot\beta^Ty)/\{E(\alpha^Tx)^2\cdot E(\beta^Ty)^2\}^{1/2}$ then $\rho_1:=\max\{\rho(\alpha,\beta);\ \alpha\neq0,\ \beta\neq0\}$. If $\xi_1:=\alpha_1^Tx$ and $\eta_1:=\beta_1^Ty$ give the optimal correlation, then (ξ_1,η_1) are called the first canonical variables. Analogously, the second canonical variables (ξ_2,η_2) and second canonical correlation coefficient ρ_2 are obtained by maximizing $\rho(\alpha,\beta)$ under the requirement that α^Tx and β^Ty are uncorrelated with ξ_1 and η_1 respectively, i.e., $E(\alpha^Tx\xi_1)=0$ and $E(\beta^Ty\eta_1)=0$. The i-th canonical variables (ξ_i,η_i) and i-the canonical correlation coefficient ρ_i are obtained in a similar way, $i\leq\min\{n_1,n_2\}$.

Imposing the auxiliary condition $E(\alpha^Tx)^2=E(\beta^Ty)^2=1$ and using the notation introduced before it is easily seen that $\rho(\alpha,\beta)=\bar{\alpha}^T\Sigma\bar{\beta}$, where $\bar{\alpha}:=U^TS_{xx}^{1/2}\alpha$ and $\bar{\beta}:=V^TS_{yy}^{1/2}\beta$. It hence follows that the i-th canonical variables are obtained by taking $\bar{\alpha}$ and $\bar{\beta}$ to be the i-th unit vectors in $\mathbb{R}^{n_1}$ and $\mathbb{R}^{n_2}$ respectively. Hence $\rho_i=\sigma_i$, $\xi_i=(S_{xx}^{-1/2}Ue_i)^Tx=\langle a_1^{(i)},x\rangle$, and $\eta_i=(S_{yy}^{-1/2}Ve_i)^Ty=\langle a_2^{(i)},y\rangle$, $i\leq\min\{n_1,n_2\}$. □

Interpretation. For the model M_k^* of minimal predictive misfit the *predictions are based on the main k canonical variables of the empirical covariance matrix*. So the predictions are based on the relations of maximal correlation. Moreover, the predictive misfit can be expressed in terms of the main k canonical correlation coefficients. In particular the first predictive misfit is equal to $(1-\sigma_k^2)^{1/2}$ where σ_k is the k-th canonical correlation coefficient. □

Remark. This gives a *deterministic* interpretation of canonical correlation analysis in terms of prediction by means of deterministic models. □

2.2.3. Algorithms for predictive modelling

Using the results on canonical correlation in section 2.2.2 we give simple algorithms for predictive modelling of static data. These algorithms form the basis of the algorithms for predictive modelling of time series as described in section 4.3.

The next propositions imply algorithms for the predictive procedures $P^P_{c_{tol}}$, corresponding to minimizing lexicographically the predictive misfit under the condition $\dim(M(0))\le c_{tol}$, and $P^P_{\varepsilon_{tol}}$, corresponding to minimizing complexity under the condition $\varepsilon^P(d,M)<\varepsilon_{tol}$. We also refer to sections II.2.2 and II.2.3.

Notation. We summarize some notation which was introduced in section 2.2.2. Assume for the data $d=\{(\tilde{x}_i,\tilde{y}_i);\ i\in[1,N]\}\in(\mathbb{R}^{n_1+n_2})^N$ that $N\ge\max\{n_1,n_2\}$ and that the data are generic in the sense that $\text{span}\{\tilde{x}_1,\ldots,\tilde{x}_N\}\in\mathbb{R}^{n_1}$ and $\text{span}\{\tilde{y}_1,\ldots,\tilde{y}_N\}\in\mathbb{R}^{n_2}$. Let $\begin{pmatrix} S_{xx} & S_{xy} \\ S_{yx} & S_{yy}\end{pmatrix}:=\frac{1}{N}\Sigma_{i=1}^N\begin{pmatrix}\tilde{x}_i\\ \tilde{y}_i\end{pmatrix}\begin{pmatrix}\tilde{x}_i\\ \tilde{y}_i\end{pmatrix}^T\in\mathbb{R}^{(n_1+n_2)\times(n_1+n_2)}$ and let $S_{xx}^{-1/2}S_{xy}S_{yy}^{-1/2}$ have (SVD) $U\Sigma V^T$, with $U\in\mathbb{R}^{n_1\times n_1}$ and $V\in\mathbb{R}^{n_2\times n_2}$ both orthogonal matrices and $\Sigma=\begin{pmatrix}\Sigma' & 0\\ 0 & 0\end{pmatrix}\in\mathbb{R}^{n_1\times n_2}$, $\Sigma'=\text{diag}(\sigma_1,\ldots,\sigma_r)$, $\sigma_1\ge\ldots\ge\sigma_r>0$. There holds that $\sigma_1\le 1$ and $r=\text{rank}(S_{xy})$. Let r^* denote the number of singular values equal to 1. Denote the columns of $S_{xx}^{-1/2}U$ by $a_1^{(i)}$, $i=1,\ldots,n_1$, and those of $S_{yy}^{-1/2}V$ by $a_2^{(i)}$, $i=1,\ldots,n_2$. For $k=1,\ldots,r$ define $M_k^*:=\{(x,y);\ \langle a_2^{(i)},y\rangle=\sigma_i\langle a_1^{(i)},x\rangle,\ i=1,\ldots,k\}$ and let $M_{n_2}^*:=M_r^*\cap\{(x,y);\ \langle a_2^{(i)},y\rangle=0,\ i=r+1,\ldots,n_2\}$. Finally, let $M(\sigma):=\{(x,y);\ \langle a_2^{(i)},y\rangle=\sigma\langle a_1^{(i)},x\rangle$ for all i with $\sigma_i=\sigma\}$. □

Proposition 2-21 For generic data $d=\{(\tilde{x}_i,\tilde{y}_i);\ i=1,\ldots,N\}$ and given tolerated complexity c_{tol}, $P^P_{c_{tol}}$ is given by

(*i*) $P^P_{c_{tol}}(d)=\{M\in\mathbb{M};\ M\subset M_r^*,\ \dim(M_2^\perp)=n_2-c_{tol}\}$ if $c_{tol}<n_2-r$;

(*ii*) $P^P_{c_{tol}}(d)=M^*_{r^*}$ if $c_{tol}\ge n_2-r^*$;

(*iii*) $P^P_{c_{tol}}(d)=M^*_{n_2-c_{tol}}$ if $r^*<n_2-c_{tol}\le r$ and $\sigma_{n_2-c_{tol}}>\sigma_{n_2-c_{tol}+1}$;

(*iv*) if $\sigma_1\ge\ldots\ge\sigma_{c_1}>\sigma_{c_1+1}=\ldots=\sigma_{n_2-c_{tol}}=\sigma_{n_2-c_{tol}+1}=\ldots=\sigma_{c_2}>\sigma_{c_2+1}\ge$ $\ge\ldots\ge\sigma_r>0$, then $P^P_{c_{tol}}(d)=\{M^*_{c_1}\cap L;\ L\supset M(\sigma_{n_2-c_{tol}}),\ c^P(L)=c_{tol}+c_1\}$.

Proof. See the appendix.

Proposition 2-22 Let data $d=\{(\tilde{x}_i,\tilde{y}_i);\ i=1,\ldots,N\}$ be generic. Assume moreover that a maximal tolerated misfit level is given with $\varepsilon_{tol}=\varepsilon_1^{tol}.(1,\ldots,1)$, $\varepsilon_1^{tol}>0$, so the misfit restriction concerns only the worst prediction made by a model. Then

(*i*) $P^P_{\varepsilon_{tol}}(d)=M^*_{n_2}$ if $\varepsilon_1^{tol}>(1-\sigma^2_{n_2})^{1/2}$;

(*ii*) $P^P_{\varepsilon_{tol}}(d)=\mathbb{R}^{n_1+n_2}$ if $\varepsilon_1^{tol}\leq(1-\sigma_1^2)^{1/2}$;

(*iii*) $P^P_{\varepsilon_{tol}}(d)=M^*_r$ if $r<n_2$ and $(1-\sigma_r^2)^{1/2}<\varepsilon_1^{tol}\leq 1$;

(*iv*) if $(1-\sigma_1^2)^{1/2}<\varepsilon_1^{tol}\leq(1-\sigma_r^2)^{1/2}$, then $P^P_{\varepsilon_{tol}}(d)=M^*_k$ where k is such that $(1-\sigma_k^2)^{1/2}<\varepsilon_1^{tol}\leq(1-\sigma^2_{k+1})^{1/2}$.

Proof. See the appendix.

Remark. The algorithms hence amount to singular value decomposition of the (cross-correlation) matrix $S_{xx}^{-1/2}S_{xy}S_{yy}^{-1/2}$ derived from the empirical covariance matrix. □

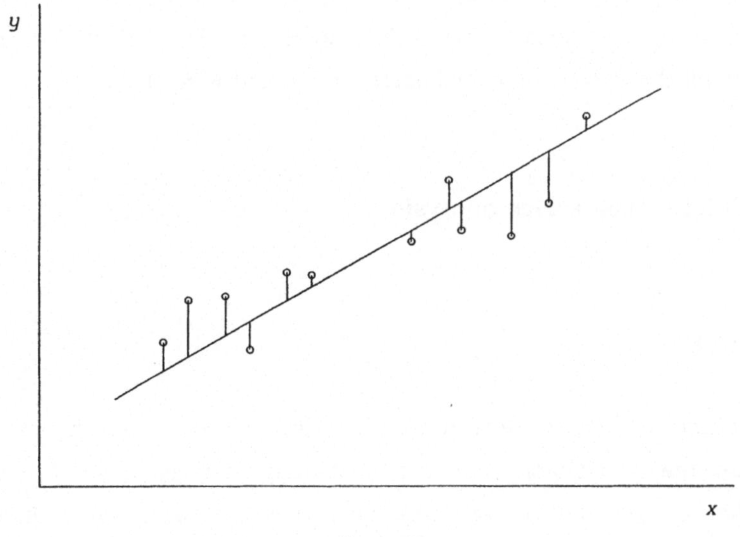

figure 10

Remark. The procedure $\bar{\bar{P}}^P_{\varepsilon_{tol}}$ corresponding to the requirement $\varepsilon_1^P(d,M)\leq\varepsilon_1^{tol}$ with $\varepsilon_1^{tol}\geq 0$ can be described in a way analogous to proposition 2-22. So the identified model is $M^*_{n_2}$ if $\varepsilon_1^{tol}\geq(1-\sigma^2_{n_2})^{1/2}$, it is $\mathbb{R}^{n_1+n_2}$ if $\varepsilon_1^{tol}<(1-\sigma_1^2)^{1/2}$, M^*_r if $r<n_2$ and $(1-\sigma_r^2)^{1/2}\leq\varepsilon_1^{tol}<1$, and M^*_k if $(1-\sigma_1^2)^{1/2}\leq(1-\sigma_k^2)^{1/2}\leq\varepsilon_1^{tol}<(1-\sigma^2_{k+1})^{1/2}\leq(1-\sigma_r^2)^{1/2}$. □

Remark. For $n_2=1$ and $c_{tol}=0$ the procedure $P^P_{c_{tol}}$ reduces to ordinary least-squares fitting. See figure 10. The special, vertical way of measuring the error reflects the purpose of predicting y on the basis of x. □

Remark. The results on canonical correlation in section 2.2.2 also can be used to construct an alternative descriptive procedure. For $a=(a_1,a_2)\in\mathbb{R}^{n_1}\times\mathbb{R}^{n_2}$ and generic data define $\tilde{e}(d,a):=\sigma_{a_1a_2}/\{\sigma_{a_1}.\sigma_{a_2}\}^{1/2}$, where $\sigma_{a_1a_2}:=\frac{1}{N}\Sigma^N_{i=1}(<a_1,\tilde{x}_i>+<a_2,\tilde{y}_i>)^2$, $\sigma_{a_1}:=\frac{1}{N}\Sigma^N_{i=1}<a_1,\tilde{x}_i>^2$ and $\sigma_{a_2}:=\frac{1}{N}\Sigma^N_{i=1}<a_2,\tilde{y}_i>^2$. Define $\tilde{\varepsilon}(d,M)$ analogous to definition 2-17, requiring in step k that $a_1^{(k)}\perp_{(1)}\text{span}\{a_1^{(1)},\dots,a_1^{(k-1)}\}$ and $a_2^{(k)}\perp_{(2)}\text{span}\{a_2^{(1)},\dots,a_2^{(k-1)}\}$, where $\perp_{(1)}$ denotes orthogonality with respect to S_{xx} and $\perp_{(2)}$ with respect to S_{yy}, cf. section 2.2.1. Defining $(\alpha_1,\alpha_2):=(U^TS^{1/2}_{xx}a_1,\ V^TS^{1/2}_{yy}a_2)$ it follows that $\tilde{e}(d,a)=\frac{\|\alpha_1\|}{\|\alpha_2\|}+\frac{\|\alpha_2\|}{\|\alpha_1\|}+2\frac{\alpha_1^T\Sigma\alpha_2}{\|\alpha_1\|.\|\alpha_2\|}$ and it is easily seen that an optimal model claiming k independent relationships is given by $\tilde{M}_k:=\{(x,y);\ <a_1^{(i)},x>=<a_2^{(i)},y>,\ i\in[1,k]\}$ with error $\tilde{\varepsilon}(d,\tilde{M}_k)=(2(1-\sigma_k),\dots,2(1-\sigma_1),0,\dots,0)$. This gives a deterministic descriptive interpretation of canonical correlation analysis. The identified model is independent of the choice of coordinates in $\mathbb{R}^{n_1}$ and $\mathbb{R}^{n_2}$. □

3. Deterministic time series analysis

3.1. Introduction

In this section we define procedures for deterministic time series analysis. In this case the specification of the modelling problem is as follows. It is assumed that q real-valued variables have been specified which have to be included in the model and that data on these variables are available in the form of a finite time series. We denote the variables by $w:=\text{col}(w_1,\dots,w_q)$, the time interval of observation by $\mathcal{T}:=[t_0,t_1]$ for some $-\infty<t_0\leq t_1<+\infty$, and the data by $\tilde{w}:=(\tilde{w}(t);t\in\mathcal{T})\in(\mathbb{R}^q)^{\mathcal{T}}$, i.e., $\tilde{w}$ is an ordered sequence of observations.

For reasons discussed in section I.1.2 and II.1.3, in identification we make *no stochastic assumptions* concerning a data generating mechanism. The aim is to identify a deterministic model for the time series which is optimal for the given modelling objectives.

In the notation of section II.2.1 the data set is $\mathcal{D}=\bigcup\{(\mathbb{R}^q)^n;\ n\in\mathbb{N}\}$, so the data consists of a time series of some (finite) length n in $\mathbb{R}^q$. As model class we take the class $\mathbb{B}$ of dynamical systems in $(\mathbb{R}^q)^{\mathbb{Z}}$ as defined in section II.3.1.2. So the time series has to be modelled by means of a linear, time invariant, complete system. The interpretation is that the model gives a local description, cf. section II.3.1.2. As modelling objectives π we consider description and prediction.

The modelling problem now amounts to choosing an identification procedure $P_\pi:\mathcal{D}\to 2^{\mathbb{B}}$, corresponding to a utility u_π reflecting the objectives π of modelling. We follow the approximate modelling approach described in section II.2.1. Therefore we will define complexity maps $c_\pi:\mathbb{B}\to C_\pi$ and misfit maps $\varepsilon_\pi:\mathcal{D}\times\mathbb{B}\to E_\pi$ and impose orderings on C_π and E_π. The resulting identification problem is depicted in figure 11.

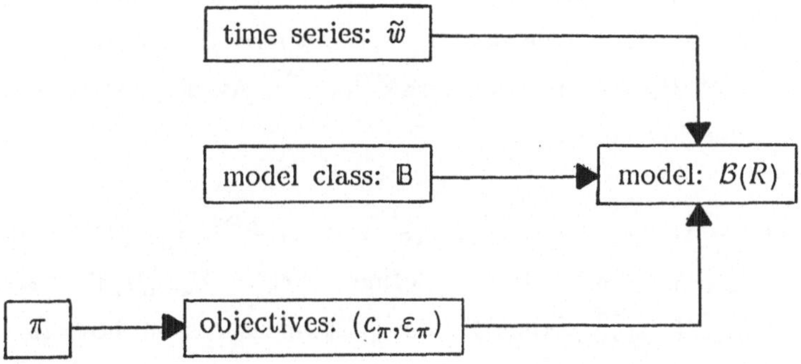

figure 11: modelling a time series

The complexity of models in $\mathbb{B}$ is defined as in section IV.2.1 with the lexicographic ordering $\geq$ of definition IV.2-4. In section 3.2 we define the descriptive and predictive misfit of models in $\mathbb{B}$ with respect to an observed time series. These misfits are defined in accordance with the corresponding static misfit measures of section 2.

The procedures for deterministic time series analysis correspond to utilities which express the usefulness of models for modelling a given time series. These utilities are based on complexity and misfit and correspond to the utilities defined in sections II.2.2 and II.2.3. This is described in section 3.3. The utilities both have a natural interpretation and allow explicit numerical algorithms as described in section 4.

Remark. The complexity and misfit measures have two desirable properties. First, these measures are *defined* intrinsically in terms of models, considered as sets of trajectories, and not artificially in terms of parametrizations. Second, the measures can be numerically *expressed* in terms of canonical parametrizations which leads to explicit numerical algorithms for identifying optimal models. □

3.2. Descriptive and predictive misfit

3.2.1. Descriptive misfit

In this section we define the misfit of a model $\mathcal{B}\in\mathbb{B}$ in describing data consisting of a finite time series $\tilde{w}:=(\tilde{w}(t);\ t\in\mathcal{T})$ on an interval $\mathcal{T}=[t_0,t_1]$. As in section 2.1.1 we first consider the case where $\mathcal{B}$ imposes one restriction, in the sense that $\mathcal{B}=\mathcal{B}(r)$ for some $r\in\mathbb{R}^{1\times q}[s,s^{-1}]$. As descriptive misfit we take the average equation error.

Notation. Let $n\in\mathbb{Z}$, $d\in\mathbb{Z}_+$, $r=\Sigma_{k=n}^{n+d}\ r_k s^k$ with $r_k\in\mathbb{R}^{1\times q}$, $r_n\neq0\neq r_{n+d}$. It is assumed that $d(r):=d\leq t_1-t_0$. We then define $\|r\|^2:=\Sigma_{k=n}^{n+d}\|r_k\|^2$ and $\|r\tilde{w}\|^2:=\frac{1}{t_1-t_0-d+1}\cdot\Sigma_{t=t_0-n}^{t_1-n-d}\{\Sigma_{k=n}^{n+d}r_k\tilde{w}(t+k)\}^2$. So $\|r\tilde{w}\|$ measures in how far $\tilde{w}$ satisfies the restriction imposed by $\mathcal{B}(r)$ that $(r\tilde{w})(t)=0$ for $t=t_0-n,\ldots,t_1-n-d$. □

Definition 3-1 The *descriptive misfit* of $r\in\mathbb{R}^{1\times q}[s,s^{-1}]$ with respect to data $\tilde{w}\in(\mathbb{R}^q)^{\mathcal{T}}$ is defined as the mean *equation error*, i.e., $e^D(\tilde{w},r):=\|r\tilde{w}\|/\|r\|$. If $d(r)\geq\#(\mathcal{T})$ then $e^D(\tilde{w},r):=0$.

We define the misfit of $\mathcal{B}(r)$ as $\varepsilon_{d,1}^D(\tilde{w},\mathcal{B}(r)):=e^D(\tilde{w},r)$.

Next consider $\mathcal{B}\in\mathbb{B}$ with $\dim(\mathcal{B}^{\perp})\geq 2$, where $\mathcal{B}^{\perp}$ denotes the class of laws which are satisfied by $\mathcal{B}$, cf. section II.3.2.2. For $r\in\mathcal{B}^{\perp}$ we measure the descriptive misfit by $e^{D}(\tilde{w},r)$. The problem is to define the misfit of $\mathcal{B}$, which imposes an infinite number of laws on the phenomenon. We define the misfit of $\mathcal{B}$ by choosing a canonical basis in $\mathcal{B}^{\perp}$, using the canonical descriptive form (CDF) which is defined in section II.3.2.5.

Remark. The idea is to define a sequence of misfits measuring the quality of laws of different order claimed by $\mathcal{B}$. Note that using (CDF) guarantees that laws of different order are orthogonal, so loosely stated these quality measures become more or less *independent*. By this we mean that e.g. a first order law should not be judged as being of small misfit if this is due to the fact that this first order law is ("near" to being) implied by good zero order laws. This is made explicit by the orthogonality conditions in (CDF). We refer to sections II.3.2.4 and II.3.2.5. As a simple example consider the case that $\tilde{w}\in(\mathbb{R}^{2})^{\mathcal{T}}$ with $\tilde{w}_1(t)=\tilde{w}_2(t)+\tilde{e}(t)$ where $\|\tilde{e}\|$ is small. Let $r:=(\delta\sigma+1,-1)$, then $e^{D}(\tilde{w},r)\approx\frac{1}{2}\sqrt{2}\|\tilde{e}\|=e^{D}(\tilde{w},(1,-1))$ for $|\delta|$ sufficiently small. However, if we accept both the law $(1,-1)$ and r, then the time series would be modelled by $\mathcal{B}(R)$ with $R=\begin{bmatrix}(1\ -1)\\ r\end{bmatrix}$ and $\mathcal{B}(R)=\{0\}$ for all $\delta\neq 0$. Declaring $\tilde{w}$ to be zero clearly may not be a reasonable model. □

Notation. For $\mathcal{B}\in\mathbb{B}$ let $L_t^D\subset\mathcal{B}^{\perp}$ denote the space of truly t-th order descriptive laws of $\mathcal{B}$ as defined in section II.3.2.5. Let v_t be as defined in definition II.3-9 and let $n_t:=\dim(v_t(L_t^D))$, then $n_t=e_t$ where $(e_t;\ t\in\mathbb{Z}_+)$ is the tightest equation structure of AR-representations of $\mathcal{B}$, see proposition II.3-10 in section II.3.2.4. □

Definition 3-2 For data $\tilde{w}\in(\mathbb{R}^{q})^{\mathcal{T}}$, the *main* t-th *order descriptive misfit* of a model $\mathcal{B}\in\mathbb{B}$ is defined as $\varepsilon_{t,1}^{D}(\tilde{w},\mathcal{B}):=\max\{e^{D}(\tilde{w},r);\ r\in L_t^D\}$ if $n_t>0$, else $\varepsilon_{t,1}^{D}(\tilde{w},\mathcal{B}):=0$.

Interpretation. The main t-th order descriptive misfit of a model $\mathcal{B}$ measures the worst fit of the truly t-th order laws claimed by $\mathcal{B}$. □

Remark. The misfit of the spaces L_t^D is defined in accordance with definition

2–4. Hence if $n_t>1$ then we define $\varepsilon^D_{t,2}(\tilde{w},\mathcal{B})$ as the misfit of the worst–but–one t–th order law, i.e., if $\varepsilon_{t,1}(\tilde{w},\mathcal{B})=e^D(\tilde{w},r_1)$, $r_1\in L^D_t$, then $\varepsilon^D_{t,2}(\tilde{w},\mathcal{B}):=\max\{e^D(\tilde{w},r);\ r\in v_t^{-1}\{v_t(L^D_t)\cap[v_t(r_1)]^\perp\}\}$. For $k=2,...,n_t$, $\varepsilon^D_{t,k}(\tilde{w},\mathcal{B})$ is inductively defined as the worst–but–$(k-1)$ t–th order misfit, as follows. If $\varepsilon^D_{t,j}(\tilde{w},\mathcal{B})=e^D(\tilde{w},r_j)$, $r_j\in v_t^{-1}\{\ v_t(L^D_t)\cap[\text{span}(v_t(r_1),...,v_t(r_{j-1}))]^\perp\}$ for $j=1,2,...,k-1$, then $\varepsilon^D_{t,k}(\tilde{w},\mathcal{B}):=\max\{e^D(\tilde{w},r);\ r\in v_t^{-1}\{\ v_t(L^D_t)\cap[\text{span}(v_t(r_1),...,v_t(r_{k-1}))]^\perp\}\}$. For $k=n_t+1,...,q$, $\varepsilon^D_{t,k}(\tilde{w},\mathcal{B}):=0$. It follows from proposition 2–11 that $\varepsilon^D_{t,k}$ is well–defined in this way, i.e., independent of the maximizing arguments $r_j\in L^D_t$. □

Definition 3–3 The *descriptive misfit* is a map $\varepsilon^D:(\mathbb{R}^q)^{\mathcal{T}}\times\mathbb{B}\to(\mathbb{R}_+^{1\times q})^{\mathbb{Z}_+}$, where $\varepsilon^D_{t,k}(\tilde{w},B)$ is the descriptive misfit for $\tilde{w}$ of the worst–but–$(k-1)$ law of the truly t–th order descriptive laws in L^D_t claimed by $\mathcal{B}$, $t\in\mathbb{Z}_+$, $k=1,...,q$.

Interpretation. Note that there are at most $\Sigma^\infty_{t=n}\ e_t\leq q$ misfit numbers unequal to zero. These numbers give the equation errors of a suitably chosen basis for all the equations which are claimed by the model. The numbers $\{\varepsilon^D_{t,k};\ k=1,...,q\}$ measure the quality of the t–th order equations which are orthogonal to the lower order ones. □

Remark. Using the bilateral row properness of (CDF) implied by definition II.3–11 and proposition II.3–8, it follows that for $\tilde{w}\in(\mathbb{R}^q)^{\mathcal{T}}$ there holds that $\{\varepsilon^D(\tilde{w},\mathcal{B})=0\}\Leftrightarrow\{\tilde{w}\in\mathcal{B}|_{\mathcal{T}}\}$. □

Remark. The descriptive misfit is defined in terms of the spaces L^D_t which are uniquely defined by $\mathcal{B}$. Hence the misfit is defined independent from autoregressive parametrization. Using proposition 2–11 the misfit can be numerically expressed in terms of a (CDF) representation. According to proposition II.3–14 such a representation is unique up to a choice of basis of L^D_t. The misfit clearly is defined independent from this choice. A convenient basis for L^D_t is $\{r_1,...,r_{n_t}\}$ as defined above. With this choice of basis the descriptive misfit of $\mathcal{B}$ consists of the misfits of the rows of the corresponding (CDF) representation of $\mathcal{B}$. □

We use the following lexicographic ordering of misfits.

Definition 3-4 $\{\varepsilon'=(\varepsilon'_{t,k})\geq\varepsilon''=(\varepsilon''_{t,k})\}:\Leftrightarrow\{\varepsilon'=\varepsilon''$; or there exists $t_0\in\mathbb{Z}_+$, $2\leq k_0\leq q$ such that $\varepsilon'_{t_0,k_0}>\varepsilon''_{t_0,k_0}$ and $\varepsilon'_{t,k}=\varepsilon''_{t,k}$ for all $t<t_0$, $k=1,\dots,q$ and for $t=t_0$, $k=1,\dots,k_0-1$; or there exists $t_0\in\mathbb{Z}_+$ such that $\varepsilon'_{t_0,1}>\varepsilon''_{t_0,1}$ and $\varepsilon'_{t,k}=\varepsilon''_{t,k}$ for all $t<t_0$, $k=1,\dots,q\}$.

Remark. Note that the misfit of $\mathcal{B}_1$ in general will be larger than the misfit of $\mathcal{B}_2$ if $\mathcal{B}_1$ has laws of lower order than $\mathcal{B}_2$. On the other hand the complexity of $\mathcal{B}_1$ then is smaller than that of $\mathcal{B}_2$, as the complexity also is ordered lexicographically, see definition IV.2-4 and corollary IV.2-5. In section 3.3 we describe two procedures to balance the desires for low misfit and for low complexity by fixing a maximal tolerated level for one of the objectives and optimizing with respect to the other one. These procedures correspond to the utilities defined in sections II.2.2.1 and II.2.3.1. □

3.2.2. Predictive misfit

The (one-step-ahead) predictive misfit of a dynamical system in predicting a time series is based on the prediction error defined in section 2.2.1 for static prediction. Now the data consists of a finite time series $\tilde{w}=(\tilde{w}(t);\ t\in\mathcal{T}=[t_0,t_1])\in(\mathbb{R}^q)^{\mathcal{T}}$ and the model class consists of the class $\mathbb{B}$ of linear, time invariant, complete systems.

Again we first consider the case where $\mathcal{B}=\mathcal{B}(r)$ with $r\in\mathbb{R}^{1\times q}[s,s^{-1}]$.

Notation. Let $n\in\mathbb{Z}$, $d\in\mathbb{Z}_+$, $r=\sum_{k=n}^{n+d} r_k s^k$ with $r_k\in\mathbb{R}^{1\times q}$, $r_n\neq 0\neq r_{n+d}=:r^*$. It is assumed that $d\leq t_1-t_0-q+1$. Now $\mathcal{B}(r)$ predicts that $r^*w(t)=-\sum_{k=0}^{d-1} r_k w(t-d+k)$. Let $r^*\tilde{w}(t)=-\sum_{k=0}^{d-1} r_k\tilde{w}(t-d+k)+e(t)$ for $t=t_0+d,\dots,t_1$. So $e(t)$ is the error made at time t in the prediction of $r^*w(t)$. Let $\|e\|^2:=\frac{1}{t_1-t_0-d+1}\sum_{t=t_0+d}^{t_1} e^2(t)$ denote the average prediction error and let $\|r^*\tilde{w}\|_d^2:=\frac{1}{t_1-t_0-d+1}\sum_{t=t_0+d}^{t_1}\{r^*\tilde{w}(t)\}^2$ denote the average magnitude of the predicted functional. □

Definition 3-5 The *predictive misfit* of $r\in\mathbb{R}^{1\times q}[s,s^{-1}]$, with $1\leq d(r)\leq t_1-t_0-q+1$ and with leading coefficient vector $r^*\in\mathbb{R}^{1\times q}$, with respect to data $\tilde{w}\in(\mathbb{R}^q)^{\mathcal{T}}$ is defined as the *relative mean prediction error*, i.e., $e^P(\tilde{w},r):=\|r\tilde{w}\|/\|r^*\tilde{w}\|_d=\|e\|/\|r^*\tilde{w}\|_d$.

Remark. The predictive misfit is well-defined provided that the data are generic in the sense that $\Sigma_{t=t_1-q+1}^{t_1}\tilde{w}(t)\tilde{w}(t)^T\in\mathbb{R}^{q\times q}$ has full rank. □

We define the predictive misfit of $\mathcal{B}(r)$ by $\varepsilon_{d,1}^P(\tilde{w},\mathcal{B}(r)):=e^P(\tilde{w},r)$.

Next we define the misfit for models with $\dim(\mathcal{B}^\perp)\geq 2$. We measure the predictive quality of a model by means of a sequence of numbers which measure the quality of predictive laws of different order. The quality assessment for laws of different orders is made *independently* by using the canonical predictive form (CPF) which is defined in section II.3.2.6.

Remark. Note that in a (CPF) representation of $\mathcal{B}$ the laws of order t are truly t-th order, i.e., they are not impleid by lower order laws. Further, predicted functionals of different orders are required to be orthogonal and prediction polynomials should be orthogonal to predictive laws of lower order, cf. definition II.3-15. These requirements are essential to guarantee that good quality of one predictive law is not due to good quality of another predictive law of lower order. As a simple and extreme example, suppose that the data consists of a time series generated by a system $\mathcal{B}(r)$ which is observed under small noise. Let e.g. $r=r_0+r_1s^{-1}\in\mathbb{R}^{1\times q}[s^{-1}]$ with $q>1$, let $w\in\mathbb{B}(r)$ and $\tilde{w}=w|_{\mathcal{T}}+\tilde{e}$ with $\|r\tilde{e}\|/\|r_0\tilde{w}\|$ small. Let $\delta>0$ be fixed, let e_i denote the i-th unit vector in $\mathbb{R}^q$, and let $r_i:=r+\delta e_i^T s^{-2}$, $i\in[1,q]$. Then for $|\delta|$ succifiently small $e^P(\tilde{w},r_i)\approx e^P(\tilde{w},r)=\|r\tilde{e}\|/\|r_0\tilde{w}\|$. Let $R:=\mathrm{col}(r_1,\ldots,r_q)$, then it is easily seen that in $\mathcal{B}(R)$ especially $w_i=w_j$ for all $i,j\in[1,q]$. This clearly is in general not a reasonable model for $\tilde{w}$. Hence we should not accept all laws of small predictive misfit but only those laws which have small misfit and which are independent of other laws of small misfit. This is made explicit by the orthogonality conditions of (CPF). □

Notation. For $\mathcal{B}\in\mathbb{B}$ let L_t^P denote the space of truly t-th order predictive laws of $\mathcal{B}$ as defined in section II.3.2.6 and let $n_t:=\dim(v_t(L_t^P))$. □

Definition 3-6 For data $\tilde{w}\in(\mathbb{R}^q)^{\mathcal{T}}$ and $t\geq 1$, the *main* t-th *order predictive misfit* of a model $\mathcal{B}\in\mathbb{B}$ is defined as $\varepsilon_{t,1}^P(\tilde{w},\mathcal{B}):=\max\{e^P(\tilde{w},r);\ r\in L_t^P\}$ if $n_t>0$, else $\varepsilon^P_{t,1}(\tilde{w},\mathcal{B}):=0$.

Interpretation. The main t-th order predictive misfit of a model $\mathcal{B}$ expresses the worst predictive quality of truly t-th order predictive laws claimed by $\mathcal{B}$. □

Remark. These misfits are well-defined for generic data, provided that $\max\{t;\ n_t\neq 0\}\leq t_1-t_0-q+1$. □

Remark. For $t=0$ we define $\varepsilon^P_{0,1}(\tilde{w},\mathcal{B}):=\varepsilon^D_{0,1}(\tilde{w},\mathcal{B})$, as for $d(r)=0$ $e^P(\tilde{w},r)=1$ for generic $\tilde{w}$, so the predictive misfit makes no sense for these static laws. In this case we measure the misfit simply by $\|e\|/\|r\|$. □

The predictive misfit is defined in analogy with definition 3–3 and with the same motivation. The misfit of the spaces L^P_t is defined in accordance with definition 2–17. So $\varepsilon^P_{t,k}$ measures the predicive misfit of the worst-but-$(k-1)$ law of the truly t-th order predictive laws in L^P_t claimed by $\mathcal{B}$.

Notation. Formally, let $Q_t:=\frac{1}{t_1-t_0-t+1}\Sigma_{i=t_0+t}^{t_1}\tilde{w}(i)\tilde{w}(i)^T$ and let $\perp_{(t)}$ denote orthogonality with respect to Q_t, i.e. $\{a\perp_{(t)}b\}:\Leftrightarrow\{a^TQ_tb=0\}$. If $t\geq 1$ and $n_t>1$, then $\varepsilon^P_{t,k}(\tilde{w},\mathcal{B})$, for $k=2,\ldots,n_t$ is inductively defined as follows. For $r\in\mathbb{R}^{1\times q}[s,s^{-1}]$ let $r^*\in\mathbb{R}^{1\times q}$ denote the leading coefficient vector of r. If $\varepsilon^P_{t,j}(\tilde{w},\mathcal{B})=e^P(\tilde{w},r_j)$ with $r_j\in L^P_t$ and $r^*_j\perp_{(t)}\text{span}\{r^*_1,\ldots,r^*_{j-1}\}$ for $j=1,\ldots,k-1$, then $\varepsilon^P_{t,k}(\tilde{w},\mathcal{B}):=\max\{e^P(\tilde{w},r);\ r\in L^P_t,\ r^*\perp_{(t)}\text{span}\{r^*_1,\ldots,r^*_{k-1}\}\}$. For $k=n_t+1,\ldots,q$ we define $\varepsilon^P_{t,k}(\tilde{w},\mathcal{B}):=0$. □

Remark. The orthogonality is taken with respect to Q_t for reasons discussed in section 2.2.1. The interpretation of L^P_t is that it imposes n_t restrictions on the values of w one-step-ahead. The quality of the corresponding predictions is measured independent of a choice of basis in $\mathbb{R}^q$ by taking $\perp_{(t)}$, cf. the interpretation of definition 2–17 in section 2.2.1. □

Remark. Proposition 2–18 and the remark following it imply that $\varepsilon^P_{t,k}(\tilde{w},\mathcal{B})$ is well-defined provided that the predictive relations in L^P_t are observed at least q times, i.e., $t_1-t_0-t+1\geq q$, and provided that the data are generic in the sense that span$\{\tilde{w}(i);\ i\in[t_0+t,t_1]\}=\mathbb{R}^q$. In fact it suffices to assume that $\tilde{w}$ is such that $\{r\in L_t^P,\ r^*\perp\tilde{w}(i)$ for all $i\in[t_0+t,t_1]\}\Rightarrow\{r^*=0\}$. □

We summarize the foregoing remarks and notation by the following definition of predictive misfit.

Definition 3-7 Let $\tilde{w}\in(\mathbb{R}^q)^{[t_0,t_1]}$ and let $\mathcal{B}\in\mathbb{B}$ with $t^*:=\max\{t;\ e_t^*(\mathcal{B})\neq 0\}\leq t_1-t_0+1-q$. If the data are generic in the sense that span$\{\tilde{w}(i);\ i\in[t_0+t^*,t_1]\}=\mathbb{R}^q$ then the *predictive misfit* of $\mathcal{B}$ for $\tilde{w}$ is $\varepsilon^P(\tilde{w},\mathcal{B})\in(\mathbb{R}_+^{1\times q})^{\mathbb{Z}_+}$, where $\varepsilon_{0,k}^P(\tilde{w},\mathcal{B}):=\varepsilon_{0,k}^D(\tilde{w},\mathcal{B})$ and for $t\geq 1$ $\varepsilon_{t,k}^P(\tilde{w},\mathcal{B})$ is the predictive misfit of the worst–but–$(k-1)$ law of the truly t–th order predictive laws in L_t^P claimed by $\mathcal{B}$, $k=1,\ldots,q$.

Interpretation. The predictive misfit of a model $\mathcal{B}$ consists of the relative mean prediction errors of a suitably chosen basis for all the predictive relations which are claimed by $\mathcal{B}$. □

Remark. Using definition II.3–15 and proposition II.3–8 it follows that for $\tilde{w}\in(\mathbb{R}^q)^{\mathcal{T}}$ with $\varepsilon^P(\tilde{w},\mathcal{B})$ well–defined there holds that $\{\varepsilon^P(\tilde{w},\mathcal{B})=0\}\Leftrightarrow\{\tilde{w}\in\mathcal{B}|_{\mathcal{T}}\}$. □

Remark. The predictive misfit is defined in terms of the spaces L_t^P which are uniquely determined by $\mathcal{B}$, hence the definition is independent from autoregressive parametrization. From the proof of proposition 2–18 it follows that the predictive misfit can be numerically expressed in terms of a (CPF) representation. □

We order the predictive misfit in the same way as the descriptive misfit, i.e., lexicographically as described in definition 3–4.

3.3. Procedures for deterministic time series analysis

3.3.1. Specification of the model class

Given an observed time series of finite length, the set of laws for which the quality can be reasonably assessed is restricted. In general terms we should not allow identification of laws for which the order is too large in comparison with the length of the observed time series. We make this explicit in this section. For given time interval of observation $\mathcal{T}$ we specify a class of models $\mathbb{B}(\mathcal{T})\subset\mathbb{B}$ consisting of those models in $\mathbb{B}$ which can be reasonably used in identification on the basis of an observed time series on $\mathcal{T}$.

Notation. Let the data consist of a finite time series $\tilde{w}\in(\mathbb{R}^q)^{\mathcal{T}}$ with $\mathcal{T}=[t_0,t_1]$. Let $0\leq d\leq t_1-t_0$ and $\tau(\mathcal{T},d):=t_1-t_0-d+1$, then for $r\in\mathbb{R}^{1\times q}[s]$, $r=\Sigma_{k=0}^{d}r_k s^k$, $r_k\in\mathbb{R}^{1\times q}$, $r_0\neq 0\neq r_d$, there holds $\|r\tilde{w}\|^2:=\frac{1}{\tau(\mathcal{T},d)}.\Sigma_{t=t_0}^{t_1-d}\{\Sigma_{k=0}^{d}r_k\tilde{w}(t+k)\}^2=v_d(r).S(\tilde{w},d).v_d(r)^T$, where $S(\tilde{w},d):=\frac{1}{\tau(\mathcal{T},d)}.\Sigma_{t=t_0}^{t_1-d}(\tilde{w}(t)^T,\ldots,\tilde{w}(t+d)^T)^T.(\tilde{w}(t)^T,\ldots,\tilde{w}(t+d)^T)$ is the empirical covariance matrix of order d and where $v_d(r)$ is as defined in definition II.3-9. □

Remark. Note that $\text{rank}(S(\tilde{w},d))\leq\min\{t_1-t_0-d+1,\ q(d+1)\}$. If $t_1-t_0-d+1<q(d+1)$ then for any $\tilde{w}\in(\mathbb{R}^q)^{\mathcal{T}}$ there exists an r with $d(r)\leq d$ and $\|r\tilde{w}\|=0$, hence with $e^D(\tilde{w},r)=0$ and $e^P(\tilde{w},r)=0$. Such laws clearly are *not corroborated* by the data. In order to prevent *overparametrization* it is necessary at least to require that $t_1-t_0-d+1\geq q(d+1)$, i.e., that $d\leq\bar{d}(\mathcal{T}):=(t_1-t_0+1-q)/(q+1)$. Then over-parametrization is prevented in the sense that for generic data $\tilde{w}\in(\mathbb{R}^q)^{\mathcal{T}}$ $S(\tilde{w},d)$ is nonsingular. This can be seen by explicit construction. Define $a\in(\mathbb{R}^q)^{t_1-t_0+1}$ by $a_{q-i}(d+i(d+1)):=1$, $i\in[0,q-1]$, and zero elsewhere. It is a matter of easy verification to show that

$$\text{rank}\left(\begin{bmatrix} a(0) & a(1) & \ldots & a(t_1-t_0-d) \\ a(1) & a(2) & \ldots & a(t_1-t_0-d+1) \\ \vdots & \vdots & & \vdots \\ a(d) & a(d+1) & \ldots & a(t_1-t_0) \end{bmatrix}\right)=q(d+1)$$

and that for $\tilde{w}:=a$ $\text{rank}(S(\tilde{w},d))=q(d+1)$, hence also for generic data $\tilde{w}$. □

Notation. For $\mathcal{B}\in\mathbb{B}$ let $e^*(\mathcal{B})$ denote the tightest equation structure of autoregressive parametrizations of $\mathcal{B}$, cf. section II.3.2.4. □

Definition 3-8 The model class $\mathbb{B}(\mathcal{T})$ is defined as $\mathbb{B}(\mathcal{T}):=\{\mathcal{B}\in\mathbb{B};\ \max\{t;\ e_t^*(\mathcal{B})\neq 0\}\leq\bar{d}(\mathcal{T})\}$, where $\bar{d}(\mathcal{T}):=(\#(\mathcal{T})-q)/(q+1)$.

Interpretation. For given interval of observation $\mathcal{T}$ the appropriate model class in $\mathbb{B}$ is given by $\mathbb{B}(\mathcal{T})$. It makes no sense to identify a model $\mathcal{B}\in\mathbb{B}\backslash\mathbb{B}(\mathcal{T})$, as such a model contains laws of orders which are too large for the given number of observations, i.e., such a model is overparametrized. □

3.3.2. Modelling under a complexity constraint

In this section we describe a descriptive and a predictive procedure for deterministic time series analysis in case a tolerated complexity is specified. Under this restriction the misfit is minimized.

So suppose that a maximal tolerated complexity $c_{tol}:=(c_t^{tol};\ t\in\mathbb{Z}_+)$ is given. Fixing c_{tol} is interpreted as requiring that allowable models should satisfy $c_t(\mathcal{B})\leq c_t^{tol}$ for all $t\in\mathbb{Z}_+$.

Remark. So $c(\mathcal{B})\overset{(p)}{\leq}c^{tol}$ in the partial ordering defined in section IV.2.2, not only in the lexicographic ordering of definition IV.2–4. According to proposition IV.2–2 the requirement $c(\mathcal{B})\overset{(p)}{\leq}c_{tol}$ is equivalent to requiring $\Sigma_{k=0}^{t}(t+1-k)e_k^*(\mathcal{B})\geq(t+1).(q-c_t^{tol})$ for all $t\in\mathbb{Z}_+$, where $(e_t^*(\mathcal{B});\ t\in\mathbb{Z}_+)$ is the equation structure of a tightest equation representation of $\mathcal{B}$, see section II.3.2.4. □

Interpretation. Fixing a maximal tolerated complexity amounts to requiring that $\mathcal{B}$ imposes a minimal tolerated number of t–th order restrictions on the behaviour. □

Under the requirement $c(\mathcal{B})\overset{(p)}{\leq}c_{tol}$ the misfit in $(\mathbb{R}^{1\times q})^{\mathbb{Z}_+}$ of $\mathcal{B}$ with respect to a given time series $\tilde{w}\in(\mathbb{R}^q)^{\mathcal{T}}$ will be minimized lexicographically. The utility $u_{c_{tol}}$ corresponding to definition II.2–3 is defined as follows.

Definition 3–9 For given $c_{tol}\in(\mathbb{R}_+)^{\mathbb{Z}_+}$ the utility $u:=u_{c_{tol}}$ is defined by the following total ordering of $(\mathbb{R}_+)^{\mathbb{Z}_+}\times(\mathbb{R}^{1\times q})^{\mathbb{Z}_+}$:

(*i*) $\{u(c^{(1)},\varepsilon^{(1)})=u(c^{(2)},\varepsilon^{(2)})\}:\Leftrightarrow\{\exists t_i\in\mathbb{Z}_+$ such that $c_{t_i}^{(i)}>c_{t_i}^{tol}$, $i=1,2$; or $(c^{(1)},\varepsilon^{(1)})=(c^{(2)},\varepsilon^{(2)})\}$;

(*ii*) $\{u(c^{(1)},\varepsilon^{(1)})<u(c^{(2)},\varepsilon^{(2)})\}:\Leftrightarrow\{\exists t_0\in\mathbb{Z}_+$ such that $c_{t_0}^{(1)}>c_{t_0}^{tol}$ and $\forall t\in\mathbb{Z}_+$ $c_t^{(2)}\leq c_t^{tol}$; or $\forall t\in\mathbb{Z}_+$ $c_t^{(1)},c_t^{(2)}\leq c_t^{tol}$, and $\exists t_0\in\mathbb{Z}_+$ such that $\varepsilon_{t_0}^{(1)}>\varepsilon_{t_0}^{(2)}$ and $\varepsilon_t^{(1)}=\varepsilon_t^{(2)}$ for all $t<t_0$; or $\forall t\in\mathbb{Z}_+$ $c_t^{(1)},c_t^{(2)}\leq c_t^{tol}$, $\varepsilon^{(1)}=\varepsilon^{(2)}$, and $\exists t_0\in\mathbb{Z}_+$ such that $c_{t_0}^{(1)}>c_{t_0}^{(2)}$ and $c_t^{(1)}=c_t^{(2)}$ for all $t<t_0\}$. Here the

vectors $\varepsilon_t \in \mathbb{R}^{1\times q}$ are ordered lexicographically.

Interpretation. A complexity larger than c_{tol} is not tolerated, small misfit is preferred and for equal misfit small complexity is preferred. □

The procedures $P^D_{c_{tol}}$ and $P^P_{c_{tol}}$ are defined as the procedures corresponding to $u_{c_{tol}}$, with c the complexity of a model and ε the descriptive or predictive misfit.

Definition 3-10 The procedure $P^D_{c_{tol}}:\mathcal{D} \to 2^{\mathbb{B}}$ is defined as follows. If $\tilde{w} \in (\mathbb{R}^q)^{\mathcal{T}}$ then $P^D_{c_{tol}}(\tilde{w}):=\text{argmax}\{u_{c_{tol}}(c(\mathcal{B}),\ \varepsilon^D(\tilde{w},\mathcal{B}));\ \mathcal{B}\in\mathbb{B}(\mathcal{T})\}$.

Remark. Propositions 2-11 and 2-12 imply that $P^D_{c_{tol}}$ is well-defined. □

Interpretation. Under the complexity constraint first the misfit of the zero order laws (in L^D_0) is minimized, then the misfit of the truly first order laws (in L^D_1), and so on, cf. definitions 3-3 and 3-4. Moreover, the procedure only identifies laws of order not larger than the critical level $\bar{d}(\mathcal{T})$, i.e., it prevents accepting laws which have good fit due to overparametrization, cf. section 3.3.1. □

Remark. In order to define $P^P_{c_{tol}}$ in terms of $u_{c_{tol}}$ we need to guarantee that we only consider data $\tilde{w}$ and models $\mathcal{B}\in\mathbb{B}$ for which the predictive misfit is well-defined. For given time series $\tilde{w}\in(\mathbb{R}^q)^{\mathcal{T}}$ the model class is restricted to $\mathbb{B}(\mathcal{T})$ in order to prevent overparametrization. As for $\mathcal{B}\in\mathbb{B}(\mathcal{T})$ there holds that $L^P_t=\{0\}$ for $t>\bar{d}(\mathcal{T})$ it follows from the remark preceding definition 3-7 that $\varepsilon^P(\tilde{w},\mathcal{B})$ is well-defined provided that $\text{span}\{\tilde{w}(t);\ t\in[t_0+\bar{d}(\mathcal{T}),t_1]\}=\mathbb{R}^q$ for $\mathcal{T}=[t_0,t_1]$. As $t_1-t_0-\bar{d}(\mathcal{T})+1=q(\bar{d}(\mathcal{T})+1)\geq q$ it hence follows that $\varepsilon^P(\tilde{w},\mathcal{B})$ is well-defined on $\mathbb{B}(\mathcal{T})$ for generic data. □

Definition 3-11 For generic data $\tilde{w}\in(\mathbb{R}^q)^{\mathcal{T}}$ $P^P_{c_{tol}}(\tilde{w}):=\text{argmax}\{u_{c_{tol}}(c(\mathcal{B}),\ \varepsilon^P(\tilde{w},\mathcal{B}));\ \mathcal{B}\in\mathbb{B}(\mathcal{T})\}$.

Remark. Propositions 2-18 and 2-21 imply that $P^P_{c_{tol}}$ is well-defined for generic data. □

Remark. Both for $P^D_{c_{tol}}$ and $P^P_{c_{tol}}$ it is crucial that the misfit of laws of different order is measured independently, as discussed in section 3.2. □

Remark. The procedures $P^D_{c_{tol}}$ and $P^P_{c_{tol}}$ can be numerically implemented by relatively simple algorithms, described in sections 4.2.1 and 4.3.1. □

Remark. For univariate time series the descriptive and predictive procedures are equivalent. That is, for arbitrary c_{tol} and generic $\tilde{w}\in\mathbb{R}^{\mathcal{T}}$ there holds that $P^D_{c_{tol}}(\tilde{w})=P^P_{c_{tol}}(\tilde{w})$. □

3.3.3. Modelling under a misfit constraint

Next we describe a descriptive and a predictive procedure for deterministic time series analysis in case the misfit should remain below a specified toleration level. Under this restriction the complexity is minimized.

We assume that $\bar{\varepsilon}:=(\bar{\varepsilon}_t^{tol};\ t\in\mathbb{Z}_+)\in\mathbb{R}^{\mathbb{Z}_+}$ is given and that it is required that the misfit of (truly) t–th order laws is smaller than $\bar{\varepsilon}_t^{tol}$.

Remark. This corresponds to $\varepsilon_{tol}:=(\varepsilon_t^{tol};\ t\in\mathbb{Z}_+)$ with $\varepsilon_t^{tol}=\bar{\varepsilon}_t^{tol}.(1,\dots,1)\in\mathbb{R}^{1\times q}$, and to a pointwise interpretation of e.g. $\varepsilon^D(\tilde{w},\mathcal{B})<\varepsilon_{tol}$. Note that $\varepsilon^D_{t,k}(\tilde{w},\mathcal{B})\leq\varepsilon^D_{t,l}(\tilde{w},\mathcal{B})$ for $k\geq l$, so indeed a model $\mathcal{B}\in\mathbb{B}$ then is tolerated if and only if $\varepsilon^D_{t,1}(\tilde{w},\mathcal{B})<\bar{\varepsilon}_t^{tol}$ for all $t\in\mathbb{Z}_+$. All laws of order t and hence especially the worst one should have misfit less than $\bar{\varepsilon}_t^{tol}$. □

Under this requirement of fit the complexity has to be minimized with respect to the lexicographic ordering of definition IV.2–4.

Interpretation. Under the misfit restriction the equation structure $(e_t^*(\mathcal{B});\ t\in\mathbb{Z}_+)$ has to be maximized lexicographically, see corollary IV.2–5. So the purpose is to find as many relations of small order as possible. For an interpreation we refer to section IV.2.2. □

Remark. The utility function $u^*:=u^*_{\varepsilon_{tol}}$, corresponding to minimizing lexicographically the complexity under a misfit constraint, is defined by the ordering $\{u^*(c^{(1)},\varepsilon^{(1)})=u^*(c^{(2)},\varepsilon^{(2)})\}:\Leftrightarrow\{\ \exists t_i\in\mathbb{Z}_+$ such that $\varepsilon^{(i)}_{t_i,1}\geq\bar{\varepsilon}^{tol}_{t_i}$,

$i=1,2$; or $(c^{(1)},\varepsilon^{(1)})=(c^{(2)},\varepsilon^{(2)})\}$, and $\{u^*(c^{(1)},\varepsilon^{(1)})<u^*(c^{(2)},\varepsilon^{(2)})\}:\Leftrightarrow\{\exists t_0\in\mathbb{Z}_+$ such that $\varepsilon_{t_0,1}^{(1)}\geq\bar{\varepsilon}_{t_0}^{tol}$ and $\forall t\in\mathbb{Z}_+$ $\varepsilon_{t,1}^{(2)}<\bar{\varepsilon}_t^{tol}$; or $\forall t\in\mathbb{Z}_+$ $\varepsilon_{t,1}^{(1)}, \varepsilon_{t,1}^{(2)}<\bar{\varepsilon}_t^{tol}$, and $\exists t_0\in\mathbb{Z}_+$ such that $c_{t_0}^{(1)}>c_{t_0}^{(2)}$ and $c_t^{(1)}=c_t^{(2)}$ for all $t<t_0$; or $\forall t\in\mathbb{Z}_+$ $\varepsilon_{t,1}^{(1)}, \varepsilon_{t,1}^{(2)}<\bar{\varepsilon}_t^{tol}$, $c^{(1)}=c^{(2)}$, and $\varepsilon^{(1)}>\varepsilon^{(2)}$ in lexicographic ordering$\}$.

This utility corresponds to the one of definition II.2–4 in section II.2.3.1. It turns out that the corresponding descriptive and predictive procedures $P^{*D}_{\varepsilon_{tol}}$ and $P^{*P}_{\varepsilon_{tol}}$ are difficult to implement algorithmically. In general optimal models even do not exist. If they exist they are difficult to compute. We will consider a slight variation $u_{\varepsilon_{tol}}$ of $u^*_{\varepsilon_{tol}}$. The corresponding procedures $P^D_{\varepsilon_{tol}}$ and $P^P_{\varepsilon_{tol}}$ can be implemented by relatively simple algorithms as described in sections 4.2.2 and 4.3.2. We will illustrate the difference between $u^*_{\varepsilon_{tol}}$ and $u_{\varepsilon_{tol}}$ by means of a simple example in section 6.5. □

Notation. For given $c\in(\mathbb{R}_+)^{\mathbb{Z}_+}$ we define the equation structure e corresponding to c as the smallest equation structure of models $\mathcal{B}\in\mathbb{B}$ with $c(\mathcal{B})\overset{(p)}{\leq}c$, i.e., (*i*) $\exists\mathcal{B}\in\mathbb{B}$ with $c(\mathcal{B})\overset{(p)}{\leq}c$ and $e^*(\mathcal{B})=e$; (*ii*) $\{c(\mathcal{B}')\overset{(p)}{\leq}c\}\Rightarrow\{e^*(\mathcal{B}')\geq e$ in lexicographic ordering$\}$. So in particular if $c=c(\mathcal{B})$ for some $\mathcal{B}\in\mathbb{B}$ then $e=e^*(\mathcal{B})$, see corollary IV.2–5. □

Definition 3-12 For given $\varepsilon_{tol}\in(\mathbb{R}^{1\times q})^{\mathbb{Z}_+}$ the utility $u:=u_{\varepsilon_{tol}}$ is defined by the following total ordering of $(\mathbb{R}_+)^{\mathbb{Z}_+}\times(\mathbb{R}^{1\times q})^{\mathbb{Z}_+}$:

(*i*) $\{u(c^{(1)},\varepsilon^{(1)})=u(c^{(2)},\varepsilon^{(2)})\}:\Leftrightarrow\{\exists t_i\in\mathbb{Z}_+$ such that $\varepsilon_{t_i,1}^{(i)}\geq\bar{\varepsilon}_{t_i}^{tol}$, $i=1,2$; or $(c^{(1)},\varepsilon^{(1)})=(c^{(2)},\varepsilon^{(2)})\}$;

(*ii*) $\{u(c^{(1)},\varepsilon^{(1)}) < u(c^{(2)},\varepsilon^{(2)})\}:\Leftrightarrow\{\exists t_0\in\mathbb{Z}_+$ such that $\varepsilon_{t_0,1}^{(1)}\geq\bar{\varepsilon}_{t_0}^{tol}$ and $\forall t\in\mathbb{Z}_+$ $\varepsilon_{t,1}^{(2)}<\bar{\varepsilon}_t^{tol}$; or $\forall t\in\mathbb{Z}_+$ $\varepsilon_{t,1}^{(1)}, \varepsilon_{t,1}^{(2)}<\bar{\varepsilon}_t^{tol}$ and $(c_0^{(1)}, \varepsilon_{0,1}^{(1)}, \ldots, \varepsilon_{0,e_0^{(1)}}^{(1)}, c_1^{(1)}, \varepsilon_{1,1}^{(1)}, \ldots, \varepsilon_{1,e_1^{(1)}}^{(1)}, c_2^{(1)}, \varepsilon_{2,1}^{(1)}, \ldots, \varepsilon_{2,e_2^{(1)}}^{(1)}, c_3^{(1)}, \ldots) > (c_0^{(2)}, \varepsilon_{0,1}^{(2)}, \ldots, \varepsilon_{0,e_0^{(2)}}^{(2)}, c_1^{(2)}, \varepsilon_{1,1}^{(2)}, \ldots, \varepsilon_{1,e_1^{(2)}}^{(2)}, c_2^{(2)}, \varepsilon_{2,1}^{(2)}, \ldots, \varepsilon_{2,e_2^{(2)}}^{(2)}, c_3^{(2)}, \ldots)$ in the lexicographic ordering, where $e^{(i)}$ is the equation structure corresponding to $c^{(i)}$, $i=1,2\}$.

The procedures $P^D_{\varepsilon_{tol}}$ and $P^P_{\varepsilon_{tol}}$ are defined as the procedures corresponding to $u_{\varepsilon_{tol}}$, with c the model complexity and ε the descriptive or predictive misfit.

Definition 3-13 The procedure $P^D_{\varepsilon_{tol}}:\mathcal{D}\to 2^{\mathbb{B}}$ is defined as follows. If $\tilde{w}\in(\mathbb{R}^q)^{\mathcal{T}}$ then $P^D_{\varepsilon_{tol}}(\tilde{w}):=\text{argmax}\{u_{\varepsilon_{tol}}(c(\mathcal{B}),\ \varepsilon^D(\tilde{w},\mathcal{B}));\ \mathcal{B}\in\mathbb{B}(\mathcal{T})\}$.

Interpretation. According to definition 3–12 this procedure corresponds to first finding a maximal number of zero order relations under the misfit constraint. Among solutions, which in general are highly non-unique, the ones with minimal zero order misfit are preferred. Subsequently the number of first order relations is maximized, and then the first order misfit is minimized, and so on. □

Remark. Propositions 2–11 and 2–13 imply that $P^D_{\varepsilon_{tol}}$ is well-defined. □

Remark. Suppose that $\varepsilon^{tol}_{t,k}\le 0$ for some $t\in\mathbb{Z}_+$ and $k\in[1,q]$. Let $\mathcal{T}$ be such that $t\le\bar{d}(\mathcal{T})$, then for any $\tilde{w}\in(\mathbb{R}^q)^{\mathcal{T}}$ $P^D_{\varepsilon_{tol}}(\tilde{w})=\emptyset$. Indeed, the remark preceding definition 3–3 in section 3.2.1 shows that in this case in step t not even accepting no law is tolerated, while any law has misfit at least zero. So in this case no model is identified. Hence for given time interval of observation $\mathcal{T}$ it is reasonable to require $\varepsilon^{tol}_{t,k}>0$ for all $t\in[0,\bar{d}(\mathcal{T})]$, $k\in[1,q]$. □

Remark. In order to define $P^P_{\varepsilon_{tol}}$ in terms of $u_{\varepsilon_{tol}}$ we recall from section 3.3.2 that for data $\tilde{w}\in(\mathbb{R}^q)^{\mathcal{T}}$ on $\mathcal{T}=[t_0,t_1]$ $\varepsilon^P(\tilde{w},\mathcal{B})$ is well-defined for all $\mathcal{B}\in\mathbb{B}(\mathcal{T})$, provided that $\text{span}\{\tilde{w}(t);\ t\in[t_0+\bar{d}(\mathcal{T}),t_1]\}=\mathbb{R}^q$. This condition is satisfied generically on $(\mathbb{R}^q)^{\mathcal{T}}$. □

Definition 3-14 For generic data $\tilde{w}\in(\mathbb{R}^q)^{\mathcal{T}}$ $P^P_{\varepsilon_{tol}}(\tilde{w}):=\text{argmax}\{u_{\varepsilon_{tol}}(c(\mathcal{B}),\ \varepsilon^P(\tilde{w},\mathcal{B}));\ \mathcal{B}\in\mathbb{B}(\mathcal{T})\}$.

Remark. Propositions 2–18 and 2–22 imply that $P^P_{\varepsilon_{tol}}$ is well-defined for generic data. For the same reasons as given before it is reasonable for given time interval of observation $\mathcal{T}$ to require that $\varepsilon^{tol}_{t,k}>0$ for all $t\in[0,\bar{d}(\mathcal{T})]$, $k\in[1,q]$. The definition of predictive misfit in definition 3–5 implies that moreover it is reasonable to take $\varepsilon^{tol}_{t,k}\le 1$ for $t\in[1,\bar{d}(\mathcal{T})]$, $k\in[1,q]$. □

Remark. Corollary IV.2–5 indicates a close relationship between $P^D_{\varepsilon_{tol}}$ and $P^{*D}_{\varepsilon_{tol}}$ as well as between $P^P_{\varepsilon_{tol}}$ and $P^{*P}_{\varepsilon_{tol}}$. However, $P^D_{\varepsilon_{tol}}$ and $P^P_{\varepsilon_{tol}}$ need not always minimize the complexity with respect to the lexicographic ordering on $(c_t(\mathcal{B});\ t\in\mathbb{Z}_+)$, as will be illustrated by means of a simulation in section 6.5. This is due to the auxiliary minimization of misfits, which is essential for obtaining simple (recursive) algorithms. □

Remark. The procedures amount to sequentially minimizing complexity and misfit. It is crucial that the misfit of laws of different order is measured independently in the way discussed in section 3.2. □

By $\bar{\bar{P}}^D_{\varepsilon_{tol}}(\tilde{w})$ we denote the procedure which is defined in analogy with $P^D_{\varepsilon_{tol}}$, but requiring $\varepsilon^D_{t,1}(\tilde{w},\mathcal{B})\leq\bar{\varepsilon}^{tol}_t$, in contrast with $P^D_{\varepsilon_{tol}}$ which requires $\varepsilon^D_{t,1}(\tilde{w},\mathcal{B})<\bar{\varepsilon}^{tol}_t$. We define $\bar{\bar{P}}^P_{\varepsilon_{tol}}$ in analogy with $P^P_{\varepsilon_{tol}}$, replacing the constraints $\varepsilon^P_{t,1}(\tilde{w},\mathcal{B})<\bar{\varepsilon}^{tol}_t$ by $\varepsilon^P_{t,1}(\tilde{w},\mathcal{B})\leq\bar{\varepsilon}^{tol}_t$.

Remark. For given time interval of observation $\mathcal{T}$ it is reasonable to require that $\varepsilon^{tol}_{t,k}\geq 0$ for all $t\in[0,\bar{d}(\mathcal{T})]$, $k\in[1,q]$. Otherwise no model is identified. □

Remark. The procedures for deterministic time series analysis presented in this and the foregoing section need not always identify a unique model. However, in the next section we will see that non-unique identification only occurs for specifications of c_{tol} or ε_{tol} which are not reasonable for the given data, cf. sections 4.1.1 and 4.4. □

4. Algorithms for deterministic time series analysis

4.1. Introduction

4.1.1. Overview

In this section we present algorithms for the deterministic approximate modelling procedures of section 3.3. These algorithms basically consist of recursive implementation of the static procedures of section 2. This sequential optimization is illustrated in section 4.1.3 by describing $P^D_{c_{tol}}$ in general terms.

We give algorithms for the descriptive procedures $P^D_{c_{tol}}$ and $P^D_{\varepsilon_{tol}}$ in section 4.2 and for the predictive procedures $P^P_{c_{tol}}$ and $P^P_{\varepsilon_{tol}}$ in section 4.3. The algorithms generate optimal models provided that the specifications of c_{tol} and ε_{tol} are reasonable for the available data. The conditions on c_{tol} and ε_{tol} are made explicit in sections 4.2 and 4.3. In section 4.4 we show that these conditions amount to requiring that the modelling problem is well–specified.

Remark. The algorithms consist of sequential construction of complementary spaces $\{V_t;\, t\in\mathbb{Z}_+\}$, cf. section II.3.2.4. The identified model $\mathcal{B}$ then is defined in terms of $L_t:=v_t^{-1}(V_t)$ by $\mathcal{B}:=\{w\in(\mathbb{R}^q)^{\mathbb{Z}};\, r(\sigma)w=0 \text{ for all } r\in L_t,\, t\in\mathbb{Z}_+\}$. The spaces V_t are constructed by sequential application of the results stated in propositions 2–12(*iii*) and 2–13(*iii*) in section 2.1.3 and propositions 2–21(*iii*) and 2–22(*iv*) in section 2.2.3. It is assumed that c_{tol} and ε_{tol} are such that the available data satisfy the relevant conditions in terms of singular values as stated in these propositions. If these conditions are not satisfied this indicates misspecification of c_{tol} or ε_{tol}, see section 4.4. □

In the next section we describe a reasonable requirement for the specification of c_{tol}.

4.1.2. Sensibility

In section 3.3.1 we derived that for data $\tilde{w}\in(\mathbb{R}^q)^{\mathcal{T}}$ the quality of laws of order larger than $\bar{d}(\mathcal{T})$ cannot be reasonably assessed. Hence we have reason to require that c_{tol} is specified in such a way that acceptance of laws of order larger than $\bar{d}(\mathcal{T})$ is not even considered. This implies a restriction on the specification of c_{tol} to be sensible.

Notation. For given $c_{tol}\in(\mathbb{R}_+)^{\mathbb{Z}_+}$ let $E(c_{tol})$ be the class of tightest equation structures of allowable models $\mathcal{B}\in\mathbb{B}$, i.e., $E(c_{tol}):=\{(e_t^*;\, t\in\mathbb{Z}_+);\ \exists\mathcal{B}\in\mathbb{B},\ c_t(\mathcal{B})\leq c_t^{tol}$ for all $t\in\mathbb{Z}_+$, such that $(e_t^*;\, t\in\mathbb{Z}_+)$ is the tightest equation structure of $\mathcal{B}\}$, cf. section II.3.2.4. Equip $E(c_{tol})$ with the lexicographic ordering and let $e(c_{tol})$ be the corresponding minimal element of $E(c_{tol})$. We call $e(c_{tol})$ the equation structure corresponding to c_{tol}. □

Definition 4-1 For given tolerated complexity c_{tol}, the *equation structure corresponding to* c_{tol} is the minimal achievable tightest equation structure of tolerated models in $\mathbb{B}$ with respect to the lexicographic ordering.

Definition 4-2 Let $\tilde{w}\in(\mathbb{R}^q)^{\mathcal{T}}$ and $c_{tol}\in(\mathbb{R}_+)^{\mathbb{Z}_+}$ be given. Then c_{tol} is called *sensible* (for $\tilde{w}$) if $\max\{t;\ e_t(c_{tol})\neq 0\}\leq\bar{d}(\mathcal{T}):=(\#(\mathcal{T})-q)/(q+1)$.

Interpretation. A specification makes sense only if overparametrization is precluded. □

Remark. For data $\tilde{w}\in(\mathbb{R}^q)^{\mathcal{T}}$ the identification procedure $P^D_{c_{tol}}$ of section 3.3 is restricted to the model class $\mathbb{B}(\mathcal{T})$. Hence if c_{tol} is not sensible then there exists a sensible c'_{tol} such that $P^D_{c'_{tol}}(\tilde{w})=P^D_{c_{tol}}(\tilde{w})$. The same holds true for $P^P_{c_{tol}}$. □

Remark. In section 4.1.3 we illustrate that for given c_{tol} the equation structure $e(c_{tol})$ is of relevance. In sections 4.2.1 and 4.3.1 we show that for given data $\tilde{w}$ and for sensible c_{tol} both $P^D_{c_{tol}}(\tilde{w})$ and $P^P_{c_{tol}}(\tilde{w})$ are singletons with tightest equation structure $e(c_{tol})$, provided that the modelling problem is well-specified for $\tilde{w}$. □

4.1.3. Sequential optimization

In this section we illustrate the main ideas for the algorithms in sections 4.2 and 4.3 by describing the procedure $P^D_{c_{tol}}$ in general terms. We indicate that for given sensible c_{tol} and under some assumptions on the data $\tilde{w}$ the optimal model $P^D_{c_{tol}}(\tilde{w})$ can be determined by sequential solution of static modelling problems. This is essentially due to the lexicographic ordering of misfits as defined in definition 3-4. We moreover indicate that under these assumptions the optimal model is unique and has tightest equation structure $e_{tol}:=e(c_{tol})$ of definition 4-1.

Step 0. The condition $c(\mathcal{B})\leq c_{tol}$ implies that we have to accept at least e_0^{tol} zero order laws, see definition 4–1. Let e_0 denote the number of independent equations of order 0 which are exactly satisfied by the data. It is reasonable to suppose that $e_0^{tol}\geq e_0$. In this case any optimal model $\mathcal{B}$ satisfies $e_0^*(\mathcal{B})=e_0^{tol}$. Indeed, models $\mathcal{B}$ with $e_0^*(\mathcal{B})<e_0^{tol}$ are not allowed while models $\mathcal{B}$ with $e_0^*(\mathcal{B})>e_0^{tol}\geq e_0$ can be improved, which is seen as follows. Suppose that $e_0=0$, then $\varepsilon_{0,k}^D(\tilde{w},\mathcal{B})>0$ for all $k\in[1,e_0^*(\mathcal{B})]$ and hence a model $\mathcal{B}$ with $e_0^*(\mathcal{B})>e_0^{tol}$ can be improved by simply deleting $e_0^*(\mathcal{B})-e_0^{tol}$ zero order equations, cf. definitions 3–3 and 3–4. A similar argument holds true if $e_0>0$.

Due to definition 3–9 and the lexicographic ordering on ε^D we hence first have to identify e_0^{tol} zero order equations of minimal misfit. Under some assumptions on the data this problem has a unique solution, cf. proposition 2–12(*iii*). Let the solution be L_0 and define $\mathcal{B}_0^\perp:=L_0$.

Step t $(t\leq\bar{d}(\mathcal{T}))$. Given that $e_k^*(\mathcal{B})=e_k^{tol}$ for $k\in[0,t-1]$, the condition $c(\mathcal{B})\leq c_{tol}$ implies a minimal required number e_t^{tol} of truly t–th order laws in the space $v_t^{-1}\{[v_t(\mathcal{B}_{t-1}^\perp+s\mathcal{B}_{t-1}^\perp)]^\perp\}$. Let e_t denote the number of independent t–th order equations in this space which are exactly satisfied by the data. Under the reasonable assumption that $e_t^{tol}\geq e_t$ it follows that for optimal models $e_t^*(\mathcal{B})=e_t^{tol}$, cf. step 0.

Due to the lexicographic ordering on ε^D we then have to identify e_t^{tol} equations of minimal misfit in $v_t^{-1}\{V_t\}$, where $V_t:=[v_t(\mathcal{B}_{t-1}^\perp+s\mathcal{B}_{t-1}^\perp)]^\perp$. This can be done by applying proposition 2–12 for transformed data, as follows. Let $q_t:=\dim(V_t)$ and let $P_t\in\mathbb{R}^{q_t\times q(t+1)}$ have orthonormal rows which span V_t. Define transformed data by $\tilde{x}_i:=P_t\cdot\mathrm{col}(\tilde{w}(t_0+i),\ldots,\tilde{w}(t_0+i+t))$, $i\in[0,t_1-t_0-t]$. Note that for $r\in\mathbb{R}_t^{1\times q}[s]$ $v_t(r)\in V_t$ if and only if there is an $a\in\mathbb{R}^{q_t}$ with $v_t(r)=a^TP$ and that then $\|r\tilde{w}\|^2=v_t(r)S(\tilde{w},t)v_t(r)^T=a^T\left[\frac{1}{t_1-t_0-t+1}\cdot\Sigma_{i=0}^{t_1-t_0-t}\tilde{x}_i\tilde{x}_i^T\right]a$. We assume that the data are such that the optimal set of e_t^{tol} laws in $v_t^{-1}(V_t)$ is uniquely determined by proposition 2–12(*iii*) for data $\{\tilde{x}_i;\, i\in[0,t_1-t_0-t]\}$, i.e., we assume that these transformed data have a covariance matrix which satisfies the condition on the singular values stated in proposition 2–12(*iii*). Let the solution be L_t and define $\mathcal{B}_t^\perp:=\mathcal{B}_{t-1}^\perp+s\mathcal{B}_{t-1}^\perp+L_t$.

Remark. In the algorithm for $P_{c_{tol}}^D$ in section 4.2.1 we assume that the data are such that in all steps $t\leq\bar{d}(\mathcal{T})$ proposition 2–12(*iii*) is applicable, i.e., that the relevant singular values do not coincide, see assumption 4–3(*ii*). □

Remark. The resulting model is defined by $\mathcal{B}:=\{w\in(\mathbb{R}^q)^{\mathbb{Z}};\ r(\sigma)w=0$ for all $r\in L_t$, $t\in[0,\bar{d}(\mathcal{T})]\}$. Under assumptions on the data $\tilde{w}$ this sequential optimization algorithm generates the optimal model $P^D_{c_{tol}}(\tilde{w})$, which in this case is a singleton, i.e., the optimal model is unique. For $\mathcal{B}$ there holds $L_t=L_t^D$ of (CDF), cf. section II.3.2.5. Moreover $e^*(\mathcal{B})=e(c_{tol})$. This is due to the lexicographic ordering of misfits which, roughly stated, implies that it is preferable to accept as few low order equations as possible, given the complexity constraint. In this way low order errors are minimized. We refer to definitions 3–3 and 3–4 and the remark preceding definition 3–3. □

Remark. Hence under assumptions on the data the optimal model can be determined by sequential optimal choice of e_t^{tol} laws of minimal descriptive misfit by applying proposition 2–12(*iii*). □

Remark. In the next sections we describe computational details of this algorithm and the other ones. We specify input, initialization, recursive part, termination and output of the algorithms. Moreover, we explicitly state the optimality properties of the resulting models in terms of assumptions on the data. We comment on these assumptions in section 4.4. For the algorithms we also refer to Willems [73, section 25], and Heij and Willems [30, section 7]. □

4.2. Algorithms for the descriptive procedures

4.2.1. Algorithm for descriptive modelling under a complexity constraint

In this section we describe an algorithm which, under conditions on the data $\tilde{w}\in(\mathbb{R}^q)^{\mathcal{T}}$, generates the model $\{\mathcal{B}\}=P^D_{c_{tol}}(\tilde{w})$ as defined in section 3.3.2. We explicitly state conditions on the data which imply optimality of the model calculated by the algorithm.

Algorithm for $P^D_{c_{tol}}$.

1. *Input.*

1.1. Data $\tilde{w}=(\tilde{w}(t);\ t\in\mathcal{T}=[t_0,t_1])\in(\mathbb{R}^q)^{\mathcal{T}}$.

1.2. Tolerated complexity $c_{tol}=(c_t^{tol};\ t\in\mathbb{Z}_+)\in(\mathbb{R}_+)^{\mathbb{Z}_+}$.

Let $e_{tol}:=e(c_{tol})$ denote the equation structure corresponding to c_{tol}.

2. *Initialization (step* 0).

2.1. Let $S(\tilde{w},0):=\frac{1}{t_1-t_0+1}\cdot\Sigma_{t=t_0}^{t_1}\tilde{w}(t)\tilde{w}(t)^T$, the empirical covariance matrix of order 0, have singular value decomposition (SVD) $S(\tilde{w},0)=U_0\Sigma_0U_0^T$, $\Sigma_0=\text{diag}(\sigma_1^{(0)},\ldots,\sigma_q^{(0)})$, $\sigma_1^{(0)}\geq\ldots\geq\sigma^{(0)}_{q-e_0^{tol}}\geq\sigma^{(0)}_{q-e_0^{tol}+1}\geq\ldots\geq\sigma_q^{(0)}\geq0$.

2.2. If $U_0=(u_1^{(0)},\ldots,u_q^{(0)})$, $u_k^{(0)}\in\mathbb{R}^q$, $k=1,\ldots,q$, then define $V_0:=\text{span}\{u_k^{(0)T};\ k\geq q-e_0^{tol}+1\}$ and $\mathcal{B}_0^{\perp}:=v_0^{-1}(V_0)$.

2.3. Define $q_0:=q$, $p_1:=2e_0^{tol}$, and let $\{v_k^{(1)T};\ k=1,\ldots,p_1\}$ be an orthonormal basis of $v_1(\mathcal{B}_0^{\perp}+s\mathcal{B}_0^{\perp})\subset\mathbb{R}^{1\times2q}$, e.g., $v_k^{(1)T}$ is the k-th row of $\begin{bmatrix}\bar{U}_0 & 0\\ 0 & \bar{U}_0\end{bmatrix}$ where $\bar{U}_0:=\text{col}(u_k^{(0)T};\ k=q-e_0^{tol}+1,\ldots,q)$.

3. *Recursion (step* t).

3.0. Input from step $t-1$: an orthonormal basis $\{v_k^{(t)T};\ k=1,\ldots,p_t\}$ of $v_t(\mathcal{B}_{t-1}^{\perp}+s\mathcal{B}_{t-1}^{\perp})\subset\mathbb{R}^{1\times q(t+1)}$, where $p_t=\dim(v_t(\mathcal{B}_{t-1}^{\perp}+s\mathcal{B}_{t-1}^{\perp}))=\Sigma_{k=0}^{t-1}(t+1-k).e_k^{tol}$.

(SVD): $\Sigma_{k=1}^{p_t}v_k^{(t)}v_k^{(t)T}=\bar{V}_t\bar{\Sigma}_t\bar{V}_t^T$, $\bar{\Sigma}_t=\text{diag}(\bar{\sigma}_1^{(t)},\ldots,\bar{\sigma}^{(t)}_{q(t+1)})$, $1=\bar{\sigma}_1^{(t)}=\ldots=\bar{\sigma}^{(t)}_{p_t}>\bar{\sigma}^{(t)}_{p_t+1}=\ldots=\bar{\sigma}^{(t)}_{q(t+1)}=0$, $\bar{V}_t=(v_1^{(t)},\ldots,v_{p_t}^{(t)},v_{p_t+1}^{(t)},\ldots,v_{q(t+1)}^{(t)})$. Let $q_t:=q(t+1)-p_t$ and define $P_t:=\text{col}(v_k^{(t)T};\ k=p_t+1,\ldots,q(t+1))\in\mathbb{R}^{q_t\times q(t+1)}$. So the rows of P_t form an orthonormal basis for $[v_t(\mathcal{B}_{t-1}^{\perp}+s\mathcal{B}_{t-1}^{\perp})]^{\perp}\subset\mathbb{R}^{1\times q(t+1)}$.

3.1. Let $S(\tilde{w},t):=\frac{1}{t_1-t_0-t+1}\cdot\Sigma_{k=t_0}^{t_1-t}(\tilde{w}(k)^T,\ldots,\tilde{w}(k+t)^T)^T.(\tilde{w}(k)^T,\ldots,\tilde{w}(k+t)^T)$, the empirical covariance matrix of order t, and let $P_tS(\tilde{w},t)P_t^T$ have (SVD) $P_tS(\tilde{w},t)P_t^T=U_t\Sigma_tU_t^T$, $\Sigma_t=\text{diag}(\sigma_1^{(t)},\ldots,\sigma_{q_t}^{(t)})$, $\sigma_1^{(t)}\geq\ldots\geq\sigma^{(t)}_{q_t-e_t^{tol}}\geq\sigma^{(t)}_{q_t-e_t^{tol}+1}\geq\ldots\geq\sigma_{q_t}^{(t)}\geq0$.

3.2. If $U_t=(u_1^{(t)},\ldots,u_{q_t}^{(t)})$, $u_k^{(t)}\in\mathbb{R}^{q_t}$, $k=1,\ldots,q_t$, then define $V_t:=\text{span}\{u_k^{(t)}.P_t;\ k\geq q_t-e_t^{tol}+1\}$, $L_t:=v_t^{-1}(V_t)$, and $\mathcal{B}_t^{\perp}:=\mathcal{B}_{t-1}^{\perp}+s\mathcal{B}_{t-1}^{\perp}+L_t$.

3.3. Output to step $t+1$: an orthonormal basis $\{v_k^{(t+1)T};\ k=1,\ldots,p_{t+1}\}$ of $v_{t+1}(\mathcal{B}_t^{\perp}+s\mathcal{B}_t^{\perp})$, $p_{t+1}:=\Sigma_{k=0}^{t}(t+2-k).e_k^{tol}$.

Note that $O_t:=\{v_k^{(t)T};\ k=1,\ldots,p_t\}\cup\{u_k^{(t)}.P_t;\ k=q_t-e_t^{tol}+1,\ldots,q_t\}$ forms an orthonormal basis of $v_t(\mathcal{B}_t^\perp)$, with $\dim(O_t)=\Sigma_{k=0}^t(t+1-k)e_k^{tol}$. Let $O_t^0:=\{(v,0);\ v\in O_t,\ 0\in\mathbb{R}^{1\times q}\}$ and ${}^0O_t:=\{(0,v);\ 0\in\mathbb{R}^{1\times q},\ v\in O_t\}$, then it suffices to choose $\Sigma_{k=0}^t e_k^{tol}$ orthonormal vectors in span$({}^0O_t)$, orthogonal to O_t^0.

4. *Termination (at step t^*).*
Either at $t^*=\bar{d}(\mathcal{T}):=(t_1-t_0+1-q)/(q+1)$, or at $t^*<\bar{d}(\mathcal{T})$ when $\Sigma_{t=0}^{t^*} e_t^{tol}=q$.

5. *Output.*
Bases for V_t, $t\leq t^*$, and $\mathcal{B}_{t^*}^\perp$. Define $\mathcal{B}:=\{w\in(\mathbb{R}^q)^{\mathbb{Z}};\ r(\sigma)w=0,\ r\in\mathcal{B}_{t^*}^\perp\}$.

Remark. The algorithm basically consists of *sequential* application of proposition 2–12 in section 2.1.3. In the initialization the data is $\tilde{x}_i:=\tilde{w}(t_0+i)$, $i=0,\ldots,t_1-t_0$. In step t of the recursion the data consists of $\tilde{x}_i:=P_t.\text{col}(\tilde{w}(t_0+i),\ldots,\tilde{w}(t_0+i+t))$, $i=0,\ldots,t_1-t_0-t$. The operators P_t take care of the requirement that the new laws should be orthogonal to the old ones, cf. definitions 3–3 and II.3–11. Concerning step 3.1 note that for laws $r\in\mathbb{R}_t^{1\times q}[s]$ with $d(r)=t$ there holds $v_t(r)\in[v_t(\mathcal{B}_{t-1}^\perp+s\mathcal{B}_{t-1}^\perp)]^\perp$ if and only if there is an $a\in\mathbb{R}^{q_t}$ with $v_t(r)=a^TP_t$ and that then $\|r\tilde{w}\|^2=a^TP_tS(\tilde{w},t)P_t^Ta$. □

Let data $\tilde{w}\in(\mathbb{R}^q)^{\mathcal{T}}$ and tolerated complexity $c_{tol}\in(\mathbb{R}_+)^{\mathbb{Z}_+}$ be given. The next assumption implies conditions on c_{tol} and $\tilde{w}$.

Assumption 4–3 (*i*) c_{tol} is sensible;
(*ii*) $\sigma^{(0)}_{q-e_0^{tol}} > \sigma^{(0)}_{q-e_0^{tol}+1}$; in step t $\sigma^{(t)}_{q_t-e_t^{tol}} > \sigma^{(t)}_{q_t-e_t^{tol}+1}$;
(*iii*) for step t, let $u_k^{(t)T}.P_t=(u_{k,0},\ldots,u_{k,t})$, $u_{k,j}\in\mathbb{R}^{1\times q}$, and $U_0:=\text{col}(u_{k,0};\ k\geq q_t-e_t^{tol}+1)$, $U_t:=\text{col}(u_{k,t};\ k\geq q_t-e_t^{tol}+1)$; assume $\text{rank}(U_0)=\text{rank}(U_t)=e_t^{tol}$.

Interpretation. For assumption 4–3(*i*) we refer to section 4.1.2. Assumption 4–3(*ii*) guarantees the existence of a unique solution for the problem of optimal choice of e_t^{tol} equations of order t, orthogonal to $\mathcal{B}_{t-1}^\perp+s\mathcal{B}_{t-1}^\perp$, cf. proposition 2–12(*iii*). Assumption 4–3(*iii*) corresponds to requiring that the laws identified in step t really have order t, i.e., $\{0\neq r\in L_t\}\Rightarrow\{d(r)=t\}$. □

Remark. In section 4.4 we comment on these assumptions. If these are not satisfied then the modelling problem is misspecified, i.e., c_{tol} is not a reasonable specification for the given data. □

Theorem 4-4 Suppose assumption 4-3 is satisfied, then

(*i*) $P^D_{c_{tol}}(\tilde{w})=\{\mathcal{B}\}$, the model generated by the algorithm;

(*ii*) $e^*(\mathcal{B})=e_{tol}$;

(*iii*) $\varepsilon^D_{t,k}(\tilde{w},\mathcal{B})=\{\sigma^{(t)}_{q_t-e_t^{tol}+k}\}^{1/2}$, $k=1,\ldots,e_t^{tol}$;

(*iv*) $L_t=L_t^D$ for $\mathcal{B}$, so the algorithm gives a (CDF) representation of $\mathcal{B}$.

Proof. See the appendix.

Remark. For any data the algorithm generates an allowable model, as always $e^*(\mathcal{B})=e_{tol}$ and hence $c_t(\mathcal{B})\leq c_t^{tol}$ for all $t\in\mathbb{Z}_+$. The generated model however may be suboptimal in case assumption 4-3 is not satisfied, i.e., if c_{tol} is not a reasonable specification of the modelling problem for the given data. □

4.2.2. Algorithm for descriptive modelling under a misfit constraint

Next we describe an algorithm which, under conditions on the data $\tilde{w}\in(\mathbb{R}^q)^{\mathcal{T}}$, generates the model $P^D_{\varepsilon_{tol}}(\tilde{w})$ as defined in section 3.3.3. The algorithm basically consists of sequential application of proposition 2-13. The optimality of the model generated by the algorithm is a consequence of this proposition and the special utility $u_{\varepsilon_{tol}}$ as defined in definition 3-12.

Algorithm for $P^D_{\varepsilon_{tol}}$.

1. *Input.*

1.1. Data $\tilde{w}=(\tilde{w}(t);\ t\in\mathcal{T}=[t_0,t_1])\in(\mathbb{R}^q)^{\mathcal{T}}$.

1.2. Tolerated misfit $\varepsilon_{tol}=(\varepsilon_t^{tol};\ t\in\mathbb{Z}_+)$, $\varepsilon_t^{tol}=\bar{\varepsilon}_t^{tol}.(1,\ldots,1)\in\mathbb{R}^{1\times q}$, $\bar{\varepsilon}_t^{tol}\in\mathbb{R}$.

2. *Initialization* (*step* 0).

2.1. (SVD): $S(\tilde{w},0)=U_0\Sigma_0U_0^T$, $\Sigma_0=\text{diag}(\sigma_1^{(0)},\ldots,\sigma_q^{(0)})$, $\sigma_1^{(0)}\geq\ldots\geq\sigma^{(0)}_{q-e_0}\geq(\bar{\varepsilon}_0^{tol})^2>\sigma^{(0)}_{q-e_0+1}\geq\ldots\geq\sigma_q^{(0)}\geq0$.

2.2. If $U_0=(u_1^{(0)},\ldots,u_q^{(0)})$, $u_k^{(0)}\in\mathbb{R}^q$, $k=1,\ldots,q$, then define $V_0:=\text{span}\{u_k^{(0)T}$;

$k \geq q-e_0+1\}$ and $\mathcal{B}_0^{\perp} := v_0^{-1}(V_0)$.

2.3. Define $q_0 := q$, $p_1 := 2e_0$, and let $\{v_k^{(1)T};\ k=1,\ldots,p_1\}$ be an orthonormal basis of $v_1(\mathcal{B}_0^{\perp}+s\mathcal{B}_0^{\perp}) \subset \mathbb{R}^{1\times 2q}$, e.g., $v_k^{(1)T}$ is the k–th row of $\begin{bmatrix} \bar{U}_0 & 0 \\ 0 & \bar{U}_0 \end{bmatrix}$ where $\bar{U}_0 := \mathrm{col}(u_k^{(0)T};\ k=q-e_0+1,\ldots,q)$.

3. *Recursion (step t).*

3.0. Input from step t–1: an orthonormal basis $\{v_k^{(t)T};\ k=1,\ldots,p_t\}$ of $v_t(\mathcal{B}_{t-1}^{\perp}+s\mathcal{B}_{t-1}^{\perp}) \subset \mathbb{R}^{1\times q(t+1)}$, where $p_t = \dim(v_t(\mathcal{B}_{t-1}^{\perp}+s\mathcal{B}_{t-1}^{\perp})) = \Sigma_{k=0}^{t-1}(t+1-k).e_k$ with e_k the number of accepted k–th order laws. Let $q_t := q(t+1)-p_t$, $e_t' := q-\Sigma_{k=0}^{t-1}e_k$, and define P_t as in step 3.0 of the algorithm for $P_{c_{tol}}^D$.

3.1. (SVD): $P_tS(\tilde{w},t)P_t^T = U_t\Sigma_tU_t^T$, $\Sigma_t = \mathrm{diag}(\sigma_1^{(t)},\ldots,\sigma_{q_t}^{(t)})$, $\sigma_1^{(t)} \geq \ldots \geq \sigma_{q_t-e_t''}^{(t)} \geq (\bar{\varepsilon}_t^{tol})^2 > \sigma_{q_t-e_t''+1}^{(t)} \geq \ldots \geq \sigma_{q_t}^{(t)} \geq 0$.

3.2. If $U_t = (u_1^{(t)},\ldots,u_{q_t}^{(t)})$, $u_k^{(t)} \in \mathbb{R}^{q_t}$, $k=1,\ldots,q_t$, then with $e_t := \min\{e_t', e_t''\}$ define $V_t := \mathrm{span}\{u_k^{(t)T}.P_t;\ k \geq q_t-e_t+1\}$, $L_t := v_t^{-1}(V_t)$, and $\mathcal{B}_t^{\perp} := \mathcal{B}_{t-1}^{\perp} + s\mathcal{B}_{t-1}^{\perp} + L_t$.

3.3. Output to step t+1: an orthonormal basis $\{v_k^{(t+1)T};\ k=1,\ldots,p_{t+1}\}$ of $v_{t+1}(\mathcal{B}_t^{\perp}+s\mathcal{B}_t^{\perp})$, $p_{t+1} := \Sigma_{k=0}^{t}(t+2-k).e_k$. See also step 3.3 of the algorithm for $P_{c_{tol}}^D$.

4. *Termination (at step t^*).*

Either at $t^* = \bar{d}(\mathcal{T})$, or at $t^* < \bar{d}(\mathcal{T})$ when $\Sigma_{t=0}^{t^*} e_t = q$.

5. *Output.*

Bases for V_t, $t \leq t^*$, and $\mathcal{B}_{t^*}^{\perp}$. Define $\mathcal{B} := \{w \in (\mathbb{R}^q)^{\mathbb{Z}};\ r(\sigma)w = 0,\ r \in \mathcal{B}_{t^*}^{\perp}\}$.

We make the following assumptions on $\tilde{w} \in (\mathbb{R}^q)^{\mathcal{T}}$ and $\varepsilon_{tol} \in (\mathbb{R}^{1\times q})^{\mathbb{Z}_+}$, $\varepsilon_t^{tol} = \bar{\varepsilon}_t^{tol}(1,\ldots,1) \in \mathbb{R}^{1\times q}$ for some $\bar{\varepsilon}_{tol} := (\bar{\varepsilon}_t^{tol}; t \in \mathbb{Z}_+) \in \mathbb{R}^{\mathbb{Z}_+}$.

Assumption 4-5 (*i*) $\bar{\varepsilon}_t^{tol} > 0$ for all $t \in [0, \bar{d}(\mathcal{T})]$;

(*ii*) if at t^* $e_{t^*}'' > e_{t^*}'(>0)$, then assume $\sigma_{q_t-e_{t^*}'}^{(t^*)} > \sigma_{q_t-e_{t^*}'+1}^{(t^*)}$;

(*iii*) assumption 4–3(*iii*), with e_t^{tol} replaced by e_t.

Interpretation. Assumption 4–5(*iii*) guarantees uniqueness of $P_{\varepsilon_{tol}}^D(\tilde{w})$ and (*iii*) amounts to requiring that the laws identified in step t really have order t. Finally note that for $\bar{\varepsilon}_t^{tol} \leq 0$ not even accepting no law is tolerated, hence if assumption 4–5(i) is not satisfied then $P^D_{\varepsilon_{tol}}(\tilde{w})$ is empty, i.e., no model is identified. □

Remark. We again refer to section 4–4 for comments on these assumptions. If $\tilde{w}$ does not satisfy the assumptions then $\bar{\varepsilon}_{tol}$ is misspecified for these data. □

Theorem 4–6 Suppose assumption 4–5 is satisfied, then
(i) $P^D_{\varepsilon_{tol}}(\tilde{w})=\{\mathcal{B}\}$, the model generated by the algorithm;
(ii) $e^*(\mathcal{B})=(e_t;\ t\in\mathbb{Z}_+)$;
(iii) $\varepsilon^D_{t,k}(\tilde{w},\mathcal{B})=\{\sigma^{(t)}_{q_t-e_t+k}\}^{1/2}$, $k=1,\ldots,e_t$;
(iv) $L_t=L^D_t$ for $\mathcal{B}$, so the algorithm gives a (CDF) representation of $\mathcal{B}$.

Proof. See the appendix.

Remark. We get analogous results for $\bar{\bar{P}}^D_{\varepsilon_{tol}}$, provided that $\bar{\varepsilon}_t^{tol} \geq 0$ for all $t\in[0,\bar{d}(\mathcal{T})]$. For data $\tilde{w}$ which satisfies assumption 4–5 the model $\bar{\bar{P}}^D_{\varepsilon_{tol}}(\tilde{w})$ is unique and obtained by the algorithm of this section, taking in step 2.1 $\sigma^{(0)}_{q-e_0} > (\bar{\varepsilon}_0^{tol})^2 \geq \sigma^{(0)}_{q-e_0+1}$ and in step 3.1 $\sigma^{(t)}_{q_t-e_t'} > (\bar{\varepsilon}_t^{tol})^2 \geq \sigma^{(t)}_{q_t-e_t'+1}$. □

4.3. Algorithms for the predictive procedures

4.3.1. Algorithm for predictive modelling under a complexity constraint

In this section we give an algorithm which, under conditions on the data $\tilde{w}\in(\mathbb{R}^q)^{\mathcal{T}}$, generates the model $\{\mathcal{B}\}=P^P_{c_{tol}}(\tilde{w})$ as defined in section 3.3.2. We explicitly state conditions on the data which guarantee optimality of the model calculated by the algorithm.

Algorithm for $P^P_{c_{tol}}$.

1. *Input.*
As for $P^D_{c_{tol}}$.

2. *Initialization* (*step* 0).
2.1. As for $P^D_{c_{tol}}$.

2.2. As for $P^D_{c_{tol}}$.

2.3. Define $p_0:=e_0^{tol}$, $n_0:=e_0^{tol}$, and let $\{u_k^{(0)T};\ k\geq q-e_0^{tol}+1\}$ be an orthonormal basis of $v_0(\mathcal{B}_0^{\perp})$ and of $F_0=v_0(\mathcal{B}_0^{\perp})$, where F_0 is as defined in section II.3.2.6.

3. *Recursion* (*step* t).

3.0. Input from step $t-1$: an orthonormal basis $\{v_k^{(t-1)T};\ k=1,\ldots,\ p_{t-1}\}$, $p_{t-1}:=\Sigma_{k=0}^{t-1}(t-k)e_k^{tol}$, of $v_{t-1}(\mathcal{B}_{t-1}^{\perp})\subset\mathbb{R}^{1\times qt}$, and an orthonormal basis $\{f_k^{(t-1)T};\ k=1,\ldots,n_{t-1}\}$, $n_{t-1}:=\Sigma_{k=0}^{t-1}\ e_k^{tol}$, of $F_{t-1}:=\{\tilde{r}\in\mathbb{R}^{1\times q};\ \exists r\in\mathcal{B}_{t-1}^{\perp},\ r=\Sigma_{k=0}^{t-1}\ {}_kS^k$, such that $r_{t-1}=\tilde{r}\}$.

(SVD): $\Sigma_{k=1}^{p_{t-1}}v_k^{(t-1)}v_k^{(t-1)T}=\bar{V}_{t-1}\bar{\Sigma}_{t-1}\bar{V}_{t-1}^T$, $\bar{\Sigma}_{t-1}=\mathrm{diag}(\bar{\sigma}_1^{(t-1)},\ldots,\bar{\sigma}_{q\cdot t}^{(t-1)})$, $1=\bar{\sigma}_1^{(t-1)}=\ldots=\bar{\sigma}_{p_{t-1}}^{(t-1)}>\bar{\sigma}_{p_{t-1}+1}^{(t-1)}=\ldots=\bar{\sigma}_{q.t}^{(t-1)}=0$, $\bar{V}_{t-1}=(v_1^{(t-1)},\ldots,v_{p_{t-1}}^{(t-1)},\ v_{p_{t-1}+1}^{(t-1)},\ldots,\ v_{q.t}^{(t-1)})$. Let $q_t:=q.t-p_{t-1}$ and define $P_{1t}:=\mathrm{col}(v_k^{(t-1)T};\ k=\ p_{t-1}+1,\ldots,q\cdot t)\in\mathbb{R}^{q_t\times q.t}$.

Similarly, (SVD): $\Sigma_{k=1}^{t-1}f_k^{(t-1)}f_k^{(t-1)T}=\bar{\bar{V}}_{t-1}\bar{\bar{\Sigma}}_{t-1}\bar{\bar{V}}_{t-1}^T$, $\bar{\bar{\Sigma}}_{t-1}=\mathrm{diag}(\bar{\bar{\sigma}}_1^{(t-1)},\ldots,\bar{\bar{\sigma}}_q^{(t-1)})$, $1=\bar{\bar{\sigma}}_1^{(t-1)}=\ldots=\bar{\bar{\sigma}}_{n_{t-1}}^{(t-1)}>\bar{\bar{\sigma}}_{n_{t-1}+1}^{(t-1)}=\ldots=\bar{\bar{\sigma}}_q^{(t-1)}=0$, $\bar{\bar{V}}_{t-1}=(f_1^{(t-1)},\ldots,f_q^{(t-1)})$. Define $P_{2t}:=\mathrm{col}(f_k^{(t-1)T};\ k=n_{t-1}+1,\ldots,q)\in\mathbb{R}^{(q-n_{t-1})\times q}$.

Finally let $P_t:=\begin{bmatrix}P_{1t} & 0\\ 0 & P_{2t}\end{bmatrix}$. Then the rows of P_t form an orthonormal basis for $[v_t(F_{t-1}.s^t)+v_t(\mathcal{B}_{t-1}^{\perp})]^{\perp}\subset\mathbb{R}^{1\times q(t+1)}$.

3.1 Let $P_tS(\tilde{w},t)P_t^T=\begin{bmatrix}S_-^{(t)} & S_{-+}^{(t)}\\ S_{+-}^{(t)} & S_+^{(t)}\end{bmatrix}$ with $S_-^{(t)}\in\mathbb{R}^{q_t\times q_t}$, $S_+^{(t)}\in\mathbb{R}^{(q-n_{t-1})\times(q-n_{t-1})}$, $S_{-+}^{(t)}=S_{+-}^{(t)T}\in\mathbb{R}^{q_t\times(q-n_{t-1})}$.

(SVD): $(S_-^{(t)})^{-1/2}.S_{-+}^{(t)}.(S_+^{(t)})^{-1/2}=U_t^-\Lambda_tU_t^{+T}$, $\Lambda_t=\begin{bmatrix}\Sigma_t\\ 0\end{bmatrix}\in\mathbb{R}^{q_t\times(q-n_{t-1})}$, $\Sigma_t=\mathrm{diag}(\sigma_1^{(t)},\ldots,\sigma_{q-n_{t-1}}^{(t)})$, $\sigma_1^{(t)}\geq\ldots\geq\sigma_{e_t^{tol}}^{(t)}\geq\sigma_{e_t^{tol}+1}^{(t)}\geq\ldots\geq\sigma_{q-n_{t-1}}^{(t)}\geq0$.

3.2. If $(S_-^{(t)})^{-1/2}.U_t^-=(\bar{u}_1^{(t)},\ldots,\bar{u}_{q_t}^{(t)})$ and $(S_+^{(t)})^{-1/2}.U_t^+=(\bar{\bar{u}}_1^{(t)},\ldots,\bar{\bar{u}}_{q-n_{t-1}}^{(t)})$, then for $k\leq e_t^{tol}$ let $u_k^{(t)T}:=(-\sigma_k^{(t)}.\bar{u}_k^{(t)T},\ \bar{\bar{u}}_k^{(t)T}).P_t\in\mathbb{R}^{1\times q(t+1)}$.

Define $V_t:=\mathrm{span}\{u_k^{(t)T};\ k\leq e_t^{tol}\}$, $L_t:=v_t^{-1}(V_t)$, and $\mathcal{B}_t^{\perp}:=\mathcal{B}_{t-1}^{\perp}+s\mathcal{B}_{t-1}^{\perp}+L_t$.

3.3. Output to step $t+1$: orthonormal bases $\{v_k^{(t)T};\ k=1,\ldots,p_t\}$ of $v_t(\mathcal{B}_t^{\perp})$ and $\{f_k^{(t)T};\ k=1,\ldots,n_t\}$ of F_t. Here $p_t:=p_{t-1}+\Sigma_{k=0}^te_k^{tol}$ and $n_t:=n_{t-1}+e_t^{tol}$.

Note that a basis for F_t is given by $\{f_k^{(t-1)T};\ k=1,\ldots,n_{t-1}\}\cup\{\bar{\bar{u}}_k^{(t)T}.P_{2t};\ k\leq e_t^{tol}\}$. Further, let $O_{t-1}:=\{v_k^{(t-1)T};\ k=1,\ldots,p_{t-1}\}$, $O_{t-1}^0:=\{(v,0);\ v\in O_{t-1},\ 0\in\mathbb{R}^{1\times q}\}$ and ${}^0O_{t-1}:=\{(0,v);\ 0\in\mathbb{R}^{1\times q},\ v\in O_{t-1}\}$. For $v_t(\mathcal{B}_t^{\perp})$ it then suffices to take O_{t-1}^0, an orthonormal basis for V_t, and n_{t-1} orthonormal vectors in span$({}^0O_{t-1})$, orthogonal to $O^0_{t-1}+V_t$.

4. *Termination (at step t^*).*
As for $P^D_{c_{tol}}$.

5. *Output.*
Bases for V_t, $t \leq t^*$, and $\mathcal{B}^{\perp}_{t^*}$. Define $\mathcal{B} := \{w \in (\mathbb{R}^q)^{\mathbb{Z}};\ r(\sigma)w = 0,\ r \in \mathcal{B}^{\perp}_{t^*}\}$.

Remark. The algorithm basically consists of *sequential* application of proposition 2–21 of section 2.2.3. As a rough outline, $P^P_{c_{tol}}$ models data by successively minimizing the misfit of a required number e_0^{tol} of zero order laws, then minimizing the predictive misfit of a required number e_1^{tol} of first order laws, and so on. In order to measure the misfit more or less independently, as made precise in section 3.2.2, the newly identified laws r of order t have to be elements of the space $[v_t(F_{t-1}.s^t)+v_t(\mathcal{B}^{\perp}_{t-1})]^{\perp}$, see also section II.3.2.6. The operator P_t takes care of this requirement. The resulting optimization problem of step t of the recursion is of a static nature as described in section 2.2.3. The data consists of $(\tilde{x}_i, \tilde{y}_i)$, $i = 0, \ldots, t_1 - t_0 - t$, with $\tilde{y}_i := P_{2t}\tilde{w}(t_0+t+i)$ and $\tilde{x}_i := P_{1t}.\mathrm{col}(\tilde{w}(t_0+i), \ldots, \tilde{w}(t_0+t-1+i))$. □

Let data $\tilde{w} \in (\mathbb{R}^q)^{\mathcal{T}}$ and tolerated complexity $c_{tol} \in (\mathbb{R}_+)^{\mathbb{Z}_+}$ be given.

Assumption 4–7 (*i*) assumption 4–3(*i*);
(*ii*) for step t, $S_-^{(t)}$ and $S_+^{(t)}$ have full rank;
(*iii*) $\sigma^{(0)}_{q-e_0^{tol}} > \sigma^{(0)}_{q-e_0^{tol}+1}$; in step t $\sigma^{(t)}_{e_t^{tol}} > \sigma^{(t)}_{e_t^{tol}+1}$;
(*iv*) for step t, let $u_k^{(t)T} = (u_{k,0}, \ldots, u_{k,t})$, $u_{k,j} \in \mathbb{R}^{1\times q}$, and $U_0 := \mathrm{col}(u_{k,0};\ k \leq e_t^{tol})$, $U_t := \mathrm{col}(u_{k,t};\ k \leq e_t^{tol})$; assume $\mathrm{rank}(U_0) = \mathrm{rank}(U_t) = e_t^{tol}$.

Interpretation. The algorithm is well-defined under assumption 4–7(*ii*), the identified model is unique under assumption 4–7(*iii*), and the laws identified in step t really have order t under assumption 4–7(*iv*). □

Remark. For generic data $\tilde{w} \in (\mathbb{R}^q)^{\mathcal{T}}$ $S(\tilde{w},t) > 0$ for all $t \leq \bar{d}(\mathcal{T})$. Hence under assumption (*i*) condition (*ii*) is satisfied for generic data, cf. definition 4–2 and section 3.3.1. □

Remark. See section 4.4 for comments on assumption 4–7, which amounts to assuming that c_{tol} is a reasonable specification for the data $\tilde{w}$. □

Theorem 4–8 Suppose assumption 4–7 is satisfied, then

(*i*) $P^P_{c_{tol}}(\tilde{w})=\{\mathcal{B}\}$, the model generated by the algorithm;

(*ii*) $e^*(\mathcal{B})=e_{tol}$;

(*iii*) $\varepsilon^P_{t,k}(\tilde{w},\mathcal{B})=\{1-(\sigma^{(t)}_{e_t^{tol}-k+1})^2\}^{1/2}$, $k=1,\ldots,e_t^{tol}$, $t\geq 1$;

(*iv*) $L_t=L_t^P$ for $\mathcal{B}$, so the algorithm gives a (CPF) representation of $\mathcal{B}$.

Proof. See the appendix.

4.3.2. Algorithm for predictive modelling under a misfit constraint

Finally we give an algorithm which, under conditions on the data $\tilde{w}\in(\mathbb{R}^q)^{\mathcal{T}}$, generates the model $P^P_{\varepsilon_{tol}}(\tilde{w})$ as defined in section 3.3.3. The algorithm basically consists of sequential application of proposition 2–22 of section 2.2.3. The optimality of the model generated by the algorithm is a consequence of this proposition and the special utility $u_{\varepsilon_{tol}}$ as defined in definition 3–12.

Algorithm for $P^P_{\varepsilon_{tol}}$.

1. *Input.*
   As for $P^D_{\varepsilon_{tol}}$.

2. *Initialization* (*step* 0).

2.1. As for $P^D_{\varepsilon_{tol}}$.

2.2. As for $P^D_{\varepsilon_{tol}}$.

2.3. As for $P^P_{c_{tol}}$, with e_0^{tol} replaced by e_0.

3. *Recursion* (*step* t).

3.0. As for $P^P_{c_{tol}}$, with e_k^{tol} replaced e_k, $k\leq t-1$; let $e_t':=q-\Sigma_{k=0}^{t-1}e_k$.

3.1. As for $P^P_{c_{tol}}$. Let $0\leq 1-(\sigma_1^{(t)})^2\leq\ldots\leq 1-(\sigma^{(t)}_{e_t''})^2<(\bar{\varepsilon}_t^{tol})^2\leq 1-(\sigma^{(t)}_{e_t''+1})^2\leq\ldots\leq 1-(\sigma^{(t)}_{q-n_{t-1}})^2\leq 1$.

3.2. As for $P^P_{c_{tol}}$, with e_t^{tol} replaced by $e_t:=\min\{e_t',e_t''\}$.

3.3. As for $P^P_{c_{tol}}$, with e_t^{tol} replaced by e_t.

4. *Termination* (*at step* t^*).
   As for $P^D_{\varepsilon_{tol}}$.

5. *Output.*
   Bases for V_t, $t\leq t^*$, and $\mathcal{B}^{\perp}_{t^*}$. Define $\mathcal{B}:=\{w\in(\mathbb{R}^q)^{\mathbb{Z}};\ r(\sigma)w=0,\ r\in\mathcal{B}^{\perp}_{t^*}\}$.

Assumption 4-9 (*i*) assumption 4–5(*i*);

(*ii*) assumption 4–7(*ii*);

(*iii*) if at t^* $e_t''^* > e_t'^*(>0)$, then assume $\sigma^{(t^*)}_{e'_{t^*}} > \sigma^{(t^*)}_{e'_{t^*}+1}$;

(*iv*) assumption 4–7(*iv*), with e_t^{tol} replaced by e_t.

Interpretation. The algorithm is well–defined under assumption 4–9(*ii*), the identified model exists under assumption 4–9(*i*) and is unique under assumption 4–9(*iii*), and the laws identified in step t really have order t under assumption 4–9(*iv*). For data $\tilde{w}$ the specification $\bar{\varepsilon}_{tol}$ is reasonable only if assumption 4–9 is satisfied, cf. section 4.4. □

Remark. If for some $t\in[1,\bar{d}(\mathcal{T})]$ $\bar{\varepsilon}_t^{tol}>1$ then it follows from step 3.1 that this requirement is not restrictive, i.e., e_t' laws are accepted in step t of the recursion. Hence in general it is reasonable to require that $\bar{\varepsilon}_t^{tol}\leq 1$, i.e., not to tolerate relative mean predition errors larger than one. □

Theorem 4-10 Suppose assumption 4–9 is satisfied, then

(*i*) $P^P_{\varepsilon_{tol}}(\tilde{w})=\{\mathcal{B}\}$, the model generated by the algorithm;

(*ii*) $e^*(\mathcal{B})=(e_t;\ t\in\mathbb{Z}_+)$;

(*iii*) $\varepsilon^P_{t,k}(\tilde{w},\mathcal{B})=\{1-(\sigma^{(t)}_{e_t''-k+1})^2\}^{1/2}$, $k=1,\ldots,e_t$, $t\geq 1$;

(*iv*) $L_t=L_t^P$ for $\mathcal{B}$, so the algorithm gives a (CPF) representation of $\mathcal{B}$.

Proof. See the appendix.

Remark. Taking in step 2.1 of the algorithm of section 4.2.2 $\sigma^{(0)}_{q-e_0}>(\bar{\varepsilon}_0^{tol})^2\geq \sigma^{(0)}_{q-e_0+1}$ and in step 3.1 of the algorithm of this section $1-(\sigma^{(t)}_{e_t''})^2\leq(\bar{\varepsilon}_t^{tol})^2<1-(\sigma^{(t)}_{e_t''+1})^2$, we get an algorithm which under assumption 4–9 determines $\bar{\bar{P}}^P_{\varepsilon_{tol}}(\tilde{w})$. Assumption 4-9($i$) can be replaced by $\bar{\varepsilon}_t^{tol} \geq 0$ for all $t \in [0, \bar{d}(\mathcal{T})]$. □

4.4. Comments

The algorithms described in the foregoing sections allow a simple numerical implementation of the procedures of section 3. The computational complexity is mainly determined by singular value analysis of empirical covariance matrices and, in the case of predictive modelling, determination of the square root of positive definite matrices. The algorithms have been numerically implemented and employed, e.g., for the simulations described in section 6.

The essential part of the algorithms is the construction of the complementary spaces V_t, either generating a canonical descriptive form or a canonical predictive form. The operators P_t guarantee that newly identified laws are "far" from being implied by the already identified laws. In this way the misfit is measured according to the principles of section 3.2. This is one of the main advantages of these procedures. In assessing the quality of a model, the *simultaneous* nature of AR-equations representing a system is fully taken into account. The quality is measured by means of canonical parametrizations which are not determined by (scientific) theory, but which are based upon the purpose of modelling, i.e. here, description or prediction.

Remark. The procedures of section 3 are essentially based on the utilities defined in sections II.2.2 and II.2.3. The aim is to identify models which have both small complexity and small misfit. A maximal tolerated level for one of the objectives is specified a priori and the other objective is optimized under this restriction. These procedures in some sense are extreme solutions for reconciliation of the two conflicting objectives. More general utilities could specify, e.g., which increase in misfit is tolerable to make a specified decrease in complexity acceptable. Such utilities are described, e.g., in Akaike [1], Hannan and Deistler [23], Rissanen [60], and Shibata [62]. □

Next we comment on the assumptions on the data made in sections 4.2 and 4.3. These assumptions in fact form conditions for the reasonableness of the specification of c_{tol} or ε_{tol} for the given data. If the assumptions are not satisfied this indicates *misspecification* of the modelling problem. We illustrate this by considering assumption 4–3(*ii*) and (*iii*). If in step t $\sigma^{(t)}_{q_t-e_t^{tol}}=\sigma^{(t)}_{q_t-e_t^{tol}+1}$ then it is not reasonable to require that exactly e_t^{tol} laws are accepted. This would imply that one law of misfit $(\sigma^{(t)}_{q_t-e_t^{tol}})^{1/2}$ should be accepted while another law of equal misfit should be rejected. This clearly indicates misspecification of c_{tol}, as $e(c_{tol})$ is not a reasonable specification for the given data. Similarly, if in step t, e.g., rank(U_t) $<e_t^{tol}$, then in step t we would accept a law of order $d\leq t-1$. This law then should have been accepted in step d. This again indicates misspecification of c_{tol}, as $e(c_{tol})$ is not reasonable for the given data.

The assumptions 4–5, 4–7 and 4–9 have a completely similar interpretation.

Remark. The foregoing comments illustrate the obvious fact that the specification of the modelling problem should be data dependent. In identification there is not only a problem of model selection on the basis of data, but also one of criterion or utility selection on the basis of these data. □

Remark. We state as a *conjecture* that the assumptions 4–3, 4–5, 4–7 and 4–9 are satisfied for "*generic*" data. We make this conjecture precise for assumption 4–3, as the other ones are similar.

Let $\mathcal{T}$ be given and let λ denote the Lebesgue measure on $(\mathbb{R}^q)^{\mathcal{T}}$. We call a set $V\subset(\mathbb{R}^q)^{\mathcal{T}}$ λ–generic if it contains an open subset of full Lebesgue measure, i.e., if there is an open set $O\subset V$ with $\lambda((\mathbb{R}^q)^{\mathcal{T}}\backslash O)=0$.

Let c_{tol} be given and suppose that this specification is sensible for $\mathcal{T}$, cf. definition 4–2. Moreover let $\Omega^D_{c_{tol}}\subset(\mathbb{R}^q)^{\mathcal{T}}$ denote the set of data for which assumptions 4–3(*ii*) and (*iii*) both are satisfied. We state as a conjecture that for sensible c_{tol} $\Omega^D_{c_{tol}}$ is dense in $(\mathbb{R}^q)^{\mathcal{T}}$, i.e., that assumptions 4–3(*ii*) and (*iii*) can be satisfied by an arbitrarily small disturbance of arbitrary data. This conjecture would imply that $\Omega^D_{c_{tol}}$ is λ–generic. This follows as it can be shown that $(\mathbb{R}^q)^{\mathcal{T}}=N\cup V$, with $\lambda(N)=0$ and with $V=\bigcup_{i\in\mathbb{N}} V_i$ an open set, where for each $i\in\mathbb{N}$ V_i is an open set such that the assumptions are either satisfied everywhere on V_i or nowhere on V_i.

Our intuition for the correctness of this conjecture is supported by simulation results, but its proof is still an open problem. □

Remark. The algebraic concept of genericity defined in section III.3.1.1 is stronger than the measure theoretic concept of λ–genericity, cf. Federer [12, section 2.6.5]. The concept of λ–genericity is much stronger than the topological concept of T–genericity, where a subset of a topological space T is called T–generic if it contains an open subset which is dense in T. □

Remark. The identified models may be rather sensitive for changes in c_{tol}. For changes in ε_{tol} the identified models only change at discrete critical values of ε_{tol}. This indicates that the identified models are not robust with respect to changes in the complexity (the structural form). So, in cases where we have no strong reasons to postulate the structure of a phenomenon, it seems preferable to infer approximate structure from the data by imposing a pragmatic requirement of fit. □

Remark. The effect of changes in c_{tol} and ε_{tol} concerns the robustness of identified models with respect to the specification of the modelling problem. The results on consistency in the next section show that under some conditions on the data the identified models are robust with respect to changes in the data. Stated more precise, locally around data which satisfy the conditions in sections 4–2 and 4–3 the identified models are a continuous function of the data with respect to a natural topology on the model class $\mathbb{B}$. This is made explicit in theorems 5–11 and 5–12 in section 5.4. □

5. Consistency

5.1. Definition of consistency

The procedures of section 3 have a clear optimality property as *data modelling* procedures. The identified models are optimal with respect to the utility $u_{c_{tol}}$ or $u_{\varepsilon_{tol}}$, cf. definitions 3–10, 3–11, 3–13 and 3–14. The procedures give a solution for the identification problem, i.e., for given data and model class $\mathbb{B}$, a model is chosen from the model class which is optimal with respect to a criterion, based on the objective of modelling. It need not be assumed that the data are generated by a phenomenon of a certain structure. This pure data modelling is of interest, e.g., in data compression, speech processing, econometrics, and so on.

However, in other cases we would like to construct a good model of a *phenomenon* which is supposed to generate the data. The identified model then should not only be good with respect to the particular data, but it should be good with respect to the generating system.

In this section we define a general concept of consistency, reflecting the purpose of constructing models which approximate the generating system in an optimal way. The approach is inspired by Ljung [48], [50]. We also refer to Heij and Willems [29], [30].

Intuitively, a procedure is called *consistent* if the model, identified by the procedure, converges to an *optimal approximation of the generating system* when the number of observations tends to infinity. So in the limit a consistent procedure identifies a model which, within the given model class, is as close as possible to the phenomenon. In this sense a consistent procedure gives a good model of the phenomenon, provided that the number of observations is large enough.

Notation. To define consistency we introduce some additional concepts. Let the set of conceivable data be $\mathcal{D}:=\bigcup\{(\mathbb{R}^q)^n;\ n\in\mathbb{N}\}$, so the data $\tilde{w}\in D$ consists of a finite time series $\tilde{w}=(\tilde{w}(t);\ t\in\mathcal{T}=[t_0,t_1])$ in q variables. Let $\#(\mathcal{T}):=t_1-t_0+1$ denote the number of observations. Let $\mathbb{M}$ be a class of models and $\mathbb{G}$ a class of generating systems. It is assumed that the phenomenon generating the data corresponds to a system $G\in\mathbb{G}$. This means that there is a time series $w\in(\mathbb{R}^q)^{\mathbb{Z}}$ in G from which we observe $\tilde{w}=w|_{\mathcal{T}}$. Further suppose that the objectives π have been used to construct a procedure $P:D\to 2^{\mathbb{M}}$. Moreover, assume that π induces an optimal approximation map $A:\mathbb{G}\to 2^{\mathbb{M}}$. This means that, with respect to π, $A(G)$ is the set of optimal approximations within the class $\mathbb{M}$ of the system $G\in\mathbb{G}$. Often $A(G)$ will consist of a singleton. Further, let $\to$ be a concept of convergence in $2^{\mathbb{M}}$, possibly also related to π. Finally, let n.a. denote a concept of "nearly always" for systems $G\in\mathbb{G}$. □

Remark. Such a concept of "nearly always" is crucial, as optimality of procedures can fail to hold true for very particular data which nearly never occur. □

Consistency now is defined as follows.

Definition 5-1 P is called *consistent* on $\mathbb{G}$ if for all $G \in \mathbb{G}$ and for nearly all $w \in G$, $P(w|_{\mathcal{T}}) \to A(G)$ if $\#(\mathcal{T}) \to \infty$.

Interpretation. If the length of the observed time series tends to infinity then the set of models identified by a consistent procedure converges "nearly always" to the set of optimal approximations within $\mathbb{M}$ of the generating system G. □

Remark. In the sequel we restrict attention to cases where $A(G)$ consists of a singleton, i.e., for $G \in \mathbb{G}$ there exists a unique optimal approximation $a(G) \in \mathbb{M}$, so $A(G) = \{a(G)\}$. In this case, let $\to$ be a concept of convergence in $\mathbb{M}$. Then we call $P: \mathcal{D} \to 2^{\mathbb{M}}$ *consistent* if for all $G \in \mathbb{G}$, n.a. in $w \in G$, $P(w|_{\mathcal{T}}) = \{M(w|_{\mathcal{T}})\}$, i.e., a singleton, for $\#(\mathcal{T})$ sufficiently large, and $M(w|_{\mathcal{T}}) \to a(G)$ for $\#(\mathcal{T}) \to \infty$. By slight abuse of notation we will indicate this by $P(w|_{\mathcal{T}}) \to A(G)$. □

The consistency problem is depicted in figure 12.

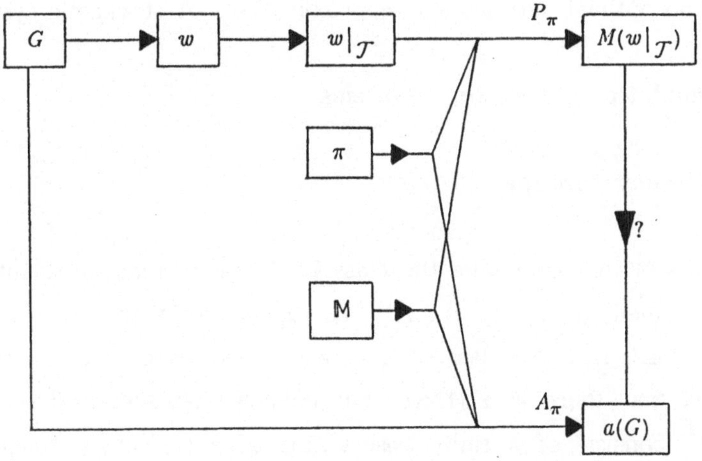

figure 12: consistency

Remark. This concept of model consistency differs in some important aspects from the concept of parameter consistency generally used in statistics, see e.g. Kendall and Stuart [41, section 17.7]. In the latter case $\mathbb{M}=\mathbb{G}=\{M(\theta);\ \theta\in\Theta\}$ for some parametrized class of models (probability distributions). The data modelling problem is formulated as an estimation problem, and a modelling procedure is a map $E:D\to\Theta$. The procedure is called consistent if (n.a.) $E(w|_{\mathcal{T}})\to\theta$ when $\#(\mathcal{T})\to\infty$, where θ parametrizes the generating system. Model consistency differs in four main respects from this parameter consistency. First, it need not be assumed that $\mathbb{M}=\mathbb{G}$, i.e., the generating system need not belong to the model class. Second, convergence is defined in terms of models, not in terms of parametrizations. Third, parameter consistency raises problems in case of non-unique parametrizations, model consistency avoids these problems. Fourth, the models need not be stochastic. □

Remark. For the case of time series analysis see e.g. Hannan and Deistler [23] for parameter consistency and e.g. Ljung and Caines [51] for model consistency. □

In the next two sections we investigate consistency of some of the procedures of section 3 for certain classes of generating systems $\mathbb{G}$. In section 5.2 we suppose $\mathbb{G}=\mathbb{B}$, i.e., the phenomenon itself is a linear, time invariant, complete (deterministic) dynamical system. In section 5.3 we consider the case where $\mathbb{G}$ consists of stochastic ARMA-systems and the purpose π is prediction. For this case we define optimal deterministic approximations of stochastic systems.

5.2. Deterministic generating systems

5.2.1. Consistency concept

As model class $\mathbb{M}$ we again take the class $\mathbb{B}$ of linear, time invariant, complete systems as defined in section II.3.1.2. We suppose that the data is generated by a system $\mathcal{B}\in\mathbb{B}$, i.e., the class of generating systems $\mathbb{G}$ is taken to be $\mathbb{B}$. So it is assumed that there is a (fixed, but unknown) system $\mathcal{B}\in\mathbb{B}$ such that the data $\tilde{w}\in(\mathbb{R}^q)^{\mathcal{T}}$ consists of a finite observation of a trajectory compatible with $\mathcal{B}$, i.e., there is a $w\in\mathcal{B}$ with $\tilde{w}=w|_{\mathcal{T}}$. In this case there exists an exact model of the phenomenon in the model class.

In order to define consistency we have to specify an optimal approximation map $A:\mathbb{G}\rightarrow\mathbb{B}$, a concept of convergence on $\mathbb{B}$, and a concept of "nearly always" for systems in $\mathbb{B}$.

Remark. To define n.a., we use the following concept of genericity on $\mathbb{B}$. Let $\mathcal{T}$ be a finite interval in $\mathbb{Z}$ and let $V\subset(\mathbb{R}^q)^{\mathcal{T}}$ be a linear subspace, then a subset $V'\subset V$ is called generic in V if there is a polynomial $p:V\rightarrow\mathbb{R}$, $p\neq 0$, such that $V'\supset V\backslash p^{-1}(0)$, cf. section III.3.1.1. For a model $\mathcal{B}\in\mathbb{B}$ we call $\mathcal{B}'\subset\mathcal{B}$ generic in $\mathcal{B}$ if $\mathcal{B}'|_{\mathcal{T}}\subset\mathcal{B}|_{\mathcal{T}}$ is generic in $\mathcal{B}|_{\mathcal{T}}$ for $\#(\mathcal{T})$ sufficiently large. A property now is said to hold true n.a. for $\mathcal{B}$ if the set of points $w\in\mathcal{B}$ where the property holds true is generic in $\mathcal{B}$. □

As $\mathbb{G}=\mathbb{B}$, an obvious choice for the optimal approximation map A is the *identity* map. Moreover, we take the *discrete* topology on $\mathbb{B}$.

Interpretation. In this setting a procedure P is consistent on $\mathcal{B}\in\mathbb{B}$ if generically in $w\in\mathcal{B}$ there holds that $P(w|_{\mathcal{T}})=\{\mathcal{B}\}$ for $\#(\mathcal{T})$ sufficiently large. In this case a consistent procedure identifies a generating system $\mathcal{B}$ *exactly*, provided that the data consists of an observation of a sufficiently large finite part of a generic time series in $\mathcal{B}$. □

5.2.2. Consistency result

We first consider the procedure of undominated unfalsified modelling P^u as defined in section III.1.2.

Remark. So $P^u:\mathcal{D}\rightarrow 2^{\mathbb{B}}$, where for $\tilde{w}\in(\mathbb{R}^q)^{\mathcal{T}}$ there holds $\mathcal{B}\in P^u(\tilde{w})$ if and only if $\mathcal{B}\in\mathbb{B}$ and $\mathcal{B}$ is unfalsified and undominated, i.e., $\tilde{w}\in\mathcal{B}|_{\mathcal{T}}$ and $\{\tilde{w}\in\mathcal{B}'|_{\mathcal{T}}, \mathcal{B}'\in\mathbb{B}, \mathcal{B}'\subset\mathcal{B}\}\Rightarrow\{\mathcal{B}'=\mathcal{B}\}$ respectively. □

Proposition 5-2 P^u is not consistent on $\mathbb{B}$.

Proof. See the appendix.

Interpretation. In general P^u accepts laws which are not corroborated by the data, cf. the comments on P^u in section III.3.1. □

Remark. For $q=1$ the procedure $\tilde{P}^*$ of definition III.3–31 is consistent. This follows directly from propositions III.3–30(*iii*) and III.3–34(*i*). □

Next we consider the procedures described in section 3.

Proposition 5–3 The procedures $P^D_{\varepsilon_{tol}}$ and $P^P_{\varepsilon_{tol}}$ are not consistent on $\mathbb{B}$. If $\varepsilon_{tol} \not\equiv 0$ or $c_{tol} < q.(1,1,1,...)$ in lexicographic ordering, then $\bar{\bar{P}}^D_{\varepsilon_{tol}}$ and $\bar{\bar{P}}^P_{\varepsilon_{tol}}$ or $P^D_{c_{tol}}$ and $P^P_{c_{tol}}$ respectively are not consistent on $\mathbb{B}$.

Proof. See the appendix.

Interpretation. If laws of positive misfit are accepted then the procedures are not consistent in the sense of exact identification for generic time series. Note that $c_{tol}<q.(1,1,1,...)$ is equivalent to $e(c_{tol})\not\equiv 0$, in which case it is required that laws of certain order are accepted, generally with positive misfit. □

We now define two exact identification procedures, cf. definition III.3–4.

Notation. For $\varepsilon_{tol}\equiv 0$ we denote the procedures $\bar{\bar{P}}^D_{\varepsilon_{tol}}$ and $\bar{\bar{P}}^P_{\varepsilon_{tol}}$ defined in section 3.3.3 by P^D_e and P^P_e respectively. Hence for $\tilde{w}\in(\mathbb{R}^q)^{\mathcal{T}}$ there holds $\mathcal{B}\in P^D_e(\tilde{w})$ if and only if $\mathcal{B}\in\mathbb{B}(\mathcal{T})$, $\varepsilon^D(\tilde{w},\mathcal{B})=0$, and $\{\mathcal{B}'\in\mathbb{B}(\mathcal{T}),\ \varepsilon^D(\tilde{w},\mathcal{B}')=0\} \Rightarrow \{c(\mathcal{B}')\geq c(\mathcal{B})$ in lexicographic ordering$\}$. P^P_e is defined in a similar way. □

Remark. We recall from the remarks following definitions 3–3 and 3–7 that for $\tilde{w}\in(\mathbb{R}^q)^{\mathcal{T}}$ and $\mathcal{B}\in\mathbb{B}$ there holds that $\mathcal{B}$ has descriptive or predictive misfit zero if and only if $\tilde{w}\in\mathcal{B}|_{\mathcal{T}}$. Hence the procedures P^D_e and P^P_e are exact. □

Interpretation. P^D_e and P^P_e accept the unfalsified laws which are significant for the given number of observations, cf. section 3.3.1. □

Remark. Note that for $\mathcal{B}\in\mathbb{B}$ $c_t(\mathcal{B})\leq q$, $t\in\mathbb{Z}_+$, cf. definition IV.2.1. Taking $c_{tol}:=$

$q.(1,1,1,\ldots)$ it follows from definitions 3–9 and 3–12 that $P_e^D = P_{c_{tol}}^D$ and $P_e^P = P_{c_{tol}}^P$ for this (unrestrictive) complexity requirement. □

Notation. Let $\mathbb{B}_c \subset \mathbb{B}$ denote the class of controllable systems as defined in section IV.3.2. □

Theorem 5–4 P_e^D and P_e^P are consistent on $\mathbb{B}_c$.

Proof. See the appendix.

Remark. For $q=1$ P_e^D and P_e^P are consistent on $\mathbb{B}$. This follows directly from proposition III.3–30(*iii*). However, for $q>1$ P_e^D and P_e^P are not consistent on $\mathbb{B}$, as the following example shows. Let $q>1$ and $\mathcal{B} := \{w \in (\mathbb{R}^q)^{\mathbb{Z}};\ \sigma w = w\}$. Then for all $\mathcal{T}$ and $w|_{\mathcal{T}} \in \mathcal{B}|_{\mathcal{T}}$ the rank of the empirical covariance matrix of any order t is at most 1 and hence $P_e^D(w|_{\mathcal{T}})$ and $P_e^P(w|_{\mathcal{T}})$ are models of dimension at most 1. If, e.g., $w(t) = c \in \mathbb{R}^q$ with $c_1 \neq 0$, then $P_e^D(w|_{\mathcal{T}}) = P_e^P(w|_{\mathcal{T}}) = \{w \in (\mathbb{R}^q)^{\mathbb{Z}}\ ;\ \sigma w_1 = w_1,\ w_i = \frac{c_i}{c_1} w_1,\ i \in [2,q]\} \neq \mathcal{B}$. So $\mathcal{B}$ is generically even not identified for any (finite or infinite) $\mathcal{T}$. □

Remark. There exist systems $\mathcal{B} \in \mathbb{B} \backslash \mathbb{B}_c$ for which P_e^D and P_e^P are consistent, cf. section 6.2.4. □

Remark. P^u and the procedures mentioned in proposition 5–3 are not consistent on $\mathbb{B}_c$, cf. the proof of propositions 5–2 and 5–3. □

Remark. A remaining question for $\varepsilon_{tol} \not\equiv 0$ is the relationship between consistency of $P_{\varepsilon_{tol}}^D$ and $P_{\varepsilon_{tol}}^P$ and a definition of n.a. in terms of "*sufficiency of excitation*". We suspect that these procedures are consistent on the class of controllable systems $\mathbb{B}_c$, provided that n.a. is defined in terms of an excitation condition on the inputs which is sufficiently strong for the given specification of ε_{tol}. Roughly stated, the inputs should satisfy no law with misfit smaller than ε_{tol}. Exact identification then is guaranteed provided that the inputs are sufficiently rich with respect to ε_{tol}. □

The procedures P_e^D and P_e^P are exact modelling procedures for multivariable time series analysis. These procedures have desirable properties as stated in the next theorem. For definitions we refer to section III.3.1. Now the model class is $\mathbb{B}_c$ instead of $\mathbb{B}_T$, i.e., we only consider controllable systems.

Theorem 5-5 On $\mathbb{B}_c$ the procedures P_e^D and P_e^P are exact, bilaterally monotone, linear, truthful, and strongly prudential. Moreover, P_e^D and P_e^P are equivalent in the sense that for every $\mathcal{B}\in\mathbb{B}_c$ generically in $w\in\mathcal{B}$ $P_e^D(w|_{\mathcal{T}})=P_e^P(w|_{\mathcal{T}})$.

Proof. See the appendix.

5.3. Stochastic generating systems

5.3.1. Introduction

In identification generally the model class consists of relatively simple models while the phenomenon is relatively complex. In order to investigate whether a modelling procedure identifies a model which is close to the data generating phenomenon we have to define optimal approximations of the phenomenon in the model class. An optimal approximate model of a phenomenon should reflect the *approximate structure* which can be modelled in the model class. The criterion for optimality will depend upon the objectives of modelling.

Remark. Hence for given objectives of modelling the question arises to specify for a given model $\mathcal{B}\in\mathbb{B}$ the class of phenomena G for which $\mathcal{B}$ reflects the structure of G in an optimal way, as far as this structure can be captured within the model class $\mathbb{B}$. This amounts to defining the concept of approximate structure. In the sequel we restrict attention to phenomena which are stochastic. Other interesting cases are those of nonlinear and time-varying phenomena. □

In the sequel we consider the purpose of prediction. We investigate the consistency of the predictive procedures $P^P_{c_{tol}}$ and $P^P_{\varepsilon_{tol}}$ in case the data consists of a finite part of a realization of a stochastic process. So the phenomenon G is supposed to correspond to a stochastic process. The approximate structure of G which can be captured in $\mathbb{B}$ is obtained by constructing the optimal predictive relationships for the given objectives of modelling as specified in $u_{c_{tol}}$ or $u_{\varepsilon_{tol}}$.

Remark. The foregoing is made explicit in section 5.3.4 where we define the optimal approximation of a stochastic process by a deterministic system, for given c_{tol} or ε_{tol}. Roughly stated, the optimal deterministic approximation is described by the predictive relationships corresponding to c_{tol} or ε_{tol} in case the stochastic process were known. Note that both deterministic and stochastic models have an interpretation in terms of (optimal) one-step-ahead prediction by means of deterministic equations. □

Remark. A similar exposition could be given for the descriptive procedures $P^D_{c_{tol}}$ and $P^D_{\varepsilon_{tol}}$. However, in general it seems difficult to give an interpretation of stochastic systems in terms of deterministic *descriptive* autoregressive laws. Therefore we restrict attention to $P^P_{c_{tol}}$ and $P^P_{\varepsilon_{tol}}$. □

In the following we introduce a concept of convergence on $\mathbb{B}$, we describe a class of generating ARMA-systems, we define optimal approximation maps $A^P_{c_{tol}}$ and $A^P_{\varepsilon_{tol}}$, and finally we state consistency results.

5.3.2. Concept of convergence

Let $\mathcal{B}_k \in \mathbb{B}$, $k \in \mathbb{N}$, and $\mathcal{B}_\infty \in \mathbb{B}$. Then $\mathcal{B}_k$ is defined to converge to $\mathcal{B}_\infty$ for $k \to \infty$ if there exist parametrizations $\mathcal{B}_k = \mathcal{B}(R_k)$, $k \in \mathbb{N}$, and $\mathcal{B}_\infty = \mathcal{B}(R_\infty)$, with the following properties. R_∞ is bilaterally row proper, $\{d(R_k); k \in \mathbb{N}\}$ is bounded, and $R_k \to R_\infty$ in Euclidean sense if $k \to \infty$. By this we mean that for k sufficiently large R_k has as many rows as R_∞, and if $R_k = \Sigma_{j=-\infty}^{\infty} R_j^{(k)} s^j$, $R_j^{(k)} = (r_{lm}^{jk}) \in \mathbb{R}^{p \times q}$, $k \in \mathbb{N} \cup \{\infty\}$, that then $\Sigma_{j=-\infty}^{\infty} \Sigma_{l=1}^{p} \Sigma_{m=1}^{q} (r_{lm}^{jk} - r_{lm}^{j\infty})^2 \to 0$ if $k \to \infty$.

Remark. This concept of convergence is analysed in Nieuwenhuis and Willems [55]. There it is shown that this convergence in terms of parametrizations is equivalent to a natural concept of convergence of systems, considered as subsets of $(\mathbb{R}^q)^{\mathbb{Z}}$. □

5.3.3. Class of generating systems

We assume that the generating system belongs to the class $\mathbb{G}$ of stochastic processes $w=\{w(t);\ t\in\mathbb{Z}\}$ which satisfy the following assumption.

Assumption 5-6 (*i*) w is second order stationary with for all $t, k\in\mathbb{Z}$ $Ew(t)=0$, $C_k:=Ew(t)w(t+k)^T$; (*ii*) almost surely for realizations w_r of w there holds for all $k\in\mathbb{Z}$ $\frac{1}{t_1-t_0+1}\cdot\Sigma_{t=t_0}^{t_1-k}\ w_r(t)w_r(t+k)^T\to C_k$ if $|t_1-t_0|\to\infty$.

Remark. A sufficient condition for this assumption to be satisfied is that w is strictly stationary and ergodic, e.g., that w is stationary Gaussian with a spectral distribution Φ which is continuous on the unit circle. We refer to Hannan [22, section IV.2 and especially theorem 4.2]. The last condition holds true for Gaussian ARMA–processes, in which case $\Phi(z):=\Sigma_{k=-\infty}^{\infty}C_kz^{-k}$ is a rational function with no poles on the unit circle. The process w then has a representation of the following form. There exist $m\in\mathbb{N}$ and polynomial matrices $N\in\mathbb{R}^{q\times m}[s]$ and $M\in\mathbb{R}^{q\times q}[s]$ with $\det(M(s))\neq 0$ on $|s|\leq 1$, such that $M(\sigma^{-1})w=N(\sigma^{-1})n$ for an m–dimensional stationary Gaussian white noise process n, i.e., $En(t)=0$, $En(t)n(t)^T=En(s)n(s)^T$ for all t,s, and $En(t)n(s)^T=0$ for all $t\neq s$. We refer to Hannan [22, section III.2], and Davis and Vinter [11, section 2.3]. □

Remark. An ARMA–process $M(\sigma^{-1})w=N(\sigma^{-1})n$ with n white noise can be considered as an errors–in–equations model. However, the model $\mathcal{B}:=\{w'\in(\mathbb{R}^q)^{\mathbb{Z}}\ ;\ M(\sigma^{-1})w'=0\}$ in general will not be an optimal approximation of the process w. In particular for the purpose of prediction the "model" $(N^{-1}M)(\sigma^{-1})w'=0$ is preferable, but this model in general does not belong to $\mathbb{B}$ as $N^{-1}M$ need not be a polynomial matrix. In the next section we define optimal approximations of stochastic processes by means of models in $\mathbb{B}$, i.e., with finite lag AR–equations. □

Remark. The consistency result of section 5.3.5 is stated in terms of subclasses of $\mathbb{G}$ which are defined in the next section. □

5.3.4. Approximation maps

In this section we construct for a given stochastic process $\underset{\sim}{w}$ optimal approximations in $\mathbb{B}$. The optimality has to be understood in the sense of a utility corresponding to the purpose of modelling. For w we define the optimal approximations $A^P_{c_{tol}}(w)$ and $A^P_{\varepsilon_{tol}}(w)$ as the models of optimal prediction of w for c_{tol} and ε_{tol} respectively, in case the generating system w were known.

Notation. The foregoing is made precise as follows. For $r \in \mathbb{R}^{1\times q}[s,s^{-1}]$ with $d(r)>0$ we define the relative expected prediction error in analogy with definition 3-5 as $e^P(w,r):=\{(E\|rw\|^2)/(E\|r^*w\|^2)\}^{1/2}$, where r^* is the leading coefficient vector of r and where $E\|rw\|^2:=E\{(r(\sigma,\sigma^{-1})w)(t)\}^2$, which does not depend on t due to stationarity. In case $E\|r^*w\|=0$ we define $e^P(w,r):=+\infty$.If $d(r)=0$ then we define $e^P(w,r):=\{E\|rw\|^2/\|r\|^2\}^{1/2}$. For $\mathcal{B}\in\mathbb{B}$ we define $\varepsilon^P(w,\mathcal{B})\in[(\mathbb{R}_+\cup\{\infty\})^{1\times q}]^{\mathbb{Z}_+}$ exactly analogous to $\varepsilon^P(\tilde{w},\mathcal{B})$ in definition 3-7. Hence $\varepsilon^P_{t,1}(w,\mathcal{B})$ measures the largest relative expected prediction error of the truly t-th order predictive laws claimed by $\mathcal{B}$, $t\in\mathbb{Z}_+$, and so on. □

Definition 5-7 For $w\in\mathbb{G}$, $A^P_{c_{tol}}(w):=\text{argmax}\{u_{c_{tol}}(c(\mathcal{B}),\ \varepsilon^P(w,\mathcal{B}));\ \mathcal{B}\in\mathbb{B}\}$ and $A^P_{\varepsilon_{tol}}(w):=\text{argmax}\{\ u_{\varepsilon_{tol}}(c(\mathcal{B}),\ \varepsilon^P(w,\mathcal{B}));\ \mathcal{B}\in\mathbb{B}\}$.

Interpretation. $A^P_{c_{tol}}$ and $A^P_{\varepsilon_{tol}}$ give deterministic approximations of stochastic processes which are optimal in terms of a utility on complexity and predictive quality of models described by (deterministic) autoregressive equations. The corresponding predictive models are optimal for specified c_{tol} and ε_{tol} respectively in case the spectral distribution of the process w were known. □

In the sequel we restrict attention to subclasses of $\mathbb{G}$ for which $A^P_{c_{tol}}$ and $A^P_{\varepsilon_{tol}}$ consist of singletons.

Notation. For $w\in\mathbb{G}$ define the t-th order covariance matrix by $S(w,t):= E[\text{col}(w(t),\dots,w(t+k)).\text{col}(w(t),\dots,w(t+k))^T]$, $t\in\mathbb{Z}_+$.

By $\varepsilon_{tol}>0$ we denote the condition that $\varepsilon_t^{tol}=\bar{\varepsilon}_t^{tol}.(1,\dots,1)$ with $\bar{\varepsilon}_t^{tol}>0$ for all $t\in\mathbb{Z}_+$. □

Remark. Note that any c_{tol} satisfies assumption 4–7(i) for $\#(\mathcal{T})$ sufficiently large, cf. definition 4–2. □

Definition 5–8 $\mathbb{G}_{c_{tol}}:=\{w\in\mathbb{G};$ assumption 4–7(ii), (iii), (iv) is satisfied for $c_{tol}\}$; for $\varepsilon_{tol}>0$ $\mathbb{G}_{\varepsilon_{tol}}:=\{w\in\mathbb{G};$ assumption 4–9(ii),(iii),(iv) is satisfied for ε_{tol}, and moreover in algorithm 4.3.2 in step 2.1 $\sigma^{(0)}_{q-e_0}>(\bar{\varepsilon}_0^{tol})^2>\sigma^{(0)}_{q-e_0+1}$ and in step 3.1 $1-(\sigma^{(t)}_{e''_t})^2<(\bar{\varepsilon}_t^{tol})^2<1-(\sigma^{(t)}_{e''_t+1})^2\}$.

Interpretation. The specification c_{tol} is only reasonable on $\mathbb{G}_{c_{tol}}$, the specification ε_{tol} only on $\mathbb{G}_{\varepsilon_{tol}}$. Otherwise the modelling problem is misspecified, cf. section 4.4. □

The next result is an immediate consequence of definitions 5–7 and 5–8 and theorems 4–8(i), (iii) and 4–10(i).

Proposition 5–9 (i) For $w\in\mathbb{G}_{c_{tol}}$ $A^P_{c_{tol}}(w)$ is a singleton, generated by the algorithm of section 4.3.1 with $S(\tilde{w},t)$ replaced by $S(w,t)$, and $e(A^P_{c_{tol}}(w))=e(c_{tol})$; ($ii$) for $\varepsilon_{tol}>0$ and $w\in\mathbb{G}_{\varepsilon_{tol}}$ $A^P_{\varepsilon_{tol}}(w)$ is singleton, generated by the algorithm of section 4.3.2 with $S(\tilde{w},t)$ replaced by $S(w,t)$.

Remark. If $\bar{\varepsilon}_t^{tol}\leq 0$ for some $t\in\mathbb{Z}_+$ then $A^P_{\varepsilon_{tol}}(w)$ is empty, cf. the remarks following definitions 3–13 and 3–14 in section 3.3.3. □

Remark. Define the set of covariance sequences $\Gamma\subset(\mathbb{R}^{q\times q})^{\mathbb{Z}}$ by $\Gamma:=\{(C_k;\ k\in\mathbb{Z});\ \exists w\in\mathbb{G}$ with $C_k=Ew(t)w(t+k)^T,\ k\in\mathbb{Z}\}$. We call a subset $\Gamma'\subset\Gamma$ λ–generic in Γ if for all $-\infty<t_0\leq t_1<+\infty$ $\Gamma'|_{[t_0,t_1]}$ is λ–generic in $\Gamma|_{[t_0,t_1]}$, i.e., $\Gamma'|_{[t_0,t_1]}$ contains an open set of full (Lebesgue) measure in $\Gamma|_{[t_0,t_1]}$. We call a class of stochastic processes $\mathbb{G}'\subset\mathbb{G}$ generic if $\Gamma':=\{(C_k;\ k\in\mathbb{Z});\ \exists w\in\mathbb{G}'$ with $C_k=Ew(t)w(t+k)^T,\ k\in\mathbb{Z}\}$ is λ–generic in Γ.

We state as a conjecture that for any c_{tol} and for any $\varepsilon_{tol}>0$ the classes $\mathbb{G}_{c_{tol}}$ and $\mathbb{G}_{\varepsilon_{tol}}$ are generic in $\mathbb{G}$ and that the Gaussian ARMA–processes in $\mathbb{G}_{c_{tol}}$ and $\mathbb{G}_{\varepsilon_{tol}}$ are generic in the class of all Gaussian ARMA–processes in $\mathbb{G}$. This conjecture is analogous to the one formulated in section 4.4. The conjecture in particular would imply that the consistency result stated in the next section holds true for generic Gaussian ARMA–processes. □

5.3.5. Consistency result

Assume that the data $\tilde{w}$ consists of an observation on a (finite) time interval $\mathcal{T}$ of a realization $w_r \in (\mathbb{R}^q)^{\mathbb{Z}}$ of a stochastic process $w \in \mathbb{G}$, i.e., w satisfies assumption 5–6. In this section we state a result on consistency of $P^P_{c_{tol}}$ and $P^P_{\varepsilon_{tol}}$. Here the concept of convergence is as defined in section 5.3.2 and the approximation maps are as defined in definition 5–7. As definition of "nearly always" for a phenomenon $w \in \mathbb{G}$ we take "almost sure" with respect to the probability measure corresponding to w.

Notation. For given ε_{tol} and $\bar{d} \in \mathbb{Z}_+$ let $P^P_{(\varepsilon_{tol},\bar{d})}$ denote the procedure which for $\tilde{w} \in (\mathbb{R}^q)^{\mathcal{T}}$ is defined by $P^P_{(\varepsilon_{tol},\bar{d})}(\tilde{w}) := \text{argmax}\{u_{\varepsilon_{tol}}(c(\mathcal{B}), \varepsilon^P(\tilde{w},\mathcal{B}));\ \mathcal{B} \in \mathbb{B}(\mathcal{T})$ and $\max\{t;\ e^*_t(\mathcal{B}) \neq 0\} \leq \bar{d}\}$. Define $\mathbb{G}_{(\varepsilon_{tol},\bar{d})} := \{w \in \mathbb{G}_{\varepsilon_{tol}}\ ;\ \mathcal{B} := A^P_{\varepsilon_{tol}}(w)$ satisfies $\max\{t;\ e^*_t(\mathcal{B}) \neq 0\} \leq \bar{d}\}$. Further for given c_{tol} let $\mathbb{G}^c_{c_{tol}} := \{w \in \mathbb{G}_{c_{tol}}\ ;\ w$ is continuous$\}$, where we call a process continuous if and only if all its finite dimensional marginal distributions are absolutely continuous with respect to the Lebesgue measure. □

Remark. The next theorem states consistency of $P^P_{c_{tol}}$ on $\mathbb{G}^c_{c_{tol}}$ and of $P^P_{(\varepsilon_{tol},\bar{d})}$ on $\mathbb{G}_{(\varepsilon_{tol},\bar{d})}$ for $\varepsilon_{tol} > 0$ and $\bar{d} \in \mathbb{Z}_+$. The only reason to assume continuity or to impose an upper bound $\bar{d}$ on the order is to prevent that for $\#(\mathcal{T}) \to \infty$ the identification procedures accept laws of order increasing to infinity. For $P^P_{(\varepsilon_{tol},\bar{d})}$ laws of order larger than $\bar{d}$ simply never are considered in identification, independent of the length of the observed time series. Note that for any $w \in \mathbb{G}_{\varepsilon_{tol}}$ there holds that $w \in \mathbb{G}_{(\varepsilon_{tol},\bar{d})}$ for $\bar{d}$ sufficiently large. For $P^P_{c_{tol}}$ and a continuous process w there holds almost surely for $\tilde{w} \in (\mathbb{R}^q)^{\mathcal{T}}$ that any law of order at most $\bar{d}(\mathcal{T})$ has positive misfit, hence accepting no law is preferred if this is tolerated for c_{tol}. □

Theorem 5–10 For every c_{tol}, $P^P_{c_{tol}}$ is consistent on $\mathbb{G}^c_{c_{tol}}$. For every $\varepsilon_{tol} > 0$ and $\bar{d} \in \mathbb{Z}_+$, $P^P_{(\varepsilon_{tol},\bar{d})}$ is consistent on $\mathbb{G}_{(\varepsilon_{tol},\bar{d})}$.

Proof. See the appendix.

Interpretation. Let w_r be a realization of a stochastic process $w \in \mathbb{G}^c_{c_{tol}}$ and let $\tilde{w}=w_r|_{\mathcal{T}}$. Let $A^P_{c_{tol}}(w)=\mathcal{B}\in\mathbb{B}$ with corresponding predictive spaces $V^P_t:=v_t(L^P_t)$, where L^P_t is as defined in section II.3.2.6. Then almost surely $P^P_{c_{tol}}(\tilde{w})$ is a singleton for $\#(\mathcal{T})$ sufficiently large. Denote the corresponding (data-dependent) predictive spaces by $V^P_t(\mathcal{T})$, the complexity by $c(\mathcal{T})$ and the predictive misfit by $\varepsilon(\mathcal{T})$. Then for $\#(\mathcal{T})\to\infty$ there holds that almost surely $c_t(\mathcal{T})\to c_t(\mathcal{B})$, $V^P_t(\mathcal{T})\to V^P_t$ in the Grassmannian topology (i.e., there exist choices of bases of $V^P_t(\mathcal{T})$ which converge to a basis of V^P_t), and $\varepsilon_{t,k}(\mathcal{T})\to\varepsilon^P_{t,k}(w,\mathcal{B})$, $k=1,\ldots,q$, $t\in\mathbb{Z}_+$. A similar result holds true for $P^P_{(\varepsilon_{tol},\bar{d})}$. The convergence $V^P_t(\mathcal{T})\to V^P_t$ implies convergence of AR–relations and hence of the corresponding models, cf. section 5.3.2. So if the number of observations tends to infinity, then the identified model almost surely converges to the optimal (prediction) model $\mathcal{B}$ which would be chosen as prediction model for the phenomenon w in case w were known. □

Remark. The proof of the theorem consists of using the ergodic properties of w and establishing continuity properties of the steps of the algorithms in sections 4.3.1 and 4.3.2 with respect to changes in $S(\tilde{w},t)$, $t\in\mathbb{Z}_+$. □

Remark. An interesting question is whether there is a function $l:\mathbb{N}\to\mathbb{N}$ with $l(n)\to\infty$ if $n\to\infty$ and such that $P^P_{c_{tol}}$ is consistent on $\mathbb{G}_{c_{tol}}$ and $P^P_{\varepsilon_{tol}}$ on $\mathbb{G}_{\varepsilon_{tol}}$, if for $\mathcal{T}$ it is required that the identified laws have order at most $l(\#(\mathcal{T}))$. □

Remark. Let $\bar{\bar{P}}^P_{(\varepsilon_{tol},\bar{d})}$ be defined in analogy with $P^P_{(\varepsilon_{tol},\bar{d})}$, requiring $\varepsilon^P_{t,k}(\tilde{w},\mathcal{B})\leq\varepsilon^{tol}_{t,k}$ instead of $\varepsilon^P_{t,k}(\tilde{w},\mathcal{B})<\varepsilon^{tol}_{t,k}$. It follows from theorem 5–10 and the assumptions stated in definition 5–8 that the procedure $\bar{\bar{P}}^P_{(\varepsilon_{tol},\bar{d})}$ is consistent on $\mathbb{G}_{(\varepsilon_{tol},\bar{d})}$, for all $\varepsilon_{tol}\geq 0$ and $\bar{d}\in\mathbb{Z}_+$. This procedure also is consistent for continuous processes $w\in\mathbb{G}_{\varepsilon_{tol}}$ with $\varepsilon_{tol}\geq 0$ such that there is a $\bar{d}\in\mathbb{Z}_+$ with $\varepsilon^{tol}_{t,k}=0$ for all $t>\bar{d}$ and $k\in[1,q]$.

Both $P^P_{\varepsilon_{tol}}$ and $\bar{\bar{P}}^P_{\varepsilon_{tol}}$ are not consistent on $\mathbb{G}_{\varepsilon_{tol}}$ for $\varepsilon_{tol}>0$. This is due to the fact that it cannot be excluded almost surely that the procedures accept laws of order increasing to infinity if $\#(\mathcal{T})\to\infty$. □

Remark. The asymptotic optimality of the models identified by $P^P_{c_{tol}}$ and $P^P_{(\varepsilon_{tol},\bar{d})}$ should not be misunderstood. Consider e.g. $P^P_{c_{tol}}$ and suppose that

$w \in \mathbb{G}^c_{c_{tol}}$ is such that $\mathcal{B}:=A^P_{c_{tol}}(w)$ satisfies $\Sigma^{\infty}_{t=0} e^*_t(\mathcal{B})=q$. Then use of $\mathcal{B}$ leads to one-step-ahead pointpredictions, which we indicate by $\hat{w}^*$. It follows from theorem 5-10 that almost surely for $\#(\mathcal{T})$ sufficiently large $P^P_{c_{tol}}(\tilde{w})$ also leads to pointpredictions, which we indicate by $\hat{w}(\mathcal{T})$. It also follows from theorem 5-10 that $E\|\hat{w}^* - \hat{w}(\mathcal{T})\| \to 0$ if $\#(\mathcal{T}) \to \infty$. In this sense the one-step-ahead predictions based on the identified model converge to the optimal predictions $\hat{w}^*$. However, in general the least-squares (causal) predictor for w does not coincide with the predictor $\hat{w}^*$ for any choice of c_{tol}. Hence in this case the predictions $\hat{w}(\mathcal{T})$ do not converge to the least-squares predictions. So the asymptotic optimality has to be understood in terms of $u_{c_{tol}}$, not in terms of asymptotic minimal mean-square prediction error. It is not unreasonable to be slightly non-optimal in predictive accuracy if the predictions can be made by much simpler models, i.e., by models of shorter lag. □

5.4. Robustness

We finally state a result concerning the robustness of the identified models with respect to changes in the data.

Notation. Let $\mathcal{T}$ be a given interval of observation. For given sensible c_{tol} and for given $\varepsilon_{tol}>0$ let $\Omega^P_{c_{tol}}$ and $\Omega^P_{\varepsilon_{tol}}$ denote the set of data for which c_{tol} and ε_{tol} respectively are not misspecified, cf. section 4.4, i.e., $\Omega^P_{c_{tol}} := \{\tilde{w} \in (\mathbb{R}^q)^{\mathcal{T}}$; assumption 4-7(*ii*),(*iii*),(*iv*) is satisfied$\}$ and $\Omega^P_{\varepsilon_{tol}} := \{\tilde{w} \in (\mathbb{R}^q)^{\mathcal{T}}$; assumption 4-9(*ii*),(*iii*),(*iv*) is satisfied, and moreover in algorithm 4.3.2 in step 2.1 $\sigma^{(0)}_{q-e_0} > (\bar{\varepsilon}^{tol}_0)^2 > \sigma^{(0)}_{q-e_0+1}$ and in step 3.1 $1-(\sigma^{(t)}_{e''_t})^2 < (\bar{\varepsilon}^{tol}_t)^2 < 1-(\sigma^{(t)}_{e''_t+1})^2\}$. We equip $(\mathbb{R}^q)^{\mathcal{T}}$ with the Euclidean topology and $\mathbb{B}$ with the topology defined in section 5.3.2. □

Theorem 5-11 For given $\mathcal{T}$, sensible c_{tol} and $\varepsilon_{tol}>0$, there holds

(*i*) $\Omega^P_{c_{tol}}$ and $\Omega^P_{\varepsilon_{tol}}$ are open in $(\mathbb{R}^q)^{\mathcal{T}}$;

(*ii*) locally on $\Omega^P_{c_{tol}}$ $P^P_{c_{tol}}$ is continuous;

(*iii*) locally on $\Omega^P_{\varepsilon_{tol}}$ $P^P_{\varepsilon_{tol}} = \bar{\bar{P}}^P_{\varepsilon_{tol}}$ and these are continuous.

Proof. See the appendix.

Interpretation. The identified models are *robust* with respect to changes in the data, provided that the modelling problem is well-specified. □

Remark. $\bar{\bar{P}}^P_{\varepsilon_{tol}}$ also is continuous on $\Omega^P_{\varepsilon_{tol}}$ for $\varepsilon_{tol}\geq 0$. □

A similar result holds true for the descriptive procedures.

Notation. For sensible c_{tol} let $\Omega^D_{c_{tol}}:=\{\tilde{w}\in(\mathbb{R}^q)^{\mathcal{T}}$; assumption 4–3(*ii*), (*iii*) is satisfied}, and for $\varepsilon_{tol}>0$ let $\Omega^D_{\varepsilon_{tol}}:=\{\tilde{w}\in(\mathbb{R}^q)^{\mathcal{T}}$; assumption 4–5(*ii*), (*iii*) is satisfied, and moreover in algorithm 4.2.2 in step 2.1 $\sigma^{(0)}_{q-e_0}>(\bar{\varepsilon}^{tol}_0)^2>\sigma^{(0)}_{q-e_0+1}$ and in step 3.1 $\sigma^{(t)}_{q_t-e''_t}>(\bar{\varepsilon}^{tol}_t)^2>\sigma^{(t)}_{q_t-e''_t+1}\}$. □

Theorem 5-12 For given $\mathcal{T}$, sensible c_{tol} and $\varepsilon_{tol}>0$, there holds

(*i*) $\Omega^D_{c_{tol}}$ and $\Omega^D_{\varepsilon_{tol}}$ are open in $(\mathbb{R}^q)^{\mathcal{T}}$;

(*ii*) locally on $\Omega^D_{c_{tol}}$ $P^D_{c_{tol}}$ is continuous;

(*iii*) locally on $\Omega^D_{\varepsilon_{tol}}$ $P^D_{\varepsilon_{tol}}=\bar{\bar{P}}^D_{\varepsilon_{tol}}$ and these are continuous.

Proof. See the appendix.

Remark. Given the conjecture that $\Omega^P_{c_{tol}}$, $\Omega^P_{\varepsilon_{tol}}$, $\Omega^D_{c_{tol}}$ and $\Omega^D_{\varepsilon_{tol}}$ are λ–generic, cf. section 4.4., it would follow that the procedures identify models which are robust for λ–generic data. □

6. Simulations

6.1. Introduction

In this section we illustrate the modelling procedures of section 3 by means of four simple numerical examples.

In section 6.2 we consider exact modelling. In this case only exactly satisfied laws are accepted. This corresponds to applying the procedures $\bar{\bar{P}}^D_{\varepsilon_{tol}}$ and $\bar{\bar{P}}^P_{\varepsilon_{tol}}$ with $\varepsilon_{tol}=0$. The data consists of an exact observation of a time series generated by a system in $\mathbb{B}$.

Section 6.3 gives an example of descriptive modelling of a time series for given maximal tolerated complexity, i.e., of the procedure $P^D_{c_{tol}}$. The data consists of a noisy observation of a signal generated by a system in $\mathbb{B}$. We compare the (non–causal) impulse response of the generating system with that of the identified model.

In section 6.4 we illustrate the difference between descriptive and predictive modelling. For a given time series we compare the models identified by the procedures $P^D_{\varepsilon_{tol}}$ and $P^P_{\varepsilon_{tol}}$.

Finally section 6.5 contains a simulation illustrating the fact that the procedures $P^D_{\varepsilon_{tol}}$ and $P^P_{\varepsilon_{tol}}$ of section 3.3 for modelling under a given maximal tolerated misfit need not generate models of minimal complexity. This indicates the difference between these procedures and $P^{*D}_{\varepsilon_{tol}}$ and $P^{*P}_{\varepsilon_{tol}}$ as defined in section 3.3.3. We also illustrate consistency of $\bar{\bar{P}}^P_{\varepsilon_{tol}}$.

6.2. Exact modelling

6.2.1. Data

In the first simulation we consider exact modelling of a signal generated by a system in $\mathbb{B}$. The signal consists of two components, each being a sum of two sinusoids. To be specific, let $f_1:=2\pi/100$, $f_2:=2\pi/120$, and $f_3:=2\pi/150$. Define $s_k(t):=\sin(f_k.t)$, $k=1,2,3$, $t\in\mathbb{R}$, and $w_1(t):=s_1(t)+s_2(t)$, $w_2(t):=s_1(t)+s_3(t)$. The data consist of observations of the signals w_1 and w_2 on times $t=1,\dots,300$, i.e., $\tilde{w}=(\begin{bmatrix} w_1(t) \\ w_2(t) \end{bmatrix};\ t=1,\dots,300)\in(\mathbb{R}^2)^{300}$. The signals are given in figure 13.

6.2.2. System

Both w_1 and w_2 are periodic, with period 600 and 300 respectively. Hence $w\in\mathcal{B}(R)$ with $R:=\begin{bmatrix} \sigma^{600}-1 & 0 \\ 0 & \sigma^{300}-1 \end{bmatrix}$. However, there are more powerful models for w. Observe that for $g(t)=\sin(f.t)$ there holds $g(t+2)+g(t)=2\cos(f).g(t+1)$, hence $g\in\mathcal{B}(r)$ with $r(s):=s^2-2\cos(f).s+1=(s-e^{if})(s-e^{-if})$. Defining $p_k(s):=(s-e^{if_k})(s-e^{-if_k})$, $k=1,2,3$, we conclude that $\tilde{w}\in\mathcal{B}(R_0)$ with $R_0:=\begin{bmatrix} p_1\cdot p_2 & 0 \\ 0 & p_1\cdot p_3 \end{bmatrix}$.

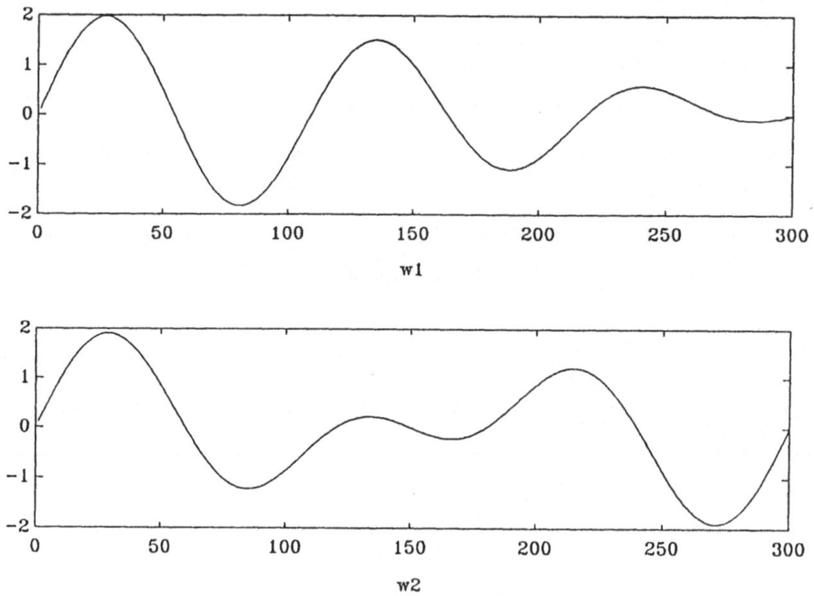

figure 13: data for simulation 6.2.

6.2.3. Model identification

Exact models for the data $\tilde{w}$ are obtained by applying the procedures $\bar{\bar{P}}^D_{\varepsilon_{tol}}$ and $\bar{\bar{P}}^P_{\varepsilon_{tol}}$ with $\varepsilon_{tol}=0$. We denote the resulting models by $\mathcal{B}(R^D):=\bar{\bar{P}}^D_0(\tilde{w})$ and $\mathcal{B}(R^P):=\bar{\bar{P}}^P_0(\tilde{w})$. These models are identified by using the algorithms of section 4 with $\varepsilon_{tol}=0$. Both models consist of one second order law and one fourth order law. Let R^D and R^P have elements r^D_{lm} and r^P_{lm} respectively, $l,m=1,2$. The identified laws are given in table 1.

6.2.4. Model validation

Two questions arise, namely, whether these AR–laws are equivalent and whether they are equivalent to R_0, i.e., if $\mathcal{B}(R^D)=\mathcal{B}(R^P)=\mathcal{B}(R_0)$.

Direct calculation shows that there exist a constant $\alpha\neq 0$ and unimodular matrices U^D and U^P such that $U^DR^D=U^PR^P=R_I:=\begin{bmatrix} p_2 & \alpha p_3 \\ p_1p_2 & 0\end{bmatrix}$. So indeed $\mathcal{B}(R^D)=\mathcal{B}(R^P)$, cf. section II.3.2.2. As $\begin{bmatrix}0 & 1\\ p_1 & -1\end{bmatrix}R_I=\begin{bmatrix}1 & 0\\ 0 & \alpha\end{bmatrix}R_0$ it follows that $\mathcal{B}(R_I)\subset\mathcal{B}(R_0)$, but $\mathcal{B}(R_I)\neq\mathcal{B}(R_0)$. So the identified laws R^D and R^P are equivalent, but not equivalent to R_0. This is due to the fact that $\mathcal{B}(R_0)$ is not the most

laws:	coefficients of: σ^0	σ^1	σ^2	σ^3	σ^4
r_{11}^D	0.5007	-1.0000	0.5007	0	0
r_{12}^D	-0.2754	0.5502	-0.2754	0	0
r_{21}^D	0.4637	-0.9568	0.5746	-0.1319	0.0507
r_{22}^D	-0.0352	-0.3517	1.0000	-0.8055	0.1920
r_{11}^P	1.2392	-2.4750	1.2392	0	0
r_{12}^P	-0.6815	1.3618	-0.6815	0	0
r_{21}^P	0.6815	-2.7224	4.0818	-2.7223	0.6815
r_{22}^P	1.2392	-4.9490	7.4196	-4.9489	1.2391

table 1: identified AR-laws for simulation 6.2.

powerful unfalsified model for $\tilde{w}$. Indeed, a short calculation gives that $p_2+\alpha p_3=\alpha' p_1$, where $\alpha:=\{\cos(f_1)-\cos(f_2)\}/\{\cos(f_3)-\cos(f_1)\}$ and $\alpha':=\{\cos(f_3)-\cos(f_2)\}/\{\cos(f_3)-\cos(f_1)\}$. Stated otherwise, the space of polynomials $\{s^2+c.s+1;\ c\in\mathbb{R}\}$ has dimension two. The most powerful unfalsified model for the generating system is $\mathcal{B}(R_0^*)$ with $R_0^*:=\begin{bmatrix} p_1p_2 & 0 \\ 0 & p_1p_3 \\ p_2 & \alpha p_3 \end{bmatrix}$. It easily follows that $\mathcal{B}(R^D)=\mathcal{B}(R^P)=\mathcal{B}(R_I)=\mathcal{B}(R_0^*)$.

Remark. The foregoing shows that the identified models correspond to the (most powerful unfalsified) model for the generating system. Hence the generating system is exactly identified. This illustrates the consistency investigated in section 5.2.2. We remark that $\mathcal{B}(R_0^*)$ is not controllable. □

6.3. Descriptive modelling

6.3.1. Introduction

In the second simulation we model a time series by minimizing the descriptive misfit, given a maximal tolerated complexity, i.e., we use the procedure $P^D_{c_{tol}}$. We first describe the data and the system generating it, then present the identified model, and finally compare this model with the generating system.

6.3.2. Data

The data consists of a two-dimensional time series $\tilde{w}=\begin{bmatrix} w_1 \\ w_2 \end{bmatrix} \in (\mathbb{R}^2)^{1000}$ and is depicted in figure 14.

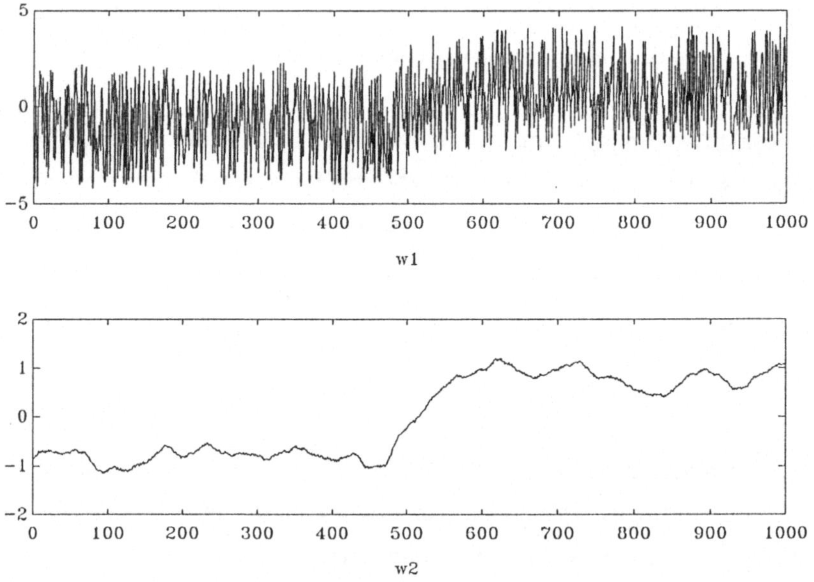

figure 14: data for simulation 6.3.

6.3.3. System

The data $\tilde{w}$ is generated by the system shown in figure 15. Here s_1 is the noise-free input, n_1 the noise on the input, and $w_1:=s_1+n_1$ the exactly observed

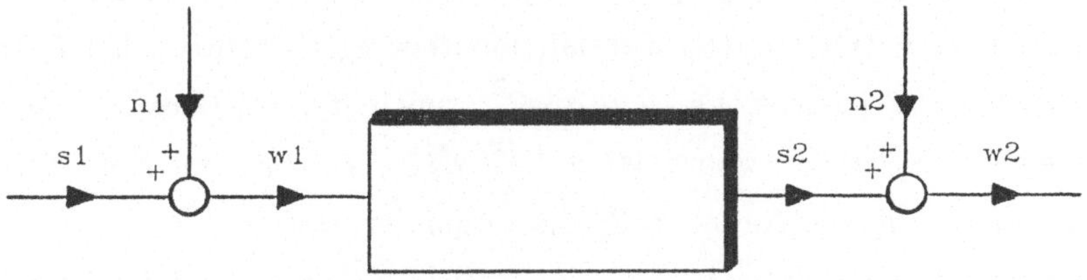

figure 15: generating system for simulation 6.3.

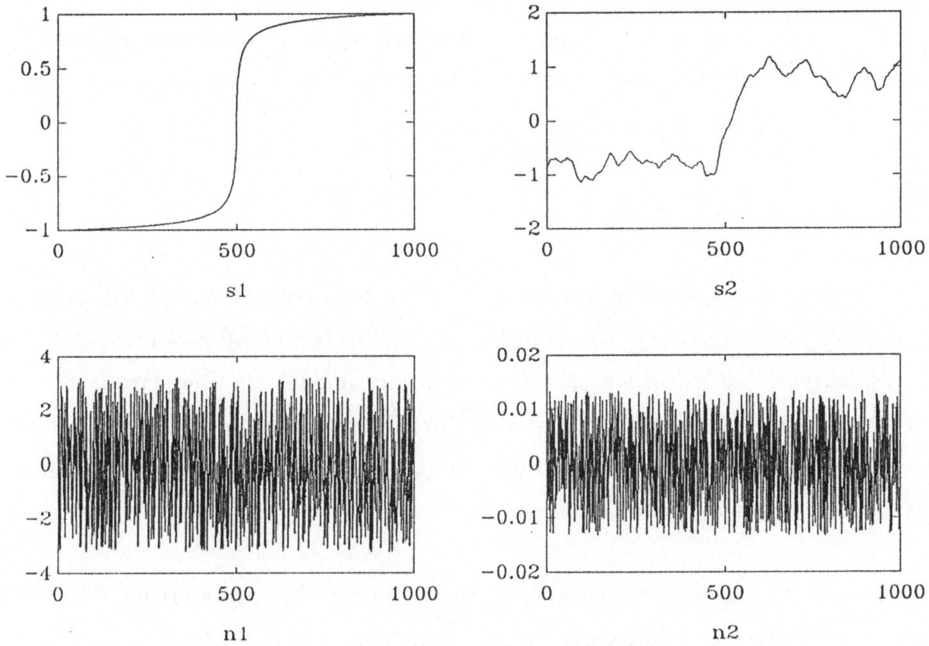

figure 16: signals and noise for simulation 6.3.

input. The signal s_2 is the output generated by the input w_1. The observed output is $w_2 := s_2 + n_2$.

The signals s_1, s_2 and the noise n_1, n_2 are given in figure 16. For a signal $s \in \mathbb{R}^T$ and noise $n \in \mathbb{R}^T$ we define the signal to noise ratio in $s+n$ as $\|s\|/\|n\| := \{\Sigma_{t=1}^T s(t)^2 / \Sigma_{t=1}^T n(t)^2\}^{1/2}$. In this simulation the signal to noise ratio for w_1 is $1/2$, for w_2 100.

The system generating s_2 from w_1 is a (symmetric) exponential smoother. For $0<\alpha<1$ we define the exponential smoother e_α as follows. Let l_∞ denote the set of bounded sequences, i.e., $l_\infty:=\{w\in\mathbb{R}^{\mathbb{Z}};\ \sup(|w(t)|;\ t\in\mathbb{Z})<\infty\}$. Then $e_\alpha: l_\infty \to l_\infty$ is defined by $e_\alpha(u):=y$, where $y(t):=\frac{1-\alpha}{1+\alpha}\cdot\Sigma_{\tau=-\infty}^{\infty}\ \alpha^{|\tau|}u(t+\tau)$. Note that for u a constant signal, $u(t)=c$ for all $t\in\mathbb{Z}$, the output is $y=u$.

We embed the graph of e_α $\mathrm{gr}(e_\alpha):=\{(u,y)\in l_\infty^2;\ y=e_\alpha(u)\}$ in a system $\mathcal{B}_\alpha\in\mathbb{B}$. In order to describe $\mathcal{B}_\alpha$, let $y=e_\alpha(u)=\frac{1-\alpha}{1+\alpha}\cdot(y_-+u+y_+)$, where $y_-(t):=\Sigma_{\tau=1}^{\infty}\ \alpha^{\tau}u(t-\tau)$ and $y_+(\tau):=\Sigma_{\tau=1}^{\infty}\ \alpha^{\tau}u(t+\tau)$. Then $(\sigma-\alpha)y_-=\alpha u$ and $(1-\alpha\sigma)y_+=\alpha\sigma u$, hence $(\sigma-\alpha)(1-\alpha\sigma)\ (y_-+u+y_+)=[(1-\alpha\sigma)\alpha+(\sigma-\alpha)(1-\alpha\sigma)+(\sigma-\alpha)\alpha\sigma]u= 1-\alpha^2)\sigma u$. Define $p_\alpha:=(s-\alpha)(1-\alpha s)$ and $q_\alpha:=\frac{1-\alpha}{1+\alpha}.(1-\alpha^2)s=(1-\alpha)^2 s$, then $\mathrm{gr}(e_\alpha)\subset\mathcal{B}_\alpha:=\mathcal{B}(R_\alpha)$ where $R_\alpha:=(-q_\alpha,p_\alpha)$.

In the simulation the signal s_2 is obtained by exponential smoothing of w_1 with $\alpha=0.95$. Hence the (most powerful unfalsified model of the) generating system is $\mathcal{B}(R_g)$ with $R_g=(-q_g,p_g):=(-q_{0.95},p_{0.95})$.

6.3.4. Model identification

Next we analyse the data $\tilde{w}$ by means of $P^D_{c_{tol}}$. We consider models of decreasing complexity, corresponding to requiring one AR-relation of order 5,4,3,2,1, and 0 respectively. For order k the resulting model is indicated by $\mathcal{B}_k:=\mathcal{B}((-q^{(k)},p^{(k)})):=\{(u,y)\in(\mathbb{R}^2)^{\mathbb{Z}};\ p^{(k)}(\sigma)y=q^{(k)}(\sigma)u\}$, $k=5,4,3,2,1,0$. See table 2. This table also contains the roots of the polynomials $p^{(k)}$, $q^{(k)}$, and the descriptive errors $\varepsilon^D_{k,1}(\tilde{w},\mathcal{B}_k)$.

The results in table 2 indicate that by reducing the order from 5 to 2 the loss in descriptive quality is small. Moreover, two of the roots of the identified polynomial p turn out to be rather invariant for different orders, while the roots of the identified polynomial q seem to be quite random, although generally one of them is close to 0. It seems reasonable to take c_{tol} such that the corresponding equation structure is $e(c_{tol})=(0,0,1,0,0,0,\ldots)$, i.e., to require one second order relation.

Remark. Note that in this identification no prior knowledge is used that w_1 is the (non-causal) input and w_2 the output. □

	coefficients of: σ^0	σ^1	σ^2	σ^3	σ^4	σ^5	roots p	q	error
order 5: $p^{(5)}$	0.4475	0.0893	-0.5333	-0.5563	0.1161	0.4295	0.9536	0.21	0.0154
$q^{(5)}$	0.0003	-0.0010	-0.0023	-0.0025	-0.0014	-0.0003	1.0548	-0.64±1.07i	
							-1.05	-1.53±0.83i	
							−0.61±0.78i		
order 4: $p^{(4)}$	0.5482	-0.3488	-0.4063	-0.3417	0.5440		0.9518	-18	0.0155
$q^{(4)}$	0.0003	-0.0014	-0.0016	-0.0017	-0.0001		1.0514	0.15	
							-0.69±0.73i	-0.56±0.88i	
order 3: $p^{(3)}$	0.5427	-0.6713	-0.2884	0.4144			0.9501	0.037	0.0159
$q^{(3)}$	0.0001	-0.0014	-0.0009	-0.0003			1.0537 -1.31	-1.31±1.55i	
order 2: $p^{(2)}$	0.4061	-0.8168	0.4099				0.9529	5.24	0.0159
$q^{(2)}$	0.0002	-0.0011	0.0002				1.0396	0.15	
order 1: $p^{(1)}$	0.7073	-0.7069					1.0006	1.20	0.0176
$q^{(1)}$	0.0011	-0.0009							
order 0: $p^{(0)}$	0.9806								0.7190
$q^{(0)}$	0.1962								

table 2: identified AR-laws for simulation 6.3.

	coefficients of: σ^0	σ^1	σ^2	roots	
system: p_g	1	-2.0026	1	0.95	1.0526
q_g	0	-0.0026	0	0	
model: p_I	0.9906	-1.9925	1	0.9529	1.0396
q_I	0.0004	-0.0028	0.0005	0.1537	5.2435

table 3: system and identified model for simulation 6.3.

6.3.5. Model validation

The identified model $\mathcal{B}((-q_I,p_I)):=\mathcal{B}_2$ is compared with the generating system $\mathcal{B}((-q_g,p_g))$ in table 3. This indicates that the AR–law of the identified system is close to the law of the generating system.

We next compare the model and the system with respect to their input–output behaviour. So we now use the prior knowledge that w_1 is an input and w_2 an output. We will compare the impulse responses of the model and the system.

For $\mathcal{B}=\{(u,y)\in(\mathbb{R}^2)^{\mathbb{Z}};\ p(\sigma)y=q(\sigma)u\}$ we define the impulse response of $\mathcal{B}$ with respect to u as $\mathcal{B}^{\delta}:=\{(u,y)\in\mathcal{B};\ u=\delta\}$, where $\delta(0):=1$ and $\delta(t):=0$ for all $t\neq 0$. It can be shown that $\mathcal{B}^{\delta}$ contains exactly one bounded element if $q\neq 0$, $p\neq 0$, and if p has no roots on the unit circle. In this case we call the time series $i\in\mathbb{R}^{\mathbb{Z}}$ such that $(\delta,i)\in\mathcal{B}^{\delta}\cap\ell_{\infty}^{2}$ the stable impulse response. The models $\mathcal{B}((-q_g,p_g))$ and $\mathcal{B}((-q_I,p_I))$ satisfy these conditions. We denote their stable impulse responses by i_g and i_I respectively. Here $i_g(t)=\frac{1-\alpha}{1+\alpha}.\alpha^{|t|}$ and i_I is determined as follows. There exist unique real numbers a_1, a_2, b_1, b_2, d, with $|a_1|<1$ and $|a_2|>1$, such that $\frac{q_I}{p_I}=\frac{b_1}{s-a_1}+\frac{b_2}{s-a_2}+d$. Define $i_I(0):=d-\frac{b_2}{a_2}$, $i_I(t):=b_1.a_1^{t-1}$ for $t>0$ and $i_I(t):=-b_2a_2^{t-1}$ for $t<0$. It then is a matter of simple calculation to verify that $p_I(\sigma)i_I=q_I(\sigma)\delta$. This corresponds to a causal interpretation of the transferfunction $\frac{b_1}{s-a_1}$ and an anticausal one of $\frac{b_2}{s-a_2}$, cf. proposition IV.3–5.

The stable impulse responses i_g of the system and i_I of the identified model are given in figure 17.

6.3.6. Remarks

We conclude this example with some remarks.

First, the stable impulse response of a system is a highly sensitive function of the AR–coefficients describing the system. For example, in the system $(\sigma-1-\varepsilon)y=u$ with $|\varepsilon|<1$ the stable impulse response is causal if $\varepsilon<0$, anticausal if $\varepsilon>0$.

Second, the result of the identification algorithm depends on scaling of the variables. In order to illustrate this, consider scaling of the output in the system $\mathcal{B}(R_g)$ by a factor $c\neq 0$. Let $\mathcal{B}_c:=\{(u,y)\in(\mathbb{R}^2)^{\mathbb{Z}};\ p_g(\sigma)y=c.q_g(\sigma)u\}$. Let

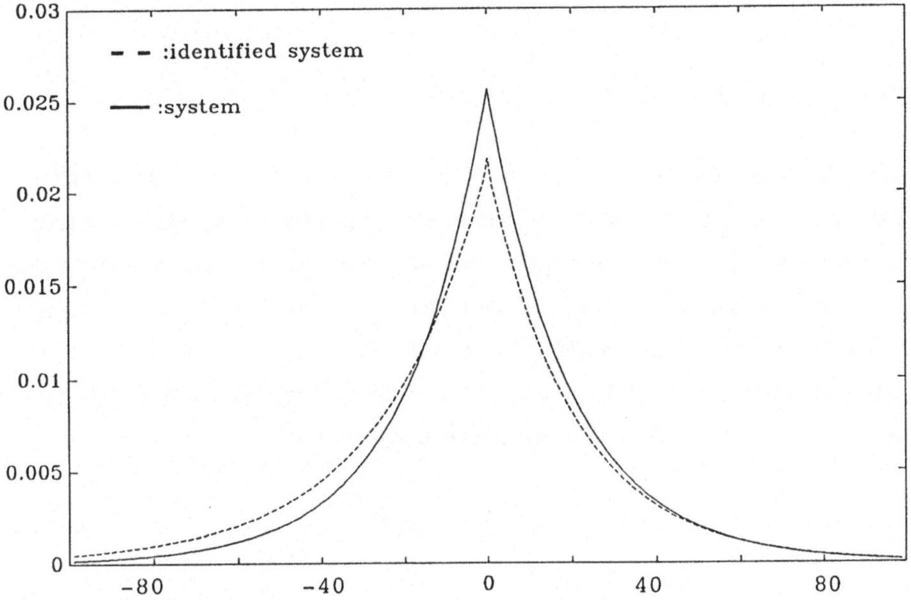

figure 17: impulse responses for simulation 6.3.

$\varepsilon := e^D(\tilde{w},(-q_I,p_I))$ denote the descriptive misfit of the identified law $(-q_I,p_I)$ with respect to the data $\tilde{w}=\begin{bmatrix} w_1 \\ w_2 \end{bmatrix}$. Denote the transformed data by $\tilde{w}_c := \begin{bmatrix} w_1 \\ cw_2 \end{bmatrix}$. From definition 3–1 it follows that $e^D(\tilde{w}_c,(-cq_I,p_I)) = \varepsilon.(\|q_I\|^2+\|p_I\|^2)^{1/2}/(\|q_I\|^2+c^{-2}.\|p_I\|^2)^{1/2}$. Using the results in table 3 it follows that the descriptive misfit of $(-cq_I,p_I)$ with respect to the scaled data $\tilde{w}_c$ is approximately $c.\varepsilon$. So, e.g., if c is sufficiently large then the law $u=0$ has smaller error. In the next section we illustrate that the predictive procedures prevent these problems of scaling. For a more general discussion of the effects of scaling for descriptive modelling of SISO systems we refer to section 6.4.5.

Finally, autoregressive modelling is subject to problems of fast sampling. Consider the case that a continuous–time input/output system is sampled at a certain sample rate Δ^{-1}. The magnitudes of the AR–coefficients of the sampled system depend on this sample rate. This affects the descriptive quality of the AR–laws, as indicated above. The constant c is related to Δ as $c=\Delta$. In section 6.4.5 we illustrate another problem of fast sampling.

6.4. Predictive modelling

6.4.1. Introduction

In the third simulation we illustrate a difference between descriptive and predictive modelling. We will see that the predictive procedures suffer less from scaling problems. On the other hand, the imposed asymmetry in time due to the one–step–*ahead* prediction criterion sometimes is artificial, in which case the descriptive procedures could be preferable.

We now first describe the data and the generating system and subsequently analyse the data by means of descriptive and predictive procedures.

6.4.2. Data

The data consists of a three–dimensional time series $\tilde{w}=\text{col}(w_1,w_{21},w_{22})\in(\mathbb{R}^3)^{200}$. We will investigate the effect of scaling. In order to illustrate this we scale w_{22} and identify models for the scaled data $\tilde{w}^{(k)}:=\text{col}(\tilde{w}_1^{(k)},\tilde{w}_2^{(k)},\tilde{w}_3^{(k)}):=\text{col}(w_1,w_{21},k.w_{22})$, $k\in\mathbb{R}_+$.

6.4.3. System

The data is generated by the system shown in figure 18. Here s_{11} is the noise–free input, n_{11} noise on the system input, $s_1:=s_{11}+n_{11}$ the input for the

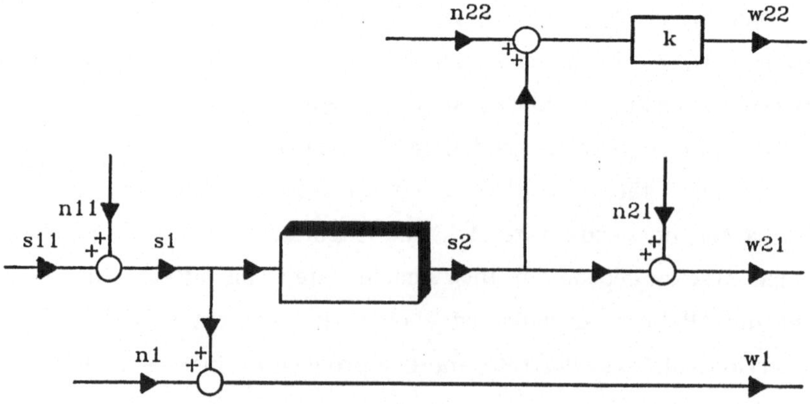

figure 18: generating system for simulation 6.4.

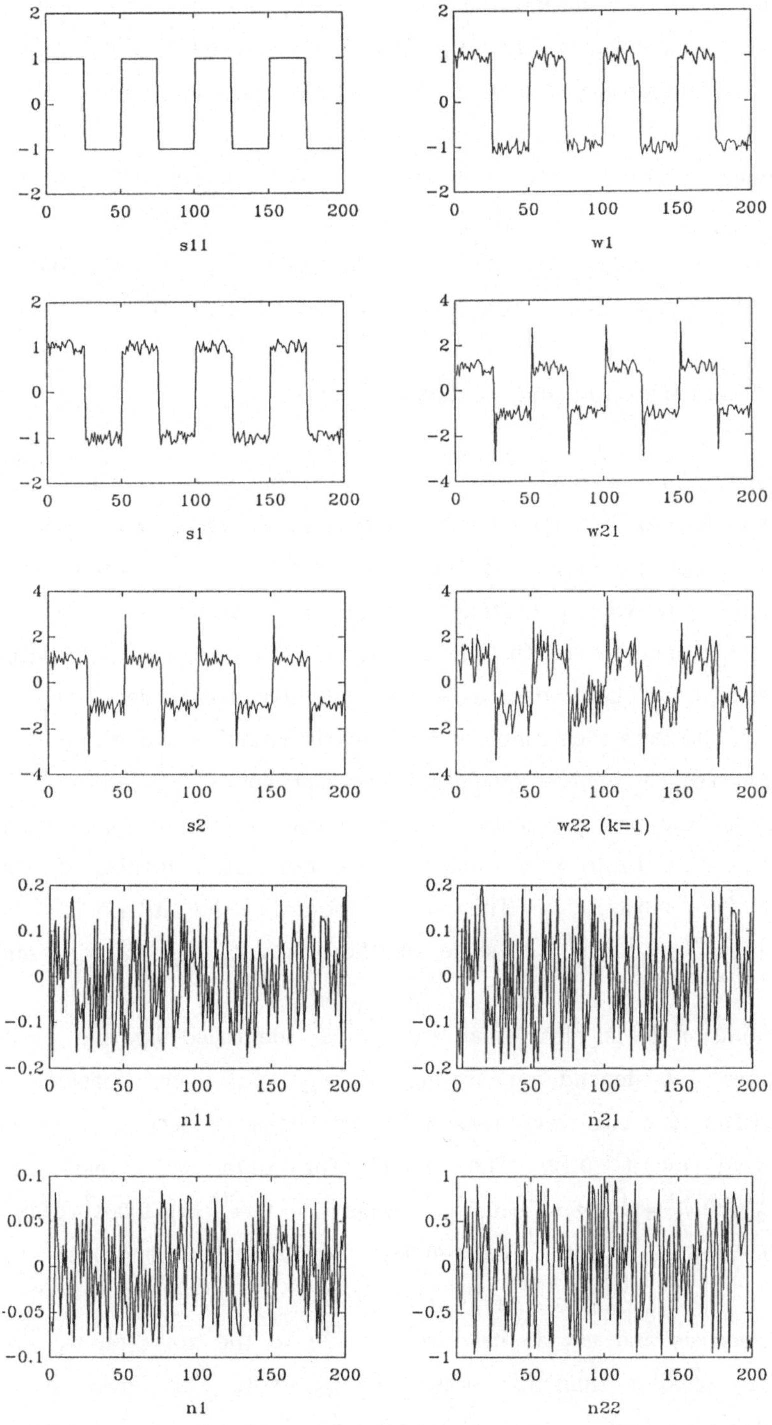

figure 19: data, signals and noise for simulation 6.4.

system, n_1 noise on the observed input, $w_1:=s_1+n_1$ the observed input, s_2 the the output of the system, n_{21} and n_{22} noise on observed outputs, $w_{21}:=s_2+n_{21}$ and $w_{22}:=s_2+n_{22}$ the observed outputs for $k=1$. The signal to noise ratios are $\|s_{11}\|/\|n_{11}\|=10$, $\|s_1\|/\|n_1\|=20$, $\|s_2\|/\|n_{21}\|=10$ and $\|s_2\|/\|n_{22}\|=2$.

The signals, observed data, and noise are given in figure 19 for the case $k=1$.

The system relating s_2 to s_1 is described by $\sigma^2 s_2=(2\sigma-1)s_1$. This corresponds to a simple linear extrapolator $s_2(t):=s_1(t-1)+\{s_1(t-1)-s_1(t-2)\}$.

6.4.4. Model identification and validation

In order to identify a model we have to reconcile the desires for low complexity and for small misfit. In the simulation we identified the laws with best descriptive and predictive fit for orders from 0 up to 4 and for data $\tilde{w}^{(k)}$ corresponding to various scaling constants k. In order to choose a model we compared the increase in fit due to increase in complexity. It turns out that the descriptive misfit decreases only slightly for orders larger than two. Moreover, the laws identified for $k>1$ nearly coincide with those for $k=1$.

The main results of the simulation are summarized in tables 4 and 5. Table 4 contains the best predictive models of orders from 0 up to 4 and for various values of k. Table 5 contains the best descriptive models of orders 0 and 2 and for various k. The tables contain the AR-coefficients in $r_1(\sigma)\tilde{w}_1^{(k)}+r_{21}(\sigma)\tilde{w}_2^{(k)}+r_{22}(\sigma)\tilde{w}_3^{(k)}=0$, some of the roots of r_1, r_{21}, r_{22}, and the misfits.

From table 4 it is clear that the model identified by the predictive procedure does not depend on scaling of w_{22}. Moreover, considering the predictive misfits it seems very reasonable to choose a second order model, with predictive misfit 0.12. The model for data $\tilde{w}^{(k)}$ then becomes $r_1^{(k)}(\sigma)w_1^{(k)}+r_{21}^{(k)}(\sigma)w_2^{(k)}+r_{22}^{(k)}(\sigma)w_3^{(k)}=0$, where $r_1^{(k)}(s)=0.08s^2-1.99s+0.96$, $r_{21}^{(k)}(s)=s^2-0.05s+0.01$, $r_{22}^{(k)}\approx k^{-1}(0.01s-0.03)$. So this law is close to the generating system $(-2\sigma+1)s_1+\sigma^2 s_2=0$. The predictive procedure identifies the relation between w_1 and w_{21} as its misfit is due to the noise on w_1 and w_{21}, which is much smaller than the noise on w_{22}. Note that, even if $\tilde{w}^{(k)}$ is observed instead of $\tilde{w}=\tilde{w}^{(1)}$, the predictive procedure for all k identifies the same AR-relation for the unscaled variables (w_1,w_{21},w_{22}).

	order 0 r_1	order 0 r_{21}	order 0 r_{22}	order 1 r_1	order 1 r_{21}	order 1 r_{22}	order 2 r_1	order 2 r_{21}	order 2 r_{22}	order 3 r_1	order 3 r_{21}	order 3 r_{22}	order 4 r_1	order 4 r_{21}	order 4 r_{22}
k=1															
coeff. σ^0	-0.60	1	-0.44	-1.82	0.48	-0.05	0.96	0.01	-0.03	0.18	0.01	-0.02	-0.18	0.05	-0.00
σ^1				0.40	1	-0.04	-1.99	-0.05	0.01	0.69	-0.02	-0.02	0.30	0.07	-0.02
σ^2							0.08	1	0.00	-1.99	0.09	0.01	0.73	-0.09	-0.02
σ^3										0.08	1	-0.00	-1.99	0.07	0.01
σ^4													0.07	1	-0.00
roots		-		4.62	-0.48	-1.25	0.49	0.02±	1.65	0.53			0.41±0.21i		
							25.2	0.11i	-10.3	-0.17			-0.43		
										24.4			26.6		
misfit		0.3250			0.2153			0.1168			0.1149			0.1134	
k=0.1															
coeff. σ^0	-0.60	1	-0.44	-1.82	0.48	-0.46	0.96	0.01	-0.28	0.18	0.01	-0.19	-0.18	0.05	-0.00
σ^1				0.40	1	-0.37	-1.99	-0.05	0.14	0.69	-0.02	-0.21	0.30	0.07	-0.18
σ^2							0.08	1	0.02	-1.99	0.09	0.14	0.73	-0.09	-0.19
σ^3										0.08	1	-0.02	-1.99	0.07	0.13
σ^4													0.07	1	-0.04
roots		-		4.62	-0.48	-1.25	0.49	0.02±	1.65	0.53			0.41±0.21i		
							25.2	0.11i	-10.3	-0.17			-0.43		
										24.4			26.6		
misfit		0.3250			0.2153			0.1168			0.1149			0.1134	
k=0.01															
coeff. σ^0	-0.60	1	-0.44	-1.82	0.48	-4.57	0.96	0.01	-2.75	0.18	0.01	-1.89	-0.18	0.05	-0.08
σ^1				0.40	1	-3.65	-1.99	-0.05	1.40	0.69	-0.02	-2.14	0.30	0.07	-1.81
σ^2							0.08	1	0.16	-1.99	0.09	1.40	0.73	-0.09	-1.93
σ^3										0.08	1	-0.17	-1.99	0.07	1.29
σ^4													0.07	1	-0.42
roots		-		4.62	-0.48	-1.25	0.49	0.02±	1.65	0.53			0.41±0.21i		
							25.2	0.11i	-10.3	-0.17			-0.43		
										24.4			26.6		
misfit		0.3250			0.2153			0.1168			0.1149			0.1134	

table 4: predictive AR-laws for simulation 6.4.

	order 0	misfit	coeff. order 2: σ^0	σ^1	σ^2	roots	misfit
k=1:							
r_1	1.36	0.3250	1.13	-1.99	0.02	0.57 ; 87.7	0.0561
r_{21}	-2.28		-0.03	-0.12	1	0.24 ; -0.12	
r_{22}	1		-0.03	0.02	-0.00	4.92 ; 1.99	
k=0.2:							
r_1	-0.00	0.1137	1.13	-1.99	0.02	0.57 ; 89.5	0.0559
r_{21}	-0.21		-0.02	-0.13	1	0.20 ; -0.08	
r_{22}	1		-0.19	0.14	-0.02	3.06 ; 2.49	
k=0.14:							
r_1	-0.01	0.0804	1.11	-1.98	0.02	0.57 ; 91.8	0.0555
r_{21}	-0.14		0.01	-0.14	1	0.07±0.09i	
r_{22}	1		-0.43	0.33	-0.08	1.98±1.09i	
k=0.12:							
r_1	-0.01	0.0691	1.08	-1.95	0.02	0.56 ; 89.6	0.0547
r_{21}	-0.12		0.06	-0.17	1	0.08±0.23i	
r_{22}	1		-0.80	0.68	-0.24	1.43±1.15i	
k=0.11:							
r_1	-0.01	0.0634	1.02	-1.88	0.02	0.55 ; 76.9	0.0535
r_{21}	-0.11		0.13	-0.22	1	0.11±0.34i	
r_{22}	1		-1.37	1.29	-0.59	1.10±1.06i	
k=0.1:							
r_1	-0.01	0.0577	0.90	-1.72	0.03	0.53 ; 49.4	0.0505
r_{21}	-0.10		0.26	-0.33	1	0.17±0.48i	
r_{22}	1		-2.54	2.71	-1.54	0.88±0.94i	
k=0.09:							
r_1	-0.01	0.0520	0.76	-1.52	0.05	0.51 ; 30.3	0.0461
r_{21}	-0.09		0.40	-0.47	1	0.24±059i	
r_{22}	1		-4.06	4.66	-2.96	0.79±0.87i	
k=0.01:							
r_1	-0.00	0.0058	-0.01	0.01	-0.00	0.40 ; 8.64	0.0052
r_{21}	-0.01		-0.01	0.01	-0.02	0.44±0.77i	
r_{22}	1		1.10	-1.39	1	0.70±0.78i	

table 5: descriptive AR-laws for simulation 6.4.

On the other hand, as shown in table 5, the model identified by the descriptive procedure strongly depends on scaling of w_{22}. Roughly stated, for values of k larger than 0.1 it seems reasonable to choose a model of order 2. These models turn out to be relatively close to the generating system. For values of k smaller than 0.1 it seems reasonable to choose a model of order 0 approximately corresponding to $w_3^{(k)}=k.w_2^{(k)}$.

In this way the simulation clearly indicates the effect of scaling of data on the model identified by the descriptive procedures. The model identified by the predictive procedures is invariant under scaling.

6.4.5. Effects of scaling for SISO-systems

We conclude this example with a few remarks on the effect of scaling on the identification of single input single output (SISO) systems.

In table 6 we give the main results of the simulation experiment which consists of modelling the data $\approx{w}^{(k)}:=$ col$(w_1,k.w_{21})$ for various k by means of the descriptive procedures. From the table of misfits it seems reasonable to accept a second order law, as the second order laws have considerably better fit than lower order laws and nearly as good fit as higher order laws. The table indicates that scaling has little influence on the model for (w_1,w_{21}), as for scaling constant k the identified AR–law $(r_1^{(k)},r_{21}^{(k)})$ is approximately equal to $(kr_1^{(1)},r_{21}^{(1)})$.

On the other hand, it turns out that by decreasing the signal to noise ratio for w_{21} the identified model becomes more sensitive to scaling. The results are similar to those in section 6.4.4. Moreover, in section 6.3.6 we concluded that for the exponential weighting system the identified model is sensitive to scaling. It hence appears that scaling sometimes has influence on the identified model, but that the effect need not always be large. Here we only give a sketch of an explanation.

Notation. For simplicity we consider a second order system $\mathcal{B}=\{(w_1,w_2);\ p(\sigma)w_2=q(\sigma)w_1\}$ with degree $d((p,q))=2$. Assume that w_2 is scaled in such a way that $\|p\|^2=\|q\|^2=\frac{1}{2}$. Let the data consist of $\tilde{w}=(\tilde{w}_1,\tilde{w}_2)$, $\tilde{w}_1=w_1+\varepsilon_1$, $\tilde{w}_2=w_2+\varepsilon_2$, with $(w_1,w_2)\in\mathcal{B}$ and where ε_1 and ε_2 are uncorrelated white noise with $\sigma_1:=\|\varepsilon_1\|$ and $\sigma_2:=\|\varepsilon_2\|$. To investigate the effect of scaling, suppose that we observe

misfit	order 0	1	2	3	4
k=100	0.4812	0.1587	0.0616	0.0564	0.0554
k=10	0.4798	0.1585	0.0616	0.0564	0.0554
k=1	0.3726	0.1370	0.0565	0.0528	0.0520
k=0.1	0.0544	0.0245	0.0134	0.0127	0.0125
k=0.01	0.0055	0.0025	0.0014	0.0013	0.0013

AR-law		coeff. of: σ^0	σ^1	σ^2	roots	
k=100:	r_1	118	-202	3.37	0.59	59.1
	r_{21}	-0.07	-0.12	1	0.33	-0.21
k=10:	r_1	11.8	-20.2	0.34	0.59	59.3
	r_{21}	-0.07	-0.12	1	0.33	-0.21
k=1:	r_1	1.15	-2.00	0.02	0.58	80.0
	r_{21}	-0.06	-0.11	1	0.31	-0.20
k=0.1:	r_1	0.10	-0.19	-0.00	0.52	-111
	r_{21}	-0.03	-0.05	1	0.19	-0.14
k=0.01:	r_1	0.01	-0.02	-0.00	0.51	-98.0
	r_{21}	-0.02	-0.05	1	0.18	-0.13
k=1: predictive:	r_1	0.97	-1.99	0.08	0.50	23.8
	r_{21}	-0.02	-0.04	1	0.17	-0.13

table 6: descriptive misfit and AR-laws for $\approx{w}^{(k)}$.

$(c_1\tilde{w}_1,c_2\tilde{w}_2)$, $c_1.c_2\neq 0$. As the identified models are invariant under a data transformation $(\pm c\tilde{w}_1,\pm c\tilde{w}_2)$, $c\neq 0$, we may consider $\tilde{w}^{(k)}:=(\tilde{w}_1,k.\tilde{w}_2)$, with $k:=|c_2/c_1|$. First let $k=1$ and let α denote the descriptive misfit of $(-q,p)$, i.e., $\alpha:=\|p\tilde{w}_2-q\tilde{w}_1\|\approx -\sqrt{2}.(\sigma_1^2+\sigma_2^2)^{1/2}$. Moreover, let β and γ denote the descriptive misfit of the best first order law for $\tilde{w}_1$ and $\tilde{w}_2$ respectively. For k let e_k^1 denote the descriptive misfit of the best first order law for $\tilde{w}^{(k)}$, and α_k the misfit of $(-kq,p)$, i.e., $\alpha_k:=e^D(\tilde{w}^{(k)},(-kq,p))=(\alpha.k\sqrt{2})/(1+k^2)^{1/2}$. □

An indication for the sensitivity to scaling is the influence of k on α_k and e_k^1. We assume that for small k $e_k^1\approx k.\gamma$ and that for large k $e_k^1\approx\beta$. This seems often to be the case. If $\alpha\sqrt{2}<\min\{\beta,\gamma\}$ we may expect small sensitivity to scaling, as it seems probable that in this case $e_k^1>\alpha_k$ for all $k\in\mathbb{R}_+$.

In the case of data $\tilde{\tilde{w}}^{(k)}:=\text{col}(w_1,kw_{21})$ in this section the underlying system is described by $p(s)=s^2$ and $q(s)=2s-1$. So for $k=1/\sqrt{5}$ we have $\|kq\|=\|p\|$. From this we get $\alpha\approx 0.04$, while for this simulation $\beta\approx 0.28$, $\gamma\approx 0.27$. So indeed $\alpha\sqrt{2}<\min\{\beta,\gamma\}$.

On the other hand, for the exponential weighting system of section 6.3 we have $\|p_g\|>>\|q_g\|$, cf. table 3. It can be calculated that for $c=850$ we have $\|cq_g\|\approx\|p_g\|$ and $\alpha\approx 9.5$, $\beta\approx 1.82$, $\gamma\approx 15.3$. So in this case $\beta<\alpha\sqrt{2}<\gamma$. For large values of k we are unable to identify the generating system. The simulation of section 6.3 corresponds to small k ($k\approx 1/850$).

Finally, if w_1 and w_2 are very smooth we always have problems in identifying the relationship between w_1 and w_2. In this case $\beta\approx e^D(\tilde{w}_1,\sigma-1)\approx\sigma_1$ and $\gamma\approx e^D(\tilde{w}_2,\sigma-1)\approx\sigma_2$, while $\alpha_k\approx(\sigma_1^2+\sigma_2^2)^{1/2}\,.\,k/(1+k^2)^{1/2}$. In this case we may expect $e_k^1<\alpha_k$ for all k.

Remark. This illustrates problems in case of fast sampling. It seems a contradiction that having a large number of observations due to fast sampling is undesirable in identification. This is due to the fact that the misfit functions defined in section 3 express only local misfit, i.e., the misfit is considered only locally in time. It would be preferable to express the distance of the observed time series, considered as a trajectory, with respect to a model. However, it appears very difficult to construct algorithms for identification on the basis of global misfit measures. □

6.5. An example illustrating non-optimality

6.5.1. Introduction

In the fourth simulation we illustrate the fact that the procedures $P^D_{\varepsilon_{tol}}$ and $P^P_{\varepsilon_{tol}}$ for modelling under a given maximal tolerated misfit need not generate models of minimal complexity. This then shows that the procedures $P^D_{\varepsilon_{tol}}$ and $P^P_{\varepsilon_{tol}}$ differ from the (optimal) procedures $P^{*D}_{\varepsilon_{tol}}$ and $P^{*P}_{\varepsilon_{tol}}$ respectively, as indicated in section 3.3.3.

We first describe the data and the generating system, then analyse the data by means of the procedures $\bar{\bar{P}}^D_{\varepsilon_{tol}}$ and $\bar{\bar{P}}^P_{\varepsilon_{tol}}$, and comment on the identified models. We finally illustrate the consistency of $\bar{\bar{P}}^P_{\varepsilon_{tol}}$.

6.5.2. Data and system

The data $\tilde{w}=\mathrm{col}(\tilde{w}_1,\tilde{w}_2,\tilde{w}_3)\in(\mathbb{R}^3)^{400}$ is generated by an ARMA-system $M(\sigma^{-1})w=N(\sigma^{-1})n$, where $n=\mathrm{col}(n_1,n_2,n_3)$ consists of three uncorrelated white noise processes with $En_k=0$, $En_k^2=1$, $k=1,2,3$. The matrices M and N are given by

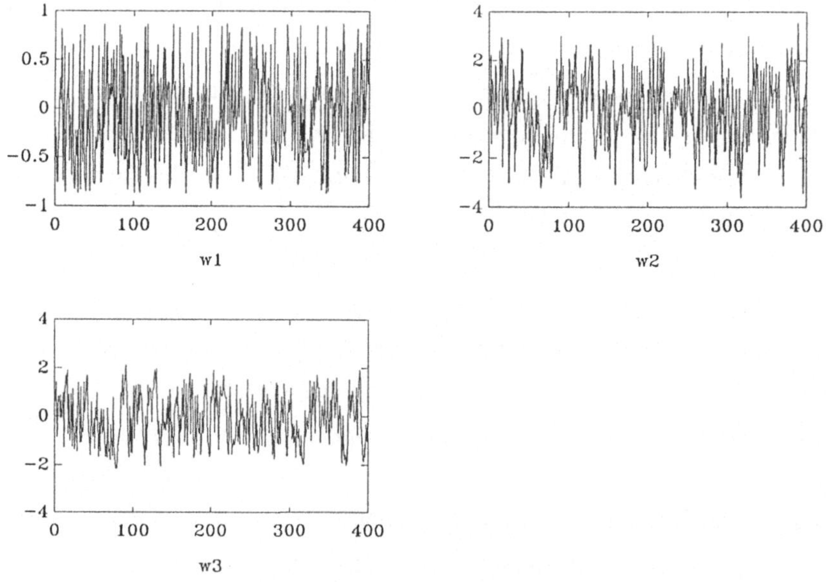

figure 20: data for simulation 6.5.

$M=\begin{bmatrix}1 & 0 & 0\\ 0 & 1 & -1\\ 0 & 0 & 1-\alpha\sigma\end{bmatrix}^{-1}$ and $N=\begin{bmatrix}½ & 0 & 0\\ 0 & \beta & 0\\ 0 & 0 & 1\end{bmatrix}$ with $\alpha:=1/\sqrt{11}$ and $\beta:=\sqrt{1.1}$. This corresponds to $w_1=½.n_1$, $\sigma w_3=\alpha w_3+\sigma n_3$, $w_2=w_3+\beta n_2$. Figure 20 shows the data $\tilde{w}$, generated by a realization of n.

6.5.3. Model identification

We identify a model for $\tilde{w}$ by means of descriptive and predictive procedures with (unfavourable) given tolerated misfits.

First we consider $\bar{\bar{P}}^D_{\varepsilon_{tol}}$ with $\varepsilon_{tol}=(\bar{\varepsilon}^{tol}_t.(1,1,1);\ t\in\mathbb{Z}_+)$, $\bar{\varepsilon}^{tol}_0:=e^D_0:=1.6$, $\bar{\varepsilon}^{tol}_1:=e^D_1:=1.2$, and $\bar{\varepsilon}^{tol}_t:=0$ for $t>1$. This practically means that only zero order and first order laws may be used in the identification of a model. The identified model is given in table 7, along with the best (not allowable) first order law.

Next we consider $\bar{\bar{P}}^P_{\varepsilon_{tol}}$ with $\varepsilon_{tol}=(\bar{\varepsilon}^{tol}_t.(1,1,1);\ t\in\mathbb{Z}_+)$, $\bar{\varepsilon}^{tol}_0:=e^P_0:=1.6$,

	identified model w_1	w_2	w_3	misfit	model B^* w_1	w_2	w_3	misfit
descr. AR order 0	0.9978	-0.0364	0.0552	0.4992	1	0	0	0.5000
	-0.0661	-0.5347	0.8425	0.6562	0	1	0	1.4938
				1.7197				
order 1: σ^0	-0.0012	-0.8443	-0.5359	1.4470	0	0	$-\alpha$	0.9574
σ^1	0.0012	0.8439	0.5356		0	0	1	
pred. AR order 0	0.9978	-0.0364	0.0552	0.4992	1	0	0	0.5000
	-0.0661	-0.5347	0.8425	0.6562	0	1	0	1.4938
				1.7197				
order 1: σ^0	-0.0004	-0.2937	-0.1865	0.9559	0	0	$-\alpha$	0.9301
σ^1	0.0014	1	0.6348		0	0	1	

table 7: descriptive and predictive AR-laws for simulation 6.5.

$\bar{\varepsilon}_1^{tol}$:=$e_1^P$:=0.95, and $\bar{\varepsilon}_t^{tol}$:=0 for $t>1$. The identified model is given in table 7, along with the best (not allowable) first order law.

6.5.4. Model validation

The identified models are not of minimal complexity, given the maximal tolerated misfit. This is also indicated in table 7. It turns out that both for the descriptive and for the predictive tolerated misfit as given before the model $\mathcal{B}^*:=\{w\in(\mathbb{R}^3)^{\mathbb{Z}};\ w_1=0, w_2=0,\ (\sigma-\alpha)w_3=0\}$ satisfies the misfit constraint. This model has complexity $c(\mathcal{B}^*)=(1,\frac{1}{2},\frac{1}{3},\frac{1}{4},...)$, which is smaller than the complexity of the identified models which is (1,1,1,1,...). It easily follows that $c(\mathcal{B}^*)$ is the lowest achievable complexity, given the misfit constraints.

Among the models of lowest achievable complexity there exist models with minimal misfit, but they seem difficult to compute. Their identification involves the question what is the lowest possible zero order misfit such that there exists a first order relation which satisfies the misfit constraint and the orthogonality conditions of the (descriptive or predictive) canonical form.

The procedures $\bar{\bar{P}}^D_{\varepsilon_{tol}}$ and $\bar{\bar{P}}^P_{\varepsilon_{tol}}$ first determine as many zero order laws as possible. Requiring three of those laws results in a zero order misfit (1.7197, 0.6562, 0.4992), which is more than tolerated. Hence two zero order laws are accepted. Moreover, the best two laws are chosen. This implies conditions on first order laws, due to the canonical forms (CDF) and (CPF), cf. definitions 3–3 and 3–7. In this simulation there is no allowable first order law satisfying these conditions. The model $\mathcal{B}^*$ shows that for minimization of complexity it is preferable not to take the best two zero order laws in order to get allowable first order laws, i.e., with misfit less than e_1^D or e_1^P.

6.5.5. Consistency

We finally consider the effect of an increase of the number of observations generated by the ARMA-system. In table 8 we summarize results for the procedure $\bar{\bar{P}}^P_{\varepsilon_{tol}}$ in case of T=50, 100, 400 and 800 observations. We also calculated the best first order laws. Observe that for T=50 the procedure for

	identified models T=50	T=100	T=400	T=800	$A^P_{\varepsilon_{tol}}$
order 0:					
AR-coeff.					
w_1	0.9999	0.9824	0.9978	0.9961	1
w_2	0.0019	0.1422	-0.0364	-0.0234	0
w_3	0.0161	-0.1210	0.0552	-0.0346	0
misfit	0.5620	0.5161	0.4992	0.4994	0.5000
AR-coeff.					
w_1	-0.0127	0.1797	-0.0661	-0.0547	0
w_2	-0.5286	-0.5440	-0.5347	-0.5246	-0.5257
w_3	0.8488	0.8196	0.8425	0.8471	0.8507
misfit	0.6593	0.6621	0.6562	0.6429	0.6482
AR-coeff.					
w_1	-0.0102				
w_2	0.8489				
w_3	0.5285				
misfit	1.5920	>1.6	>1.6	>1.6	1.6970
order 1:	-				
AR-coeff.					
σ^0: w_1		0.0228	-0.0004	-0.0004	0
w_2		-0.3708	-0.2937	-0.2874	-0.2182
w_3		-0.2511	-0.1865	-0.1772	-0.1348
σ^1: w_1		-0.0614	0.0014	0.0014	0
w_2		1	1	1	1
w_3		0.6771	0.6348	0.6164	0.6180
misfit		0.9296	0.9559	0.9578	0.9759

table 8: consistency of $\bar{\bar{P}}^P_{\varepsilon_{tol}}$.

this simulation would accept three zero order laws, while for T=100 it would accept a first order law. The table also contains the optimal approximation $A^{P}_{\varepsilon_{tol}}$, corresponding to the optimal predictive model for ε_{tol} in case the generating system were known, see definition 5-7. This model can be calculated from covariance matrices derived from M and N.

The results in table 8 illustrate consistency, as defined in section 5.3. Note especially that, in the limit, the best first order law which satisfies the orthogonality conditions of the canonical predictive form has predictive misfit $0.9759 > e^{P}_{1} = 0.95$. Hence almost surely for a sufficiently large number of observations the procedure $\bar{\bar{P}}^{P}_{\varepsilon_{tol}}$ will only accept two zero order laws.

7. Conclusion

In this chapter we described procedures and algorithms for deterministic approximate modelling based on the objectives of low complexity and small misfit and the corresponding utilities described in section II.2.

Both for the purpose of description and for that of prediction we presented deterministic procedures for approximate static modelling. These procedures correspond to total least squares and canonical correlation analysis respectively. They form the basis for procedures for approximate modelling of time series. In our approach to deterministic time series analysis the utility of models is defined in terms of a complexity of dynamical systems and a measure of descriptive or predictive misfit. These misfit measures can be numerically expressed in terms of the canonical parametrizations of dynamical systems as defined in sections II.3.2.5 and II.3.2.6. Both for the purpose of description and for that of prediction we defined procedures which minimize the misfit for a specified maximal tolerable complexity or which minimize the complexity for a specified maximal tolerable misfit. We described corresponding algorithms for deterministic time series analysis which identify models of optimal utility in case the toleration levels are well-specified for the given data.

The procedures have a clear optimality property as data modelling procedures in terms of the corresponding utility. A procedure has also an optimal performance as a method of modelling phenomena if it is consistent. This means that nearly optimal models of the phenomenon are identified if the number of observations generated by the phenomenon is sufficiently large. This has been investigated for deterministic and for stochastic data generating phenomena.

We presented an exact modelling procedure for multivariate time series analysis which has many desirable properties, and we showed that the approximate identification procedures are robust with respect to changes in the data.

The procedures for deterministic time series analysis have been illustrated by means of some numerical simulations.

Remark. Some of the main remaining topics of interest connected with this deterministic approach to time series analysis are the following.

(*i*) Construction of utilities and algorithms when the objective of modelling is (adaptive) control;

(*ii*) construction of algorithms for utilities other than $u_{c_{tol}}$ and $u_{\varepsilon_{tol}}$, especially for minimizing the number of unexplained variables (inputs) under a misfit constraint;

(*iii*) incorporation of prior model constraints as imposed, e.g., by a relevant scientific theory;

(*iv*) definition of the amount of confidence in identified models and construction of confidence regions;

(*v*) residual analysis, choice of variables, norms and toleration levels;

(*vi*) definition of approximate structure of a phenomenon and corresponding interpretation of stochastic systems, and related questions of consistency and sufficiency of excitation. □

CONCLUSIONS

We presented a deterministic approach to identification of dynamical systems. An identification procedure identifies models of optimal utility. This utility is expressed in terms of a complexity measure of models and a misfit measure of models with respect to data. The choice of these measures depends on the objectives of modelling.

For the case of exact modelling we formulated a corroboration concept. We derived a procedure which has desirable properties and which is inspired by objectives of simplicty and corroboration. This procedure also gives a new solution for the partial realization problem.

We defined appealing measures of complexity and distance for dynamical systems and derived explicit numerical expressions for these measures. We described a heuristic and simple procedure for approximating a given system by one of less complexity.

Finally we presented procedures and algorithms for deterministic time series analysis. The misfit of a dynamical system with respect to a given observed time series is defined independent from parametrization. In assessing the quality of a model the simultaneous nature of the laws describing the model is taken into account. We investigated consistency and robustness of the procedures.

In order to formulate, analyse and implement procedures for exact modelling, model approximation and approximate time series modelling, we used various representations of dynamical systems, in particular autoregressive parametrizations and state space realizations.

The results summarized before are related to identification methodologies in systems theory, statistics, and econometrics. The distinguishing features of our approach are the following. In exact modelling and for the partial realization problem a concept of corroboration is taken into account. In model approximation a system is considered as the set of all trajectories which are compatible with the laws of the system and not as an entity producing outputs in response to given inputs. In our deterministic approach to time series analysis no assumptions are made concerning the stochastic nature of a data generating mechanism. In contrast to the dominantly stochastic approach to time series analysis in statistics and econometrics these deterministic procedures are not subject to problems of parameter identifiability or structure and order estimation.

Our contribution indicates that a deterministic approach to data modelling can be formulated and implemented, without imposing prior conditions concerning stochastics or causality. For some of the main remaining topics of research connected with this approach we refer to the remarks in the conclusions in sections III.4, IV.6, and V.7.

APPENDIX : PROOFS

CHAPTER II

Proof of theorem 3-5

It easily follows that $\mathbb{B}(AR)\subset\mathbb{B}$. Now let $\mathcal{B}\in\mathbb{B}$, then according to proposition 3–3 there is a $\Delta\geq 0$ such that $\{w\in\mathcal{B}\} \Leftrightarrow \{w|_{[t,t+\Delta]}\in\mathcal{B}|_{[0,\Delta]}$ for all $t\in\mathbb{Z}\}$. Let $(\mathbb{R}^q)^{\Delta+1}$ be equipped with the Euclidean inner product and let L be a matrix the rows of which span $(\mathcal{B}|_{[0,\Delta]})^{\perp}$. Then $\{w\in\mathcal{B}\} \Leftrightarrow \{w|_{[t,t+\Delta]}\in\ker(L)$ for all $t\in\mathbb{Z}\}$. This shows $\mathcal{B}\in\mathbb{B}(AR)$. ■

Proof of proposition 3-6

In the proof we make use of two facts about polynomial matrices.

First, every submodule of $\mathbb{R}^{1\times q}[s,s^{-1}]$ is finitely generated, i.e., if $M\subset\mathbb{R}^{1\times q}[s,s^{-1}]$ is linear and $sM=M$, then there exist $g\in\mathbb{N}$ and $r_i\in\mathbb{R}^{1\times q}[s,s^{-1}]$, $i=1,\dots,g$, such that with $R:=\mathrm{col}(r_1,\dots,r_g)$ there holds $M=M(R)$. This can easily be derived e.g. from Northcott [56, proposition 1.9 and theorem 1.10].

Second, every $R\in\mathbb{R}^{g\times q}[s,s^{-1}]$ can be decomposed as $R=UDV$ with $U\in\mathbb{R}^{g\times g}[s,s^{-1}]$ and $V\in\mathbb{R}^{q\times q}[s,s^{-1}]$ both unimodular and $D=\begin{pmatrix}\Delta & 0\\ 0 & 0\end{pmatrix}$ with $\Delta=\mathrm{diag}(d_1,\dots,d_n)$, $d_i\in\mathbb{R}[s,s^{-1}]$, $i=1,\dots,n$, where $n=\mathrm{rank}(R)$ and d_i divides d_{i+1}, $i=1,\dots,n-1$. D is called the Smith form of R. We refer to e.g. Kailath [33, section 6.3.3].

(*i*) First, if $\mathcal{B}\in\mathbb{B}$, then $\mathcal{B}^{\perp}$ is a module, $\mathcal{B}^{\perp}\subset\mathbb{R}^{1\times q}[s,s^{-1}]$, hence it is finitely generated, so $\mathcal{B}^{\perp}\in\mathbb{B}^{\perp}$.

Second, if $\mathcal{B}=\mathcal{B}(R)$, then $\mathcal{B}^{\perp}=M(R)$. This is seen as follows. If $\mathcal{B}=\mathcal{B}(R)$, then $M(R)\subset\mathcal{B}^{\perp}$ is evident. On the other hand, let $r\in\mathcal{B}^{\perp}$, then we have to prove that $r\in M(R)$. Let $R=UDV$ with U and V unimodular and D the Smith form of R, $D=\begin{pmatrix}\Delta & 0\\ 0 & 0\end{pmatrix}$, $\Delta=\mathrm{diag}(d_1,\dots,d_n)$. As $r\in\mathcal{B}^{\perp}$, there holds $\{w\in\ker(R(\sigma,\sigma^{-1}))\}\Rightarrow\{w\in\ker(r(\sigma,\sigma^{-1}))\}$. Let $\tilde{w}:=Vw$ and $\tilde{r}:=rV^{-1}=(\tilde{r}_1,\dots,\tilde{r}_q)$, $\tilde{r}_i\in\mathbb{R}[s,s^{-1}]$, $i=1,\dots,q$. Then by using the fact that U and V are unimodular it follows that $\{\tilde{w}\in\ker(D(\sigma,\sigma^{-1}))\}\Rightarrow\{\tilde{w}\in\ker(\tilde{r}(\sigma,\sigma^{-1}))\}$. For $p,q\in\mathbb{R}[s,s^{-1}]$ there holds

$\{\ker(p(\sigma,\sigma^{-1}))\subset\ker(q(\sigma,\sigma^{-1}))\} \Leftrightarrow \{\exists\alpha\in\mathbb{R}[s,s^{-1}]$ such that $q=\alpha p\}$. It hence follows that $\tilde{r}_i=0$ for $i=n+1,...,q$ and that there exist $\alpha_i\in\mathbb{R}[s,s^{-1}]$ such that $\tilde{r}_i=\alpha_i d_i$, $i=1,...,n$. Hence $r=(\alpha_1,...,\alpha_n,0,...,0)U^{-1}\cdot R$, and as U is unimodular $r\in M(R)$.

Third, if for $\mathcal{B}\in\mathbb{B}$ there holds $\mathcal{B}^{\perp}=M(R)$, then $\mathcal{B}=\mathcal{B}(R)$, which is seen as follows. As $\mathcal{B}\in\mathbb{B}$, according to theorem 3–5 there exists a polynomial matrix R^* such that $\mathcal{B}=\mathcal{B}(R^*)$. It has just been shown that then $\mathcal{B}^{\perp}=M(R^*)$, hence $M(R)=M(R^*)$. This implies $\{R^*(\sigma,\sigma^{-1})\,w=0\} \Leftrightarrow \{R(\sigma,\sigma^{-1})\,w=0\}$. Hence $\mathcal{B}(R^*)=\mathcal{B}(R)$.

Finally we show that $f:\mathbb{B}\to\mathbb{B}^{\perp}:\mathcal{B}\to\mathcal{B}^{\perp}$ is a bijection onto $\mathbb{B}^{\perp}$. Let $M\in\mathbb{B}^{\perp}$, so M is finitely generated, say $M=M(R)$, then with $\mathcal{B}:=\mathcal{B}(R)\in\mathbb{B}$ there holds $\mathcal{B}^{\perp}=M(R)$, so f is surjective. Injectivity follows from the fact that if $\mathcal{B}_1,\mathcal{B}_2\in\mathbb{B}$, then $\{\mathcal{B}_1^{\perp}=\mathcal{B}_2^{\perp}=M(R)\} \Rightarrow \{\mathcal{B}_1=\mathcal{B}(R)=\mathcal{B}_2\}$.

(*ii*) If $\dim(\mathcal{B}^{\perp})=p$, then there exist p elements $r_1,...,r_p$ in $\mathbb{R}^{1\times q}[s,s^{-1}]$ such that with $R:=\mathrm{col}(r_1,...,r_p)$ there holds $\mathcal{B}^{\perp}=M(R)$. Moreover, R has full row rank p over the polynomials. According tó (*i*) $\mathcal{B}=\mathcal{B}(R)$. Now suppose $\tilde{R}$ also has p rows and $\mathcal{B}=\mathcal{B}(\tilde{R})$. As $\dim(\mathcal{B}^{\perp})=p$, $\tilde{R}$ has full row rank. According to (*i*) $\mathcal{B}^{\perp}=M(R)=M(\tilde{R})$. This implies that there exist $F,\tilde{F}\in\mathbb{R}^{p\times p}[s,s^{-1}]$ such that $R=F\tilde{R}$ and $\tilde{R}=\tilde{F}R$. Hence $(I-F\tilde{F})\,R=0=(I-\tilde{F}F)\tilde{R}$, and as R and $\tilde{R}$ have full row rank $F\tilde{F}=\tilde{F}F=I$. So $\tilde{R}=\tilde{F}R$ with $\tilde{F}$ unimodular. ■

Proof of proposition 3-12

Let $R\in\mathbb{A}$, $\mathcal{B}:=\mathcal{B}(R)$. Define $V_t^D:=v_t(L_t^D)=[v_t(\mathcal{B}_{t-1}^{\perp}+s\mathcal{B}_{t-1}^{\perp})]^{\perp}\cap[v_t(\mathcal{B}_t^{\perp})]$, $t\geq0$. Then clearly $\{V_t^D;\ t\geq0\}$ forms a set of complementary spaces for $\mathcal{B}$. Let $\{v_i^{(t)};\ i=1,...,n_t\}$ be an arbitrary basis of V_t^D, $t\geq0$, and $d:=\max\{t;\ n_t\neq0\}$. Define $r_i^{(t)}=v_t^{-1}(v_i^{(t)})$, $i=1,...,n_t$, $t=0,...,d$, and let $\tilde{R}$ be a matrix with rows $r_i^{(t)}$, $i=1,...,n_t$, $t=0,...,d$. According to proposition 3–10 $\tilde{R}$ is a tightest equation representation of $\mathcal{B}$. Moreover, in $\tilde{R}$ laws of different order are evidently orthogonal. Hence $\tilde{R}$ is in (CDF) and $R\sim\tilde{R}$. ■

Proof of proposition 3-13

First suppose that R is in (CDF). Clearly L_+ is the leading coefficient matrix of R, and as $v_t(L_t^D)\subset v_t(\mathcal{B}_t^{\perp})\cap[v_t(s\mathcal{B}_{t-1}^{\perp})]^{\perp}$ it follows that L_- is the trailing coefficient matrix of R. Let $\mathcal{B}:=\mathcal{B}(R)$, then R is a tightest equation representation of $\mathcal{B}$. According to proposition 3–8 R is bilaterally row proper, hence L_+ and L_- have full row rank. As L_- has full row rank, $R^{(t)}$ consists of the rows of R of order t, and as R is in (CDF) the rows of $R^{(t)}$ are contained in L_t^D and those of N_t in $v_d(L_t^D)$. As $L_{t-1}^D+sL_{t-1}^D \subset \mathcal{B}_t^{\perp}$ it follows by induction that the rows of $\bar{V}_t$ are contained in $v_d(\mathcal{B}_t^{\perp})$. As $v_d(L_t^D) \perp v_d(\mathcal{B}_{t-1}^{\perp}+s\mathcal{B}_{t-1}^{\perp})$ we conclude that $N_t \perp \mathrm{col}(\bar{V}_{t-1}, s\bar{V}_{t-1})$ for all $t=1,\dots,d$.

Next suppose that L_+ and L_- have full row rank and that $N_t \perp \mathrm{col}(\bar{V}_{t-1}, s\bar{V}_{t-1})$, $t=1,\dots,d$. Then R is bilaterally row proper, hence it is a tightest equation representation of $\mathcal{B}$. It remains to show that laws of order t are contained in L_t^D. As L_- has full row rank, $R^{(t)}$ consists of the rows of R of order t, and according to proposition 3–10 the number n_t of rows of $R^{(t)}$ equals the dimension of $v_d(L_t^D)$. We now show by induction for $t=0,\dots,d$ that the rows of $\bar{V}_t$ span $v_d(\mathcal{B}_t^{\perp})$ and that the rows of N_t are contained in $v_d(L_t^D)$. Then the rows of $R^{(t)}$ are contained in L_t^D, as desired.

Now $\bar{V}_0=N_0$ consists of n_0 independent elements in $v_d(\mathcal{B}_0^{\perp})=v_d(L_0^D)$, hence the rows span $v_d(L_0^D)$. Next suppose that the rows of $\bar{V}_{t-1}$ span $v_d(\mathcal{B}_{t-1}^{\perp})$, then the rows of $\mathrm{col}(\bar{V}_{t-1}, s\bar{V}_{t-1})$ span $v_d(\mathcal{B}_{t-1}^{\perp}+s\mathcal{B}_{t-1}^{\perp})$. As $N_t \perp \mathrm{col}(\bar{V}_{t-1}, s\bar{V}_{t-1})$, the rows of N_t, which are contained in $v_d(\mathcal{B}_t^{\perp})$, are orthogonal to $v_d(\mathcal{B}_{t-1}^{\perp}+s\mathcal{B}_{t-1}^{\perp})$, so they are contained in $v_d(L_t^D)$. Further, as L_- and L_+ have full row rank N_t contains n_t independent rows, and $n_t=\dim(v_d(L_t^D))$. Hence the rows of N_t span $v_d(L_t^D)$. As $\mathcal{B}_t^{\perp}=\mathcal{B}_{t-1}^{\perp}+s\mathcal{B}_{t-1}^{\perp}+L_t^D$ it follows that $\bar{V}_t:=\mathrm{col}(\bar{V}_{t-1}, s\bar{V}_{t-1}, N_t)$ spans $v_d(\mathcal{B}_t^{\perp})$, which concludes the induction part of the proof. ■

Proof of proposition 3–14

Let $\mathcal{B}=\mathcal{B}(R)$ with R in (CDF) and the rows ordered with increasing degree. If R' is in (CDF) with $\mathcal{B}(R')=\mathcal{B}$, then let Π be such that in $R''=\Pi R'$ the rows are ordered with increasing degree. For $t=0,\dots,d$ let $R^{(t)}$ and $R''^{(t)}$ denote the matrices consisting of the rows of order t in R and R'' respectively, i.e., of the rows $(\Sigma_{\tau=1}^{t-1} n_\tau)+1,\dots,\Sigma_{\tau=1}^{t} n_\tau$. Then $v_d(R^{(t)})$ and $v_d(R''(t))$ both consist of n_t independent elements in $v_d(L_t^D)$, which has dimension n_t, see proposition 3–10. Hence there exists a nonsingular matrix $A_{tt} \in \mathbb{R}^{n_t \times n_t}$ such that $v_d(R''^{(t)})=A_{tt}v_d(R^{(t)})$ and hence $R''^{(t)}=A_{tt}R^{(t)}$. So $R'=\Pi AR$, $A:=\mathrm{diag}(A_{00},\dots,A_{dd})$.

On the other hand, if R is in (CDF) with rows ordered with increasing degree, then clearly AR also is in (CDF) with $A=\mathrm{diag}(A_{00},\dots,A_{dd})$, A_{tt} nonsingular, $t=0,\dots,d$, as the rows of order t still span L_t^D. Also ΠAR is in

(CDF). As ΠA is invertible $\mathcal{B}(\Pi AR)=\mathcal{B}(R)$. ■

Proof of proposition 3-16

Let $R\in\mathbb{A}$, $\mathcal{B}:=\mathcal{B}(R)$, and define $V_t^P:=v_t(L_t^P)=[v_t(F_{t-1}s^t)+v_t(\mathcal{B}_{t-1}^\perp)]^\perp\cap[v_t(\mathcal{B}_t^\perp)]$. We claim that $\{V_t^P;\ t\geq 0\}$ gives a set of complementary spaces for $\mathcal{B}$. Assuming this to hold true, choose arbitrary bases $\{v_i^{(t)};\ i=1,\ldots,n_t\}$ of V_t^P, $r_i^{(t)}:=v_t^{-1}(v_i^{(t)})$, and let $\tilde{R}$ have rows $\{r_i^{(t)};\ i=1,\ldots,n_t,\ t=0,\ldots,d\}$, where $d:=\max\{t;\ n_t\neq 0\}$. Then according to proposition 3–10 $\tilde{R}$ is a tightest equation representation of $\mathcal{B}$, which evidently has its rows in L_t^P. Hence $\tilde{R}$ is in (CPF) and $R\sim\tilde{R}$, as desired.

To prove that $\{V_t^P;\ t\geq 0\}$ is a set of complementary spaces we have to prove that (i) $V_t^P\cap v_t(\mathcal{B}_{t-1}^\perp+s\mathcal{B}_{t-1}^\perp)=\{0\}$ and (ii) $V_t^P+v_t(\mathcal{B}_{t-1}^\perp+s\mathcal{B}_{t-1}^\perp)=v_t(\mathcal{B}_t^\perp)$.

Concerning (i), let $v\in V_t^P\cap v_t(\mathcal{B}_{t-1}^\perp+s\mathcal{B}_{t-1}^\perp)$, say $v=[r_0,\ldots,r_t]$, $r_i\in\mathbb{R}^{1\times q}$, $i=0,\ldots,t$. Let $r:=\Sigma_{k=0}^t r_k s^k$. As $r\in\mathcal{B}_{t-1}^\perp+s\mathcal{B}_{t-1}^\perp$ it follows that $r_t\in F_{t-1}$, and as $v_t(r)\in V_t^P$ it follows that $r_t\perp F_{t-1}$, hence $r_t=0$. Then $r\in\mathcal{B}_{t-1}^\perp$, but $v_t(r)\perp v_t(\mathcal{B}_{t-1}^\perp)$, hence $r=0$, so $v=0$.

Concerning (ii), note that $V_t^P+v_t(\mathcal{B}_{t-1}^\perp+s\mathcal{B}_{t-1}^\perp))\subset v_t(\mathcal{B}_t^\perp)$ is trivial. Now let $r\in\mathcal{B}_t^\perp$, then we have to show that there exist $r^{(1)},r^{(2)}\in\mathcal{B}_{t-1}^\perp$ and $v\in V_t^P$ such that $v_t(r)=v_t(r^{(1)}+sr^{(2)})+v$. Let $r=\Sigma_{k=0}^t r_k s^k$, $r_t=\bar{r}_t+\bar{\bar{r}}_t$, $\bar{r}_t\perp F_{t-1}$, $\bar{\bar{r}}_t\in F_{t-1}$. Let $r'\in\mathcal{B}_{t-1}^\perp$ be such that $r'=\Sigma_{k=0}^{t'} r'_k s^k$, $r'_{t'}=\bar{\bar{r}}_t$, where $t'\leq t-1$. Define $r^{(2)}:=s^{t-t'-1}.r'$, then $r^{(2)}\in\mathcal{B}_{t-1}^\perp$. Next define $r'':=r-sr^{(2)}-\bar{r}_t s^t$, then $d(r'')\leq t-1$. Now let $v_{t-1}(r'')=v_{t-1}(r^{(1)})+v'$ where $r^{(1)}\in\mathcal{B}_{t-1}^\perp$ and $v'\perp v_{t-1}(\mathcal{B}_{t-1}^\perp)$. Let $v:=[v',\bar{r}_t]$. Then $v_t^{-1}(v)=r''-r^{(1)}+\bar{r}_t s^t=r-sr^{(2)}-r^{(1)}\in\mathcal{B}_t^\perp$, as this is a linear space. So $v\in[v_t(F_{t-1}s^t)+v_t(\mathcal{B}_{t-1}^\perp)]^\perp\cap v_t(\mathcal{B}_t^\perp)=V_t^P$. Moreover, $v_t(r)=v_t(r''+sr^{(2)}+\bar{r}_t s^t)=v_t(r^{(1)}+sr^{(2)})+v\in v_t(\mathcal{B}_{t-1}^\perp+s\mathcal{B}_{t-1}^\perp)+V_t^P$, as desired. ■

Proof of proposition 3-17

First assume that R is in (CPF). Again L_+ clearly is the leading coefficient matrix of R, and as $v_t(L_t^P)\subset v_t(\mathcal{B}_t^\perp)\cap[v_t(F_{t-1}s^t)]^\perp$ it follows that L_- is the trailing coefficient matrix of R. Let $\mathcal{B}:=\mathcal{B}(R)$. Then R is a tightest equation representation of $\mathcal{B}$, hence L_+ and L_- have full row rank. Moreover, $R^{(t)}$ consists of the rows of R of order t. As R is in (CPF), it follows that $v_t(R^{(t)})\perp v_t(F_{t-1}s^t)$, hence especially $R_t^{(t)}\perp R_s^{(s)}$ for $s<t$, $t=0,\ldots,d$. Moreover, it easily follows by induction that the rows of $\bar{V}_t$ are contained in $v_d(\mathcal{B}_t^\perp)$. As

the rows of N_t are contained in $v_d(L_t^P) \perp v_d(\mathcal{B}_{t-1}^{\perp})$ we conclude that $N_t \perp \bar{V}_{t-1}$, $t=1,\ldots,d$.

Next suppose that conditions (*i*), (*ii*) and (*iii*) of proposition 3–17 are satisfied. We have to prove that the corresponding R is in (CPF). Condition (*i*) implies that R is a tightest equation representation of $\mathcal{B}$, see proposition 3–8. Hence the number n_t of rows of $R^{(t)}$ equals the dimension of $v_d(L_t^P)$, see proposition 3–10. We now show by induction for $t=0,\ldots,d$ that the rows of $\bar{V}_t$ span $v_d(\mathcal{B}_t^{\perp})$ and that the rows of N_t are contained in $v_d(L_t^P)$. Then the rows of $R^{(t)}$ are contained in L_t^P, as desired, and R is in (CPF).

$\bar{V}_0=N_0$ consists of n_0 independent elements in $v_d(\mathcal{B}_0^{\perp})=v_d(L_0^P)$, hence the rows span $v_d(L_0^P)$. Next suppose that the rows of $\bar{V}_{t-1}$ span $v_d(\mathcal{B}_{t-1}^{\perp})$. Condition (*ii*) implies that $N_t \perp v_d(F_{t-1}s^t)$, condition (*iii*) that $N_t \perp v_d(\mathcal{B}_{t-1}^{\perp})$. As the rows of N_t evidently are contained in $v_d(\mathcal{B}_t^{\perp})$ if follows that they are contained in $v_d(L_t^P)$. Moreover, the rows of N_t are independent, due to (*i*), and their number equals $\dim(v_d(L_t^P))$, hence they span $v_d(L_t^P)$. As $\mathcal{B}_t^{\perp}=\mathcal{B}_{t-1}^{\perp}+s\mathcal{B}_{t-1}^{\perp}+L_t^P$ it follows that $\bar{V}_t:=\mathrm{col}(\bar{V}_{t-1}, s\bar{V}_{t-1}, N_t)$ spans $v_d(\mathcal{B}_t^{\perp})$, which concludes the induction part of the proof. ■

Proof of proposition 3–22

In the proof we make use of a result from abstract realization theory, given in the next lemma. For a proof we refer to Willems [74, sections 2.2.4, 2.4.3, 4.7.4 and 4.7.5]. For $z \in (\mathbb{R}^d)^{\mathbb{Z}}$ we use the notation $z^{--} := z|_{(-\infty,-1]}$, $z^{-} := z|_{(-\infty,0]}$, $z^{+} := z|_{[0,\infty)}$, $z^{++} := z|_{[1,\infty)}$.

Lemma 3–22 Let $\mathcal{B} \in \mathbb{B}$ have realization $\mathcal{B}_s=\mathcal{B}_s(A,B,C,D) := \{(v,x,w) \in (\mathbb{R}^m \times \mathbb{R}^n \times \mathbb{R}^q)^{\mathbb{Z}};\ \begin{pmatrix} \sigma x \\ w \end{pmatrix} = \begin{pmatrix} A & B \\ C & D \end{pmatrix} \begin{pmatrix} x \\ v \end{pmatrix}\}$. Then the dimension n of the state space is minimal among all realizations of $\mathcal{B}$ if and only if the next three conditions are satisfied:

(1) the state is trim, i.e., for all $x_0 \in \mathbb{R}^n$ there exist $(v,x,w) \in \mathcal{B}_s$ with $x(0)=x_0$;

(2) the state is past induced, i.e., $\{(v,x,w) \in \mathcal{B}_s,\ w^{--}=0\} \Rightarrow \{x(0)=0\}$;

(3) the state is future induced, i.e., $\{(v,x,w) \in \mathcal{B}_s,\ w^{+}=0\} \Rightarrow \{x(0)=0\}$.

(*i*) Let $\mathcal{B} \in \mathbb{B}$ be given, and let $\mathcal{B}_s(A,B,C,D)$ be a realization of $\mathcal{B}$ with n and m

the number of state variables and driving variables respectively. Let $n^*$and m^* denote the smallest achievable n and m respectively. We have to show that these minima can be achieved simultaneously.

First we derive a lower bound for m^*. For $T\geq 0$ define $\mathcal{B}_T(0):=\{a\in(\mathbb{R}^q)^{T+1};\ \exists w\in\mathcal{B}$ such that $w^{--}=0,\ w|_{[0,T]}=a\}$ and $d_T:=\dim(\mathcal{B}_T(0))$. Now linearity and time invariance of $\mathcal{B}$ imply that $d_T\geq d_0(T+1)$. On the other hand, a realization $\mathcal{B}_s(A,B,C,D)$ of $\mathcal{B}$ implies $d_T\leq n+m(T+1)$. Hence $m^*\geq d_0$.

Next let $\mathcal{B}_s(A,B,C,D)$ be a realization with $n=n^*$. If suffices to prove that we can reduce m to d_0, as then $m=m^*$. If $\begin{bmatrix}B\\D\end{bmatrix}$ has column rank $m'<m$, then clearly there exists of realization of $\mathcal{B}$ with $n=n^*$ and $m=m'$. Hence we may assume $\begin{bmatrix}B\\D\end{bmatrix}$ to be injective. Then $n=n^*$ implies that D is injective, which is seen as follows. Let $v_0\in\ker(D)$, $(v,x,w)\in\mathcal{B}_s(A,B,C,D)$, $w^{--}=0$, $v(0)=v_0$, then lemma 3–22(2) implies that $w(0)=0$ and that $0=x(0)=x(1)=Bv(0)$, hence $v_0\in\ker(B)$. So $\ker(D)\subset\ker(B)$, which due to injectivity of $\begin{bmatrix}B\\D\end{bmatrix}$ implies that D is injective. By taking $(v,x,w)^{--}=0$ it follows from lemma 3–22(2) that $\mathrm{im}(D)=\mathcal{B}_0(0)$, hence $m=d_0$. So both m and n are minimal.

(*ii*) First suppose that (A,B,C,D) is perfectly observable, that $(A\ B)$ is surjective and that D is injective. We have to show that then n and m both are minimal.

Concerning the minimality of n, according to the lemma it suffices to check the conditions (1), (2) and (3) stated there. Perfect observability immediately implies (3). Condition (1) is implied by surjectivity of $(A\ B)$. Indeed, let $x_0\in\mathbb{R}^n$, then as $(A\ B)$ is surjective there exists $(v,x)^-$ such that $x(t+1)=Ax(t)+Bv(t)$, $t\leq -1$, with $x(0)=x_0$. Take v^{++} arbitrary and $x(t+1)=Ax(t)+Bv(t)$ for $t\geq 0$. Let $w:=Cx+Dv$. Then $(v,x,w)\in\mathcal{B}_s$ and $x(0)=x_0$. Finally we check condition (2). Let $(v,x,w)\in\mathcal{B}_s$ and $w^{--}=0$. Perfect observability implies $x|_{(-\infty,-n]}=0$. For $k=0,\ldots,n-1$ one easily proves inductively that, due to injectivity of D, there holds $v(-n+k)=0$, $x(-n+k+1)=0$. Hence $x(0)=0$, which verifies (2).

Now the minimality of m follows from the proof of (*i*), where it was shown that $m=m^*$ in case $n=n^*$ and $\begin{bmatrix}B\\D\end{bmatrix}$ is injective, so especially when $n=n^*$ and D is injective.

Next we consider the implication $\{n=n^*, m=m^*\}\Rightarrow\{(A,B,C,D)$ is perfectly observable, $(A\ B)$ is surjective, D injective$\}$.

Suppose $(A\ B)$ is not surjective. Then there exists a nonsingular $S\in\mathbb{R}^{n\times n}$

such that $S(A\ B)=\begin{pmatrix} A_1 & B_1 \\ 0 & 0 \end{pmatrix}$. Let $Sx:=\begin{pmatrix} x_1 \\ x_2 \end{pmatrix}$ be an according partition, then in $\mathcal{B}_s(A,B,C,D)$ we have $x_2=0$. This contradicts condition (1) of the lemma. Further, $m=m^*$ and hence $\begin{bmatrix} B \\ D \end{bmatrix}$ is injective. In the proof of (i) we have shown that for $n=n^*$ the injectivity of $\begin{bmatrix} B \\ D \end{bmatrix}$ implies injectivity of D.

Finally we have to show that $\{(v,x,w)\in\mathcal{B}_s,\ w|_{[0,n-1]}=0\} \Rightarrow \{x(0)=0\}$. Now let $(v,x,w)\in\mathcal{B}_s$ with $w|_{[0,n-1]}=0$. As D is injective there exists a $G\in\mathbb{R}^{n\times q}$ such that $B=GD$. Then for $0\leq t\leq n-1$ we have $0=Cx(t)+Dv(t)$ and $x(t+1)=Ax(t)+Bv(t)=(A-GC)x(t)=(A-GC)^t x_0$. So $C(A-GC)^t x_0\in\text{im}(D)$ for $0\leq t\leq n-1$. This implies that $C(A-GC)^t x_0\in\text{im}(D)$ for all $t\geq 0$. Hence by choosing $\bar{v}|_{[n,\infty)}$ appropriately we can construct $(\bar{v},\bar{x},\bar{w})\in B_s$ with $(\bar{v},\bar{x},\bar{w})|_{(-\infty,n-1]}=(v,x,w)|_{(-\infty,n-1]}$ and $w^+=0$. Condition (3) of the lemma implies $x(0)=0$. ■

Proof of proposition 3-25

In the proof we use a result on uniqueness of the state for minimal realizations. For this result we refer to Willems [74, sections 2.4.3, 3.2.5 and 4.7.5].

> **Lemma 3-25** Let $\mathcal{B}_s^i$, $i=1,2$, be two minimal realizations of a given system $\mathcal{B}\in\mathbb{B}$, and let n be the dimension of the state space. Then there exists a nonsingular $S\in\mathbb{R}^{n\times n}$ such that for any $w\in\mathcal{B}$, if $(v_i,x_i,w)\in\mathcal{B}_s^i$, $i=1,2$, then $x_2=Sx_1$.

Now let $B_s:=B_s(A,B,C,D)$ be a minimal realization of $\mathcal{B}\in\mathbb{B}$, i.e., $\mathcal{B}=\{w;\ \exists(v,x)$ such that $(v,x,w)\in\mathcal{B}_s\}$.

First we show that $\mathcal{B}_s(\tilde{A},\tilde{B},\tilde{C},\tilde{D})$ also is a minimal realization of $\mathcal{B}$ if $(\tilde{A},\tilde{B},\tilde{C},\tilde{D})=(S(A+BF)S^{-1},\ SBR,\ (C+DF)S^{-1}, DR)$ for $S\in\mathbb{R}^{n\times n}$, $R\in\mathbb{R}^{m\times m}$ both nonsingular and for any $F\in\mathbb{R}^{m\times n}$. It suffices to show that it is a realization. This is easily seen, as on the one hand for $w\in\mathcal{B}$ there is $(v,x,w)\in\mathcal{B}_s$, and with $(\tilde{v},\tilde{x}):=(R^{-1}(v-Fx), Sx)$ there holds $(\tilde{v},\tilde{x},w)\in\mathcal{B}_s(\tilde{A},\tilde{B},\tilde{C},\tilde{D})$, while on the other hand for $(\tilde{v},\tilde{x},\tilde{w})\in\mathcal{B}_s(\tilde{A},\tilde{B},\tilde{C},\tilde{D})$ there holds $(v,x,w)\in\mathcal{B}_s$ with $(v,x):=(R\tilde{v}+FS^{-1}\tilde{x},\ S^{-1}\tilde{x})$, hence $w\in\mathcal{B}$.

Next let $\mathcal{B}_s(\tilde{A},\tilde{B},\tilde{C},\tilde{D})$ be an arbitrary minimal realization of $\mathcal{B}$. We then have to construct S, R and F such that $(\tilde{A},\tilde{B},\tilde{C},\tilde{D})=(S(A+BF)S^{-1},\ SBR,\ (C+DF)S^{-1},\ DR)$. We will do this by considering $i/s/o$ realizations.

Let Π be a permutation matrix such that in $\begin{pmatrix}u\\y\end{pmatrix}:=\Pi w$ u plays the role of an input and y that of an output, see corollary 3–23. Let $\Pi C:=\begin{pmatrix}C_1\\C_2\end{pmatrix}$ and $\Pi D:=\begin{pmatrix}D_1\\D_2\end{pmatrix}$, where the partitions are according to $\begin{pmatrix}u\\y\end{pmatrix}$. Then $\mathcal{B}_s(A,B,\begin{pmatrix}C_1\\C_2\end{pmatrix},\begin{pmatrix}D_1\\D_2\end{pmatrix})$ is a minimal realization of $\Pi\mathcal{B}$. Now D_1 is invertible, which is seen as follows. Let m_u denote the number of components in the (free) variable u. Let $\mathcal{B}_0(0):=\{a\in\mathbb{R}^q;$ $\exists\begin{pmatrix}u\\y\end{pmatrix}\in\Pi\mathcal{B}$ with $\begin{pmatrix}u\\y\end{pmatrix}^{--}=0$, $\begin{pmatrix}u\\y\end{pmatrix}(0)=a\}$, then according to the proof of part (i) of proposition 3–22 there holds $\dim(\mathcal{B}_0(0)):=d_0=m^*$, while clearly $d_0\geq m_u$. The fact that $\mathcal{B}\in\mathbb{B}$ has $i/s/o$ realizations implies $m^*\leq m_u$, hence $d_0=m_u$. Moreover, due to minimality the state is past induced, from which it follows that $\mathcal{B}_0(0)=\operatorname{im}\begin{pmatrix}D_1\\D_2\end{pmatrix}$. As u is free it follows that $D_1\in\mathbb{R}^{m^*\times m^*}$ is surjective, hence invertible. Defining $(\bar A,\bar B,\bar C,\bar D):=(A-BD_1^{-1}C_1,\ BD_1^{-1},\ C_2-D_2D_1^{-1}C_1,\ D_2D_1^{-1})$ we get $\Pi\mathcal{B}=\{\begin{pmatrix}u\\y\end{pmatrix};\ \exists(v,x)$ such that $\begin{pmatrix}\sigma x\\u\\y\end{pmatrix}=\begin{pmatrix}A & B\\C_1 & D_1\\C_2 & D_2\end{pmatrix}\begin{pmatrix}x\\v\end{pmatrix}\}=\{\begin{pmatrix}u\\y\end{pmatrix};\ \exists x$ such that $\begin{pmatrix}\sigma x\\y\end{pmatrix}=\begin{pmatrix}\bar A & \bar B\\ \bar C & \bar D\end{pmatrix}\begin{pmatrix}x\\u\end{pmatrix}\}$.

For $(\tilde A,\tilde B,\tilde C,\tilde D)$ we analogously get $\Pi\mathcal{B}=\{\begin{pmatrix}u\\y\end{pmatrix};\ \exists\tilde x$ such that $\begin{pmatrix}\sigma\tilde x\\y\end{pmatrix}=\begin{pmatrix}\bar{\bar A} & \bar{\bar B}\\ \bar{\bar C} & \bar{\bar D}\end{pmatrix}\begin{pmatrix}\tilde x\\u\end{pmatrix}\}$ where $(\bar{\bar A},\bar{\bar B},\bar{\bar C},\bar{\bar D}):=(\tilde A-\tilde B\tilde D_1^{-1}\tilde C_1,\ \tilde B\tilde D_1^{-1},\ \tilde C_2-\tilde D_2\tilde D_1^{-1}\tilde C_1,\ \tilde D_2\tilde D_1^{-1})$. As these are two realizations of the same system, according to the lemma there is an S such that $\tilde x=Sx$. Hence (x,u,y) satisfies $\begin{pmatrix}\sigma x\\y\end{pmatrix}=\begin{pmatrix}\bar A & \bar B\\ \bar C & \bar D\end{pmatrix}\begin{pmatrix}x\\u\end{pmatrix}$ if and only if $\begin{pmatrix}\sigma x\\y\end{pmatrix}=\begin{pmatrix}S^{-1}\bar{\bar A}S & S^{-1}\bar{\bar B}\\ \bar{\bar C}S & \bar{\bar D}\end{pmatrix}\begin{pmatrix}x\\u\end{pmatrix}$. As the state is trim and past induced we can generate any $x_0\in\mathbb{R}^n$ and take $u(0)=0$, from which we conclude $\begin{pmatrix}S^{-1}\bar{\bar A}S\\ \bar{\bar C}S\end{pmatrix}=\begin{pmatrix}\bar A\\ \bar C\end{pmatrix}$. By taking $x_0=0$ and u arbitrary we conclude $\begin{pmatrix}S^{-1}\bar{\bar B}\\ \bar{\bar D}\end{pmatrix}=\begin{pmatrix}\bar B\\ \bar D\end{pmatrix}$. A direct calculation now shows that $(\tilde A,\tilde B,\tilde C,\tilde D)=(S(A+BF)S^{-1},\ SBR,\ (C+DF)S^{-1},\ DR)$, where $R:=D_1^{-1}\tilde D_1$ is invertible and $F:=D_1^{-1}(\tilde C_1S-C_1)$. ■

Proof of proposition 3-32

In the proof we use a result on representation of linear, time invariant, complete systems for time axis $T=\mathbb{N}$. A system $\mathcal{B}\subset(\mathbb{R}^q)^{\mathbb{N}}$ is called linear if it is a linear subspace of $(\mathbb{R}^q)^{\mathbb{N}}$, time invariant if $\mathcal{B}|_{[2,\infty)}\subset\mathcal{B}$, and complete if $\{w\in\mathcal{B}\}\Leftrightarrow\{w|_{[1,t]}\in\mathcal{B}|_{[1,t]}$ for all $t\in\mathbb{N}\}$. By $\mathbb{B}_\infty$ we denote the class of linear, time invariant, complete systems in $(\mathbb{R}^q)^{\mathbb{N}}$. By $\mathcal{B}_\infty(AR)$ we denote AR-systems in $(\mathbb{R}^q)^{\mathbb{N}}$, i.e., any set for which there exist $g\in\mathbb{N}$ and $R\in\mathbb{R}^{g\times q}[s]$ such that $\mathcal{B}=\mathcal{B}_\infty(R):=\{w\in(\mathbb{R}^q)^{\mathbb{N}};\ [R(\sigma)w](t)=0$ for all $t\in\mathbb{N}\}$. The following result is analogous to theorem 3–5 and follows from Willems [73, theorem 5].

Lemma 3–32 $\mathbb{B}_\infty=\mathbb{B}_\infty(AR)$. Moreover, for any $\mathcal{B}\in\mathbb{B}_\infty$ there exists a row proper R such that $\mathcal{B}=\mathcal{B}_\infty(R)$.

Now let $\mathcal{B}\in\mathbb{B}_T$ and define $\mathcal{B}^e:=[w\in(\mathbb{R}^q)^{\mathbb{N}};\ w|_{[t,t+T-1]}\in\mathcal{B}$ for all $t\in\mathbb{N}\}$. Shift invariance of $\mathcal{B}$ implies $\mathcal{B}=\mathcal{B}^e|_{[1,T]}$. Moreover, $\mathcal{B}^e\in\mathbb{B}_\infty$. Let $R\in\mathbb{R}^{g\times q}[s]$ be row proper such that $\mathcal{B}^e=\mathcal{B}_\infty(R)$. We will show that for any row proper R there holds $\mathcal{B}_t(R)=\mathcal{B}_\infty(R)|_{[1,t]}$, for all $t\in\mathbb{N}$. Then especially $\mathcal{B}=\mathcal{B}_\infty(R)|_{[1,T]}=\mathcal{B}_T(R)$, which shows $\mathcal{B}\in\mathbb{B}_T(AR)$.

Let $R\in\mathbb{R}^{g\times q}[s]$ be row proper with rows $r_i(s)=\Sigma_{k=0}^{d_i}r_k^{(i)}s^k$, $r_{d_i}^{(i)}\neq0$, $i\in[1,g]$. Let $L_+:=\mathrm{col}(r_{d_i}^{(i)};\ i\in[1,g])\in\mathbb{R}^{g\times q}$. Then L_+ is surjective as R is row proper.

Now first let $w\in\mathcal{B}_\infty(R)|_{[1,t]}$, say $w=\bar{w}|_{[1,t]}$ with $\bar{w}\in\mathcal{B}_\infty(R)$. Then $[r_i(\sigma)\bar{w}](\tau)=0$, $\tau\in\mathbb{N}$, and especially for rows of R with degree $d_i\leq t-1$ there holds $[r_i(\sigma)w](\tau)=0$, $\tau\in[1,\ t-d_i]$, which by definition means that $w\in\mathcal{B}_t(R)$.

Conversely, let $w\in\mathcal{B}_t(R)$. Then define $\bar{w}\in(\mathbb{R}^q)^{\mathbb{N}}$ recursively as follows. Let $\bar{w}|_{[1,t]}:=w$. If $\bar{w}$ is defined on $[1,t^*]$, then define $\bar{w}(t^*+1)$ as an arbitrary solution of $r_{d_i}^{(i)}\bar{w}(t^*+1)+r_{d_i-1}^{(i)}\bar{w}(t^*)+\ldots+r_0^{(i)}\bar{w}(t^*-d_i+1)=0$ for all i with $d_i\leq t^*$. Existence of a solution is guaranteed as L_+ is surjective. Clearly $\bar{w}\in\mathcal{B}_\infty(R)$, as all laws $[R(\sigma)\bar{w}](t)=0$, $t\in\mathbb{N}$, are satisfied. Hence $w\in\mathcal{B}_\infty(R)|_{[1,t]}$. ■

Proof of proposition 3–33

(*i*) First let $\mathcal{B}\in\mathbb{B}_T$. It follows from the proof of proposition 3–32 that there exists a row proper R with $\bar{d}(R)\leq T-1$ such that $\mathcal{B}=\mathcal{B}_T(R)$.

Conversely, let $\bar{d}(R)\leq T-1$ and R row proper. Let R have rows $r_i(s)=\Sigma_{k=0}^{d_i}r_k^{(i)}s^k$, $r_{d_i}^{(i)}\neq0$, $i\in[1,g]$, and let $L_+\in\mathbb{R}^{g\times q}$ have rows $r_{d_i}^{(i)}$, $i\in[1,g]$. To show that $\mathcal{B}_T(R)\in\mathbb{B}_T$ it suffices to consider shift invariance, i.e., $\sigma\mathcal{B}_T(R)\subset\mathcal{B}_T(R)|_{[1,T-1]}$. Now this condition is equivalent to existence of a solution $a\in\mathbb{R}^q$ of the set of equations $r_{d_i}^{(i)}a+r_{d_i-1}^{(i)}w(T)+\ldots+r_0^{(i)}w(T-d_i+1)=0$ for all $i\in[1,g]$, for fixed $w\in\mathcal{B}_T(R)$. Because R is row proper, L_+ is surjective and existence of a solution is guaranteed.

(*ii*) If R is zero order bilaterally row proper, $\bar{d}(r)\leq T-1$, then one easily shows by a similar reasoning as in (*i*) that $\mathcal{B}_T(R)\in\tilde{\mathbb{B}}_T$.

Conversely, let $\mathcal{B}\in\tilde{\mathbb{B}}_T$. Define $\mathcal{B}^{ee}:=\{w\in(\mathbb{R}^q)^{\mathbb{Z}};\ w|_{[t,t+T-1]}\in\mathcal{B}$ for all $t\in\mathbb{Z}\}$. As

$\mathcal{B}$ is translation invariant it follows that $\mathcal{B}=\mathcal{B}^{ee}|_{[1,T]}$. Moreover, $\mathcal{B}^{ee}\in\mathbb{B}$. According to theorem 3–5 and proposition 3–8 there exists a bilaterally row proper R such that $\mathcal{B}^{ee}=\mathcal{B}(R)\in\mathbb{B}$. It follows from proposition 3–10 that R can be chosen to be zero order bilaterally row proper by multiplying the rows by appropriate factors σ^{n_i}, $n_i\in\mathbb{Z}$. It remains to show that $\bar{d}(R)\leq T-1$ and that $\mathcal{B}(R)|_{[1,T]}=\mathcal{B}_T(R)$. This follows by a reasoning completely analogous to the one given in the proof of proposition 3–32, using the fact that R is bilaterally row proper. ■

Proof of lemma 3–34

In the proof we use a result for systems in $\mathbb{B}_\infty$ which is analogous to proposition 3–6(i). The proof is completely analogous to that of proposition 3–6. We use the following notation. For $\mathcal{B}\in\mathbb{B}_\infty$ let $\mathcal{B}^\perp:=\{r\in\mathbb{R}^{1\times q}[s];\ [r(\sigma)w](t)=0,$ for all $w\in\mathcal{B}$ and all $t\in\mathbb{N}\}$. If $R\in\mathbb{R}^{g\times q}[s]$ has rows $r_i\in\mathbb{R}^{1\times q}[s]$, then let $M_\infty(R):=\{r\in\mathbb{R}^{1\times q}[s];\ \exists p_i\in\mathbb{R}[s],\ i\in[1,g],$ such that $r=\Sigma_{i=1}^g p_ir_i\}$ denote the submodule of $\mathbb{R}^{1\times q}[s]$, generated by the rows of R.

Lemma 3–34–1 Let $\mathcal{B}\in\mathbb{B}_\infty$, Then $\{\mathcal{B}=\mathcal{B}_\infty(R)\} \Leftrightarrow \{\mathcal{B}^\perp=M_\infty(R)\}$.

Now let $\mathcal{B}_T(R_i)\in\mathbb{B}_T$, $i=1,2$. Define $\mathcal{B}_i^e:=\{w\in(\mathbb{R}^q)^\mathbb{N};\ w|_{[t,t+T-1]}\in\mathcal{B}_T(R_i)$ for all $t\in\mathbb{N}\}$, $i=1,2$. Shift invariance of $\mathcal{B}_T(R_i)$ implies that $\mathcal{B}_i^e|_{[1,T]}=\mathcal{B}_T(R_i)$, $i=1,2$, and that $\{\mathcal{B}_T(R_1)\subset\mathcal{B}_T(R_2)\} \Leftrightarrow \{\mathcal{B}_1^e\subset\mathcal{B}_2^e\}$.

Define $\mathcal{B}_\infty(R_i):=\{w\in(\mathbb{R}^q)^\mathbb{N};\ [R_i(\sigma)w](t)=0$ for all $t\in\mathbb{N}\}$, $i=1,2$. Because $\bar{d}(R_i)\leq T-1$ there holds $\mathcal{B}_i^e=\mathcal{B}_\infty(R_i)$, $i=1,2$, which is seen as follows. Let R_i have rows $r_j^{(i)}$, $j\in[1,g_i]$. If $w\in\mathcal{B}_\infty(R_i)$, then $[r_j^{(i)}(\sigma)w](t)=0$ for all $j\in[1,g_i]$ for all $t\in\mathbb{N}$, especially for all $t\in[\tau,\ \tau+T-d(r_j^{(i)})-1]$, for all $\tau\in\mathbb{N}$, so $w|_{[\tau,\tau+T-1]}\in\mathcal{B}_T(R_i)$ for all $\tau\in\mathbb{N}$, hence $w\in\mathcal{B}_i^e$. Conversely, if $w\in\mathcal{B}_i^e$ then with $w_t:=w|_{[t,t+T-1]}$, $t\in\mathbb{N}$, $w_t\in\mathcal{B}_T(R_i)$, so $[r_j^{(i)}(\sigma)w_t](\tau)=0$ for all $j\in[1,g_i]$, for all $\tau\in[1,T-d(r_j^{(i)})]\neq\emptyset$, especially $[r_j^{(i)}(\sigma)w_t](1)=[r_j^{(i)}(\sigma)w](t)=0$ for all $t\in\mathbb{N}$, hence $w\in\mathcal{B}_\infty(R_i)$.

So to prove lemma 3–34 it remains to prove that $\{\mathcal{B}_\infty(R_1)\subset\mathcal{B}_\infty(R_2)\} \Leftrightarrow$ {there exists an F such that $R_2=FR_1\}$. Now ($\Leftarrow$) is obvious, while ($\Rightarrow$) follows from lemma 3–34–1. ■

Proof of lemma 3–35 and proposition 3–36

In the proof we make use of the following result, which was shown in the proof of proposition 3–32.

Lemma 3–35–1 If R is row proper, then for all $T\in\mathbb{N}$ $\mathcal{B}_\infty(R)|_{[1,T]}=\mathcal{B}_T(R)$.

First we consider the results for LCLM and addition.

Let R_i be row proper, i=1,2. Define $\mathcal{B}_\infty(R_i):=\{w\in(\mathbb{R}^q)^{\mathbb{N}};\ [R_i(\sigma)w](t)=0$ for all $t\in\mathbb{N}\}$ and $\mathcal{B}:=\mathcal{B}_\infty(R_1)+\mathcal{B}_\infty(R_2)$. It follows from lemma 3–32 in the proof of proposition 3–32 that there exists a row proper R_0 such that $\mathcal{B}=\mathcal{B}_\infty(R_0)$. We will show that $R_0\in\text{LCLM}(R_1,R_2)$.

Row properness implies that for all $T\in\mathbb{N}$ $\mathcal{B}_T(R_i)=\mathcal{B}_\infty(R_i)|_{[1,T]}$, i=1,2, and that $\mathcal{B}_T(R_0)=\mathcal{B}_\infty(R_0)|_{[1,T]}$. As $[\mathcal{B}_\infty(R_1)+\mathcal{B}_\infty(R_2)]|_{[1,T]}=\mathcal{B}_\infty(R_1)|_{[1,T]}+\mathcal{B}_\infty(R_2)|_{[1,T]}$ this implies that for all $T\in\mathbb{N}$ also $\mathcal{B}_T(R_1)+\mathcal{B}_T(R_2)=\mathcal{B}_T(R_0)$.

Taking $T\geq\max\{\bar{d}(R_0),\ \bar{d}(R_1),\ \bar{d}(R_2)\}+1$, $\mathcal{B}_T(R_i)\subset\mathcal{B}_T(R_0)$ by lemma 3–34 implies that there exist F_i such that $R_0=F_iR_i$. Moreover, if for $\tilde{R}$ there exist $\tilde{F}_i$ with $\tilde{R}=\tilde{F}_iR_i$, i=1,2, then let U unimodular be such that $U\tilde{R}=\begin{pmatrix}\bar{R}\\0\end{pmatrix}$ with $\bar{R}$ row proper, cf. e.g. Wolovich [77, theorem 2.5.11]. Then $\mathcal{B}_\infty(\bar{R})=\mathcal{B}_\infty(\tilde{R})\supset\mathcal{B}_\infty(R_1)+\mathcal{B}_\infty(R_2)=\mathcal{B}_\infty(R_0)$. Due to row properness it follows that for all $T\in\mathbb{N}$ $\mathcal{B}_T(\bar{R})\supset\mathcal{B}_T(R_0)$. Taking $T\geq\max\{\bar{d}(R_0),\ \bar{d}(\bar{R})\}+1$ it follows from lemma 3–34 that there exists an $\bar{F}$ such that $\bar{R}=\bar{F}R_0$, hence $\tilde{R}=FR_0$ where $F:=U^{-1}\begin{pmatrix}\bar{F}\\0\end{pmatrix}$. This proves that $R_0\in\text{LCLM}(R_1,R_2)$ and (i) of lemma 3–35 for LCLM.

Next let $R\in\text{LCLM}(R_1,R_2)$ be row proper. Then there exists an F_0 such that $R=F_0R_0$ and an F such that $R_0=FR$. In the notation of lemma 3–34–1 hence $M_\infty(R)=M_\infty(R_0)$, so $\mathcal{B}_\infty(R)=\mathcal{B}_\infty(R_0)=\mathcal{B}_\infty(R_1)+\mathcal{B}_\infty(R_2)$. Due to row properness $\mathcal{B}_T(R)=\mathcal{B}_T(R_1)+\mathcal{B}_T(R_2)$, which proves (i) of proposition 3–36. Moreover, $R=F_0FR$ and $R_0=FF_0R_0$. Because R_0 and R are row proper they have full row rank over $\mathbb{R}[s]$, so that $F_0F=FF_0=I$, which proves (ii) of lemma 3–35 for LCLM.

Now second we consider the results for GCRD and intersection.

Let R_1 and R_2 be row proper. From lemma 3–32 in the proof of proposition 3–32 it follows that $\mathbb{B}_\infty$ is closed under intersection, as $\mathbb{B}_\infty(AR)$ clearly has this property. Moreover, according to this lemma there exists a row proper R_0 such that $\mathcal{B}_\infty(R_0)=\mathcal{B}_\infty(R_1)\cap\mathcal{B}_\infty(R_2)$. That $R_0\in\text{GCRD}(R_1,R_2)$ is proved in a way analogous to the result for LCLM, and one easily gets lemma 3–35(i) and (ii) for GCRD.

To prove proposition 3–36(ii), let $R \in \text{GCRD}(R_1,R_2)$ be row proper and let $\bar{d}(R_i) \leq T-1$, $i=1,2$. By lemma 3–35(ii) there exists a unimodular U such that $R=UR_0$, so $\mathcal{B}_\infty(R)=\mathcal{B}_\infty(R_0)$ and $\mathcal{B}_T(R)=\mathcal{B}_T(R_0)$. So it suffices to prove that $\mathcal{B}_T(R_0)=\Sigma\{\mathcal{B}_T(\tilde{R});\ \mathcal{B}_T(\tilde{R}) \subset \mathcal{B}_T(R_1) \cap \mathcal{B}_T(R_2)\}$.

Now $\mathcal{B}_\infty(R_0) \subset \mathcal{B}_\infty(R_i)$ and row properness implies $\mathcal{B}_T(R_0) \subset \mathcal{B}_T(R_i)$, $i=1,2$; hence $\mathcal{B}_T(R_0) \subset \mathcal{B}_T(R_1) \cap \mathcal{B}_T(R_2)$. So it suffices to prove that for $\tilde{R}$ row proper $\{\mathcal{B}_T(\tilde{R}) \subset \mathcal{B}_T(R_1) \cap \mathcal{B}_T(R_2)\} \Rightarrow \{\mathcal{B}_T(\tilde{R}) \subset \mathcal{B}_T(R_0)\}$. Let $\bar{R}$ consist of the rows of $\tilde{R}$ of degree at most $T-1$, then $\bar{d}(R_i) \leq T-1$, $i=1,2$, and lemma 3–34 implies that there exist F_i such that $R_i=F_i\bar{R}$, $i=1,2$; hence $\mathcal{B}_\infty(\bar{R}) \subset \mathcal{B}_\infty(R_0)$ and $\mathcal{B}_T(\tilde{R})=\mathcal{B}_T(\bar{R}) \subset \mathcal{B}_T(R_0)$. ■

CHAPTER III

Proof of proposition 3-7

Let P be bilaterally monotone on $\mathcal{B}\in\tilde{\mathbb{B}}_T$. By taking $t=2$ in the definition of bilateral monotonicity we have that generically in $w\in\mathcal{B}$ $\{\mathcal{B}_{T-1}(R)\in P_{T-1}(w|_{[2,T]}),\ \mathcal{B}'\in P_Tw\}\Rightarrow\{\sigma\mathcal{B}'\subset\mathcal{B}_{T-1}(R)\}\Rightarrow\{\mathcal{B}'\subset\mathcal{B}_T(\sigma R)\}$, which proves the shift invariance condition for $t=T$.

To prove this condition for general $t\in[2,T]$, let $\mathcal{B}_{t-1}(R)\in P_{t-1}(w|_{[2,t]})$ and $\mathcal{B}'\in P_t(w|_{[1,t]})$. Now $\mathcal{B}\in\tilde{\mathbb{B}}_T$, so by proposition II.3–30(ii) there exists $\tilde{w}\in\mathcal{B}$ with $\tilde{w}|_{[T-t+1,T]}=w|_{[1,t]}$ and $\mathcal{B}_{t-1}(R)\in P_{t-1}(\tilde{w}|_{[T-t+2,T]})$, $\mathcal{B}'\in P_t(\tilde{w}|_{[T-t+1,T]})$. Now bilateral monotonicity implies that generically in $\tilde{w}\in\mathcal{B}$ $\sigma\mathcal{B}'\subset\mathcal{B}_{t-1}(R)$, hence $\mathcal{B}'\subset\mathcal{B}_t(\sigma R)$. We have to prove that this holds generically in $w\in\mathcal{B}$. It is sufficient to construct a linear bijection $w\to\tilde{w}$. This can be done as follows.

Let $\mathcal{B}^{ee}:=\{w^{ee}\in(\mathbb{R}^q)^{\mathbb{Z}};\ w^{ee}|_{[\tau,\tau+T-1]}\in\mathcal{B}$ for all $\tau\in\mathbb{Z}\}$. Because $\mathcal{B}$ is translation invariant $\mathcal{B}^{ee}|_{[1,T]}=\mathcal{B}$. It can be shown that there exists a linear injection $L:\mathcal{B}\to\mathcal{B}^{ee}:w\to w^{ee}$ with $w^{ee}|_{[1,T]}=w$ such that for all $\tau\in\mathbb{Z}$ $L_\tau:\mathcal{B}\to\mathcal{B}$: $w\to w^{ee}|_{[\tau+1,\tau+T]}$ is a bijection. Then for $w\in\mathcal{B}$ take $\tilde{w}:=L_{t-T}w$.

The idea to construct L is as follows. Let $\mathcal{B}_{i/s/o}$ be a minimal (forward) input/state/output realization of $\mathcal{B}^{ee}$ (see corollary II.3–23) with state x and with $\begin{pmatrix}u\\y\end{pmatrix}=\Pi w$ a corresponding input/output decomposition of w. It can be shown that there exists a linear map f such that $x(\tau)=f(w|_{[\tau-T,\tau-1]})$ for all $\tau\in\mathbb{Z}$. Now take in $\mathcal{B}_{i/s/o}$ u^{ee} periodic on $\mathbb{Z}_+$ with $u^{ee}|_{[kT+1,(k+1)T]}:=u|_{[1,T]}$, $k\in\mathbb{N}$. Together with $x(T+1)$ this uniquely defines $w^{ee}|_{[0,\infty]}$. We define $w^{ee}|_{(-\infty,-1]}$ in an analogous way, using a backward realization $\tilde{\mathcal{B}}_{i/s/o}$ of $\mathcal{B}^{ee}$. This defines a linear injection $L:\mathcal{B}\to\mathcal{B}^{ee}$. To see that L_τ is a bijection, suppose that $w^{ee}|_{[\tau+1,\tau+T]}$ is given for some $\tau\in\mathbb{Z}$. From this we can reconstruct $x(\tau+T+1)$ and u, as u is periodic. From $\mathcal{B}_{i/s/o}$ and $\tilde{\mathcal{B}}_{i/s/o}$ we then can reconstruct w^{ee} on $\mathbb{Z}$, hence especially $w:=w^{ee}|_{[1,T]}$. ■

Proof of lemma 3-11

($\Leftarrow$) Let $r(n)$ be linearly independent from $r(1),\ldots,r(n-1)$ and $r(n+1)$ linearly dependent on $r(1),\ldots,r(n)$, say $r(n+1)=\Sigma_{i=1}^{n}a_i r(i)$ (defined for the columns $1,\ldots,T-n$ of $H_T(w)$). Define $w(\tau)$, $\tau>T$, recursively by $w(\tau):=\Sigma_{i=1}^{n}a_i w(\tau-n-1+i)$ and define a Hankel extension M of $H_T(w)$ by $m_{ij}:=w(i+j-1)$. Using the Hankel structure one gets $\text{rank}(M)=n$, hence $\text{rank}(H_T(w))\leq n$. To prove that $\text{rank}(H_T(w))\geq n$, let M' be an arbitrary extension of $H_T(w)$ and let $d:=\text{rank}(M')$. If $d<n$ this would imply that among the rows $1,\ldots,n$ of M' at least one, say row n', is either zero (in which case $r(n')$ is linearly dependent on $r(1),\ldots,r(n-1)$ by definition) or it is linearly dependent on the foregoing ones. This implies that $r(n')$ is linearly dependent on $r(1),\ldots,r(n'-1)$, and because of the Hankel structure of $H_T(w)$ and the fact $n'\leq n$ this means that $r(n)$ would be linearly dependent on $r(1),\ldots,r(n-1)$. So $\text{rank}(M')\geq n$ and hence $\text{rank}(H_T(w))\geq n$.

($\Rightarrow$) Let $\text{rank}(H_T(w))=n$. Then $r(n)$ cannot be linearly dependent on $r(1),\ldots,r(n-1)$, as the construction above would give $\text{rank}(H_T(w))\leq n-1$. Moreover, $r(n+1)$ cannot be linearly independent from $r(1),\ldots,r(n)$, as this would imply that any extension of $H_T(w)$ would have rank at least $n+1$. ■

Proof of lemma 3-12

A minimal rank extension which is Hankel was constructed in the proof of ($\Leftarrow$) of lemma 3-11. ■

Proof of proposition 3-13

Let $\text{rank}(H_T(w))=d$. Assume $w\in\mathcal{B}\in\mathbb{B}_T$. Shift invariance of $\mathcal{B}$ implies that there exists $w^e\in(\mathbb{R})^{\mathbb{N}}$, $w^e|_{[1,T]}=w$, $w^e|_{[\tau,\tau+T-1]}\in\mathcal{B}$ for all $\tau\in\mathbb{N}$. Define an extension M of $H_T(w)$ by $m_{ij}:=w^e(i+j-1)$, $i,j=1,\ldots,T$, $m_{ij}:=0$ elsewhere. Then $\text{rank}(M)\geq d$ and hence $w^e|_{[\tau,\tau+T-1]}$, $\tau\in[1,T]$, span a space of dimension at least d in $\mathcal{B}$, hence $c(\mathcal{B})\geq d$. Further, as rank $H_T(w)=d$ there exists $a=(a_1,\ldots,a_d)\in\mathbb{R}^d$ such that $r(d+1)=\Sigma_{i=1}^{d}a_i r(i)$. Define $R_a:=\sigma^d-\Sigma_{i=1}^{d}a_i\sigma^{i-1}$. Then clearly $w\in\mathcal{B}_T(R_a)$ and $c(\mathcal{B}_T(R_a))=d$. Using the definition of P_T^K this proves (i) and $\supset$ in (ii). To prove $\subset$ in (ii), let $\mathcal{B}_T(R)\in P_T^K w$, so $c(\mathcal{B}_T(R))=d$, which implies that R has degree

d, say $R=\sigma^d-\Sigma_{i=1}^d a_i\sigma^{i-1}$. Then in $H_T(w)$ $r(d+1)=\Sigma_{i=1}^d a_i r(i)$. ■

Proof of proposition 3–16

Let $R\neq 0$ have degree $\bar{d}(R)=d$. First assume $d\leq\text{ENT}(T/2)$, so we have to show that generically in $w\in\mathcal{B}(R)$ $\text{rank}(H_T(w))=d$.

For $w\in\mathcal{B}_T(R)$ row $d+1$ of $H_T(w)$ is linearly dependent on the foregoing ones, so $\text{rank}(H_T(w))\leq d$. To prove that gen. $\text{rank}(H_T(w))=d$, it suffices according to lemma 3–11 to show that gen. row d is linearly independent from the foregoing ones. Sufficient for this is that gen. in $w\in\mathcal{B}_T(R)$ $\text{rank}(H_{d,d}(w))=d$, where

$$H_{d,d}(w):=\begin{bmatrix} w(1) & w(2) & \dots & w(d) \\ w(2) & w(3) & \dots & w(d+1) \\ \vdots & \vdots & & \vdots \\ w(d) & w(d+1) & \dots & w(2d-1) \end{bmatrix}.$$

Note that $d\leq\text{ENT}(T/2)$, so $2d-1\leq T$ and $H_{d,d}(w)$ is well-defined.

In $\mathcal{B}_T(R)$, $w(\tau)$, $\tau\in[1,d]$, can be chosen arbitrarily while $w(\tau)$, $\tau\in[d+1,T]$ can be expressed as linear functions of $w(\tau)$, $\tau\in[1,d]$. So for $w\in\mathcal{B}_T(R)$ $\det(H_{d,d}(w))$ can be considered as a polynomial in $(w(1),\dots,w(d))\in\mathbb{R}^d$. It suffices to show that $\det(H_{d,d}(w))$ is not the zero polynomial, because then $\{w;\ \text{rank}(H_{d,d}(w))<d\}=\{w;\ \det(H_{d,d}(w))=0\}$ is a proper algebraic variety and hence $\text{rank}(H_{d,d}(w))=d$ gen. in $w\in\mathcal{B}_T(R)$.

That $\det(H_{d,d}(w))\not\equiv 0$ is seen as follows. We claim that $\det(H_{d,d}(w))$ contains $\{w(d)\}^d$ as a term with coefficient ± 1. Indeed, $\det(H_{d,d}(w))=\Sigma_{p\in P}\{\text{sign}(p)\cdot\Pi_{i=1}^d a_{ip(i)}\}$, where $H_{d,d}(w)=(a_{ij})$, P denotes the set of all permutations of $\{1,\dots,d\}$, and $\text{sign}(p)\in\{-1,+1\}$. In order to get $\{w(d)\}^d$, from every row and column in $H_{d,d}(w)$ one has to choose an element which involves $w(d)$. In the first row this is only the element $(1,d)$, so $p(1)=d$. In the second row only the elements $(2,d-1)$ and possibly $(2,d)$ contain $w(d)$, so necessarily $p(2)=d-1$. Going on in this way one gets for $\{w(d)\}^d$ the unique permutation $p:=\{d,d-1,\dots,2,1\}$. This proves our claim and hence $\det(H_{d,d}(w))\not\equiv 0$.

Next assume $R=0$ or $d>\text{ENT}(T/2)$. By a similar reasoning as before one can show that gen. in $w\in\mathcal{B}_t(R)$ $H_{\text{ENT}(T/2),\text{ENT}(T/2)}(w)$ has rank $\text{ENT}(T/2)$ and hence row $\text{ENT}(T/2)+1$ of $H_T(w)$ then is linearly dependent on the foregoing ones (its length is $\text{ENT}(T/2)$ if T is even, $\text{ENT}(T/2)-1$ if T is odd). So then $\text{rank}(H_T(w))\leq\text{ENT}(T/2)$ and hence it equals $\text{ENT}(T/2)$, as row $\text{ENT}(T/2)$ is linearly independent from the foregoing ones. ■

Proof of theorem 3–17

(*i*) Obvious.

(*ii*) Not monotone. Let $\mathcal{B}=\mathbb{R}^T$ and $t\in[3,T]$ odd. Then according to proposition 3–16 gen. in $w\in\mathcal{B}$ there holds that for $\mathcal{B}_{t-1}\in P^K_{t-1}(w|_{[1,t]})$ and $\mathcal{B}_t\in P^K_t(w|_{[1,t]})$ $\dim(\mathcal{B}_{t-1})=(t-1)/2$ and $\dim(\mathcal{B}_t)=(t+1)/2$. For $t\geq 3$ $(t+1)/2\leq t-1$ and $\dim(\mathcal{B}_t)=(t+1)/2$ implies that $\dim(\mathcal{B}_t|_{[1,t-1]})=(t+1)/2$ so gen. $\mathcal{B}_t|_{[1,t-1]}\not\subset\mathcal{B}_{t-1}$ and P^K is not monotone. We have used the fact that $\{\mathcal{B}\in\mathbb{B}_t, \dim(\mathcal{B})=d\}\Rightarrow\{\dim(\mathcal{B}|_{[1,\tau]})=d$ for all $\tau\in[d,t]\}$ which follows from $\{\mathcal{B}\in\mathbb{B}_t, \dim(\mathcal{B})=d\}\Leftrightarrow\{$there exists R of degree d such that $\mathcal{B}=\mathcal{B}_t(R)\}$.

Not shift invariant. Let $\mathcal{B}=\mathbb{R}^T$ and take $t=3$, so $\mathcal{B}|_{[2,t]}=\mathbb{R}^2$ and $\mathcal{B}|_{[1,t]}=\mathbb{R}^3$. Let $w\in\mathcal{B}$, $w|_{[1,3]}=(a,b,c)$ with $a\neq 0$, $b\neq 0$, $ac-b^2\neq 0$. Then $P^K_2(w|_{[2,3]})=\mathcal{B}_2(\sigma-(c/b))$ and $\mathcal{B}_3(\sigma^2-(c/a))\in P^K_3(w|_{[1,3]})$. Shift invariance would require that gen. $\mathcal{B}_3(\sigma^2-(c/a))\subset\mathcal{B}_3(\sigma(\sigma-(c/b)))=\mathcal{B}_3(\sigma^2-(c/b)\sigma)$, which clearly does not hold true.

Not linear. Take $T=3$, $\mathcal{B}_1:=\mathcal{B}_3(\sigma^2-1)$, $\mathcal{B}_2:=\mathcal{B}_3(\sigma+2)$. Then $\mathcal{B}_1+\mathcal{B}_2=\mathbb{R}^3$ so generically in $(w_1,w_2)\in\mathcal{B}_1\times\mathcal{B}_2$, if $\mathcal{B}\in P^K_3(w_1+w_2)$ then $\dim(\mathcal{B})=2$. Also generically $\mathcal{B}_1\in P^K_3 w_1$ and $\mathcal{B}_2=P^K_3 w_2$. Linearity would require that $\mathbb{R}^3=\mathcal{B}_1+\mathcal{B}_2\subset\mathcal{B}$ which is false.

(*iii*) Take for example $T=3$, $\mathcal{B}_0=\mathbb{R}^3$. Then gen. in $w\in\mathcal{B}_0$, if $\mathcal{B}\in P^K_T w$ then $\dim(\mathcal{B})=2$, so $\mathcal{B}_0\not\subset\mathcal{B}$.

(*iv*) We will determine $\operatorname{im}(P^K_T)$, $\mathbb{B}^w_{P^K_T}$, $\mathbb{B}^s_{P^K_T}$.

That $\operatorname{im}(P^K_T)=\mathbb{B}_T$ is seen as follows. If $\mathcal{B}=\mathbb{R}^T$ then take $w\in\mathbb{R}^T$ defined by $w(t)=0$, $t\in[1,T-1]$, $w(T)=1$, so $\operatorname{rank}(H_T(w))=T$ and by proposition 3–13(*i*) $\mathbb{R}^T=P^K_T w$. If $\mathbb{R}^T\neq\mathcal{B}\in\mathbb{B}_T$, then according to proposition II.3–33(*i*) there exists R with $d=\bar{d}(R)\leq T-1$ such that $\mathcal{B}=\mathcal{B}_T(R)$. Choose $w\in\mathcal{B}_T(R)$ with $w(\tau)=0$, $\tau\in[1,d-1]$, $w(d)=1$ and $w(\tau)$ for $\tau\in[d+1,T]$ computed by means of R. Then $\operatorname{rank}(H_T(w))=d$ and by proposition 3–13(*i*) $\mathcal{B}_T(R)\in P^K_T w$.

Next we prove $\mathbb{B}^w_{P^K_T}=\{\mathcal{B}\in\mathbb{B}_T;\ c(\mathcal{B})\leq\mathrm{ENT}(T/2)\}$. From example 1 we know already that $\mathbb{R}^T\notin\mathbb{B}^w_{P^K_T}$. Now let $\mathbb{R}^T\neq\mathcal{B}\in\mathbb{B}_T$ and let $R\neq 0$ with $c(\mathcal{B})=d=\bar{d}(R)\leq T-1$ be such that $\mathcal{B}=\mathcal{B}_T(R)$. If $d>\mathrm{ENT}(T/2)$ then from proposition 3–16, gen. in $w\in\mathcal{B}_T(R)$, if $\mathcal{B}'\in P_T w$ then $\dim(\mathcal{B})=\mathrm{ENT}(T/2)$, hence $\mathcal{B}_T(R)\notin P^K_T w$, so $\mathcal{B}_T(R)\notin\mathbb{B}^w_{P^K_T}$. If $d\leq\mathrm{ENT}(T/2)$ then gen. in $w\in\mathcal{B}_T(R)$ $\operatorname{rank}(H_T(w))=d$, so $\mathcal{B}_T(R)\in P^K_T w$ and $\mathcal{B}_T(R)\in\mathbb{B}^w_{P^K_T}$.

Finally we consider $\mathbb{B}^s_{P^K_T}$. First let T be even. Let $\mathcal{B}\in\mathbb{B}_T$ with $d:=c(\mathcal{B})\leq\mathrm{ENT}(T/2)$. Then gen. in $w\in\mathcal{B}$ the first d rows of $H_T(w)$ are linearly independent and row $d+1$ has $T-d\geq T/2\geq d$ elements and this row is linearly dependent on the foregoing ones. Using proposition 3–13(ii) and the linear independence of rows $1,\ldots,d$, this implies that gen. in $w\in\mathcal{B}$ $P_T^K w=\{\mathcal{B}\}$, i.e., the assigned model is unique. So $\mathbb{B}^s_{P_T^K}\supset\mathbb{B}^w_{P_T^K}$, hence equality holds.

Next let T be odd. If $c(\mathcal{B})\leq\text{ENT}(T/2)-1$ then by a reasoning as before we get $\mathcal{B}\in\mathbb{B}^s_{P_T^K}$. If $d:=c(\mathcal{B})=\text{ENT}(T/2)$ then in $H_T(w)$ row $d+1$ consists of $T-d=(T-1)/2<d=(T+1)/2$ elements and gen. $P_T^K w$ is not unique; hence $\mathcal{B}\notin\mathbb{B}^s_{P_T^K}$.

(v) As can be seen from the reasoning in (ii), (iii) and (iv), lacking the properties of monotonicity, linearity, truthfulness and prudence has not to do with possible nonunique assignment of models by P^K. We now show that shift invariance also cannot be obtained by choice of a selection rule S.

To get shift invariance, taking the example in (ii) with $a\neq0,b\neq0,c\neq0$, $ac-b^2\neq0$, this would require that for $\mathcal{B}\in P_3^K(a,b,c)$ $\mathcal{B}\subset\mathcal{B}_3(\sigma^2-(c/b)\sigma)$ while $\dim(\mathcal{B})=2$, so this requires $S_3(a,b,c)=\mathcal{B}_3(\sigma^2-(c/b)\sigma)$. Moreover it is required that (gen.) $S_4(d,a,b,c)\subset\mathcal{B}_4(\sigma^3-(c/b)\sigma^2)$. Now gen. if $\mathcal{B}\in P_4^K(d,a,b,c)$ then $\dim(\mathcal{B})=2$. Let $\mathcal{B}_4(\sigma^2+\alpha\sigma+\beta)\in P_4^K(d,a,b,c)$. In order that $\mathcal{B}_4(\sigma^2+\alpha\sigma+\beta)\subset\mathcal{B}_4(\sigma^3-(c/b)\sigma^2)$ according to lemma II.3–34 there has to exist a γ such that $(\sigma^2+\alpha\sigma+\beta)(\sigma+\gamma)=\sigma^3-(c/b)\sigma^2$, which implies that $(\alpha,\beta)=(0,0)$ or $(\alpha,\beta)=(-c/b,0)$. But $\mathcal{B}_4(\sigma^2)\notin P_4^K(d,a,b,c)$, as it requires $b=c=0$, and $\mathcal{B}_4(\sigma^2-(c/b)\sigma)\notin P_4^K(d,a,b,c)$, as it requires $ac-b^2=0$. So it follows that it is impossible to construct a shift invariant selection rule for P^K. ■

Proof of theorem 3–20

The proof of this theorem is quite lengthy and will be split in a number of steps. The result is proved by using a number of lemmas, some of which play a role in the proof of proposition 3–25.

Notation. First we introduce some notation. T is assumed to be fixed throughout. For $R=\Sigma_{k=0}^{T-1}a_k s^k\in\mathbb{R}[s]$ let $I(R):=\{k\in[0,T-1];\ a_k=0\}$, so $\#I(R)=T-c(R)$. Let $\mathbb{B}^*(d)$, $\mathbb{B}^*(I)$ and $\mathbb{B}^*(d,I)$ as subsets of $\mathbb{B}_T^*$ be defined as follows. $\mathbb{B}^*(d):=\{\mathcal{B}_T(R)\in\mathbb{B}_T^*;\ \bar{d}(R)=d\}$, $\mathbb{B}^*(I):=\{\mathcal{B}_T(R)\in\mathbb{B}_T^*;\ I(R)\supset I\}$, $\mathbb{B}^*(d,I):=\mathbb{B}^*(d)\cap\mathbb{B}^*(I)$. Moreover define $W(d)$, $W(I)$ and $W(d,I)$ as subsets of $\mathbb{R}^T$ by $W(d):=\bigcup\{\mathcal{B}_T(R);\ \mathcal{B}_T(R)\in\mathbb{B}^*(d)\}=\{w\in\mathbb{R}^T;\ \exists\mathcal{B}_T(R)\in\mathbb{B}^*(d)\text{ with }w\in\mathcal{B}_T(R)\}$, $W(I):=\bigcup\{\mathcal{B}_T(R);\ \mathcal{B}_T(R)\in\mathbb{B}^*(I)\}$, and $W(d,I):=\bigcup\{\mathcal{B}_T(R);\ \mathcal{B}_T(R)\in\mathbb{B}^*(d,I)\}$. Without loss of generality we restrict attention to (d,I) with $d+(T-\#(I))\leq T$, i.e., $d\leq\#(I)$. □

Let $\mathcal{B}_0\in\mathcal{B}_T$ be fixed, $w\in\mathcal{B}_0$ and $H_T(w)$ its incomplete Hankel array. We now first give an outline of the proof of theorem 3–20 by means of four lemmas and then will give the proof of these lemmas.

Lemma 3-20-1 For every (d,I) either (i) $w\notin W(d,I)$ generically in $w\in\mathcal{B}_0$, or (ii) $\mathcal{B}_0\subset W(d,I)$.

Lemma 3-20-2 $\{\mathcal{B}_0\subset W(d,I)\}\Rightarrow\{\ \exists\mathcal{B}_T(R(d,I))\in\mathbb{B}^*(d,I)$ such that $\mathcal{B}_0\subset\mathcal{B}_T(R(d,I))\}$.

This lemma states that if for every $w\in\mathcal{B}_0$ there exists a model $\mathcal{B}_w(R)\in\mathbb{B}^*(d,I)$ such that $w\in\mathcal{B}_w(R)$, then there exists such a model independent from $w\in\mathcal{B}_0$.

For $\mathcal{B}_0$ define $d_0\in[0,T]$ as follows. If $\mathcal{B}_0\not\subset W(d,I)$ for all I and $d\in[0,T-1]$ then $d_0:=T$, else $d_0:=\min\{d\in[0,T-1];\ \exists I$ such that $\mathcal{B}_0\subset W(d,I)\}$.

If $d_0=T$, then according to lemma 3–20–1 (and using the fact that the number of indexsets I is finite) generically in $w\in\mathcal{B}_0$ $w\notin W(d)$ for all $d\in[0,T-1]$. This means that gen. row $d+1$ of H_T is not linearly dependent on less than $T-d$ foregoing rows of H_T, so $P_T^0w=\mathbb{R}^T$ gen. on $\mathcal{B}_0$, and theorem 3–20 follows as obviously in this case there is no $\mathcal{B}_T(R)\in\mathbb{B}_T^*$ with $\mathcal{B}_0\subset\mathcal{B}_T(R)$ and $c(\mathcal{B}_T(R))<T$.

For $d_0\in[0,T-1]$ let J_0 be defined by $J_0:=\{I;\mathcal{B}_0\subset W(d_0,I)\}$. Because by lemma 3–20–1 gen. on $\mathcal{B}_0$ $w\notin W(d)$ for $d<d_0$ and by lemma 3–20–2 $\mathcal{B}_0\subset\mathcal{B}_T(R(d_0,I))$ for $I\in J_0$, for some $R(d_0,I)$, it follows from the definition of P_T^0 that gen. on $\mathcal{B}_0$ $\{\mathcal{B}_T(R(d_0,I));\ I\in J_0\}\subset P_T^0w\subset\mathbb{B}^*(d_0)$. Indeed, on $\mathcal{B}_0$ gen. no remarkable laws of degree $d<d_0$ are satisfied while remarkable laws of degree d_0 always exist. Because of lemma 3–20–1 we even have that gen. on $\mathcal{B}_0$ $\{\mathcal{B}_T(R(d_0,I));\ I\in J_0\}\subset P_T^0w\subset\bigcup\{\mathbb{B}^*(d_0,I);\ I\in J_0\}$.

Lemma 3-20-3 For $I\in J_0$ gen. on $\mathcal{B}_0$ $P_T^0w\cap\mathbb{B}^*(d_0,I)$ is a singleton, i.e., $\mathcal{B}_T(R(d_0,I))$.

The generic way in which P_T^0 assigns models on the basis of data from $\mathcal{B}_0$ is described in lemma 3–20–4, which is a direct consequence of lemma 3–20–3 and the preceding discussion.

Lemma 3-20-4 Generically for $w\in\mathcal{B}_0$ $P_T^0 w=\{\mathcal{B}_T(R(d_0,I));\ I\in J_0\}$.

Now from lemma 3–20–2 and lemma 3–20–4 it follows that gen. on $\mathcal{B}_0$ if $\mathcal{B}\in P_T^0 w$ then $\mathcal{B}_0\subset\mathcal{B}$ and $c(\mathcal{B})=d_0$. To conclude the proof of theorem 3–20 note that by definition of d_0 if $\mathcal{B}\in\mathbb{B}_T^*$ with $c(\mathcal{B})<d_0$, then $\mathcal{B}_0\not\subset\mathcal{B}$.. On the other hand, if $\mathcal{B}\in\mathbb{B}_T^*$, $c(\mathcal{B})=d_0$, $\mathcal{B}_0\subset\mathcal{B}$, then gen. $\mathcal{B}\in P_T^0 w$. This proves theorem 3–20.

Finally we prove the foregoing lemmas.

Proof of lemma 3-20-1 and lemma 3-20-2

Let (d,I) be given, $c:=T-\#(I)$, and let $c+d\leq T$. Suppose that $w\notin W(d,I)$ is not gen. true on $\mathcal{B}_0$. We will show that then there exists $R\in\mathbb{R}[s]$ with $\mathcal{B}_T(R)\in\mathbb{B}^*(d,I)$ and $\mathcal{B}_0\subset\mathcal{B}_T(R)$, which proves the desired results. The proof has the following structure. First we introduce some notation. Next in step (i) we prove the result under two conditions, (C_1) and (C_2), and using an auxiliary lemma, (L). In step (ii) we prove (L) under conditions (C_1) and (C_2). Finally in step (iii) we consider the case where condition (C_1) or (C_2) is not satisfied.

Notation. Let $[0,T-1]\backslash I=\{i_1,i_2,\ldots,i_{c-1},d\}$ with $0\leq i_1<i_2<\ldots<i_{c-1}<d$ and $c+d\leq T$. For $w\in\mathcal{B}_0$ define $H_I(w)\in\mathbb{R}^{c\times(T-d)}$ by

$$H_I(w):=\begin{bmatrix} w(i_1+1) & w(i_1+2) & \cdots & w(i_1+T-d) \\ w(i_2+1) & w(i_2+2) & \cdots & w(i_2+T-d) \\ \vdots & \vdots & & \vdots \\ w(i_{c-1}+1) & w(i_{c-1}+2) & \cdots & w(i_{c-1}+T-d) \\ w(d+1) & w(d+2) & \cdots & w(T) \end{bmatrix}.$$

Let $\mathcal{B}_0=\mathcal{B}_T(R_0)$ with $R_0=r_0+r_1s+\ldots+r_{n-1}s^{n-1}+s^n$, where $n:=\dim(\mathcal{B}_0)$. Note that $w\in W(d,I)$ if and only if the last row of $H_I(w)$ is linearly dependent on the foregoing ones. As it is given that this is not gen. false on $\mathcal{B}_0$ it follows that $\mathcal{B}_0\neq\mathbb{R}^T$ and hence $n\leq T-1$. Let $e_1:=(1,0,\ldots,0)\in\mathbb{R}^{1\times n}$ and define $A\in\mathbb{R}^{n\times n}$ by the so-called companion matrix of R_0, i.e.,

$$A := \begin{bmatrix} 0 & 1 & 0 & 0 & \dots & 0 \\ 0 & 0 & 1 & 0 & \dots & 0 \\ \vdots & \vdots & \vdots & & & \vdots \\ 0 & 0 & 0 & \dots & 1 & 0 \\ 0 & 0 & 0 & \dots & 0 & 1 \\ -r_0 & -r_1 & -r_2 & \dots & -r_{n-2} & -r_{n-1} \end{bmatrix} = \begin{bmatrix} 0 & I_{n-1} \\ -r_0 & (-r_1 \dots -r_{n-1}) \end{bmatrix}.$$

For $w \in \mathcal{B}_0$ let $x := \text{col}(w(1),\dots,w(n)) \in \mathbb{R}^n$. Note that $\{w \in \mathcal{B}_0\} \Leftrightarrow \{\exists x \in \mathbb{R}^n$ such that $w(t) = e_1 A^{t-1} x$ for all $t \in [1,T]\}$, so (A, e_1) is a "minimal realization" of $\mathcal{B}_0$, cf. definition II.3–20 and proposition II.3–22(*ii*). Finally for $x \in \mathbb{R}^n$ define $x(t) := e_1 A^t x$, $t \in \mathbb{Z}_+$, and define $H(x,k) \in \mathbb{R}^{c \times k}$ by

$$H(x,k) := \begin{bmatrix} x(i_1) & x(i_1+1) & \dots & x(i_1+k-1) \\ x(i_2) & x(i_2+1) & \dots & x(i_2+k-1) \\ \vdots & \vdots & & \vdots \\ x(i_{c-1}) & x(i_{c-1}+1) & \dots & x(i_{c-1}+k-1) \\ x(d) & x(d+1) & \dots & x(d+k-1) \end{bmatrix}.$$

Let $H_I(x) := H(x, T-d)$ and $H_I^e(x) := H(x, 2n-1)$. Note that $H_I(x) = H_I(w)$ for $w(t) := e_1 A^{t-1} x$, $t \in [1,T]$. □

(*i*) As a first step we prove the result under conditions (C_1) and (C_2) and using the auxiliary lemma (L).

(C_1) $\mathcal{B}_0 = \mathcal{B}_T(R_0)$, $R_0 = r_0 + r_1 s + \dots + r_{n-1} s^{n-1} + s^n$, where $r_0 \neq 0$;

(C_2) gen. on $\mathcal{B}_0$ the matrix consisting of the first $c-1$ rows of $H_I(w)$ has full row rank $c-1$;

(L) if $w \notin W(d,I)$ is not gen. true on $\mathcal{B}_0$, then under conditions (C_1) and (C_2) there holds that for generic $x \in \mathbb{R}^n$ the last row of $H_I^e(x)$ is linearly dependent on the foregoing ones.

So suppose (L) holds true. Note that for $x_0 := (0,\dots,0,1)^T$ there holds $\det([x_0 \, Ax_0 \dots A^{n-1} x_0]) \neq 0$, hence for generic $x \in \mathbb{R}^n$ $\det([x \, Ax \dots A^{n-1} x]) \neq 0$. This implies that there exists $x \in \mathbb{R}^n$ such that this condition is satisfied and moreover the last row of $H_I^e(x)$ is linearly dependent on the foregoing ones, i.e., there exist $\alpha_k \in \mathbb{R}$, $k \in [1, c-1]$, such that $e_1 A^{d+t} x = \Sigma_{k=1}^{c-1} \alpha_k e_1 A^{i_k+t} x$ for all $t \in [0, 2n-2]$. Then $\text{col}(e_1, e_1 A, \dots, e_1 A^{n-1}) A^d [x \, Ax \dots A^{n-1} x] = \text{col}(e_1, e_1 A, \dots, e_1 A^{n-1}) (\Sigma_{k=1}^{c-1} \alpha_k A^{i_k}) [x \, Ax \dots A^{n-1} x]$, and as both $\text{col}(e_1, e_1 A, \dots, e_1 A^{n-1})$ and $[x \, Ax \dots A^{n-1} x]$ are invertible it follows that $A^d = \Sigma_{k=1}^{c-1} \alpha_k A^{i_k}$. Define $R := s^d - \Sigma_{k=1}^{c-1} \alpha_k s^{i_k}$, then clearly $\mathcal{B}_T(R) \in \mathbb{B}^*(d,I)$ and $\mathcal{B}_0 \subset \mathcal{B}_T(R)$, as for $w \in \mathcal{B}_0$

and $x:=\text{col}(w(1),...,w(n))$ there holds $[R(\sigma)w](t)= e_1A^{t-1}(A^d-\Sigma_{k=1}^{c-1}\alpha_kA^{i_k})x=0$, $t\in[1,T-d]$. This proves the desired results.

(*ii*) As a second step we prove (L) under conditions (C_1) and (C_2). If $w\notin W(d,I)$ is not gen. true on $\mathcal{B}_0$, then $\text{rank}(H_I(x))=c$ is not gen. true on $\mathbb{R}^n$, and as this is a polynomial condition in x we conclude that $\text{rank}(H_I(x))\le c-1$ for all $x\in\mathbb{R}^n$. We state that gen. the first c-1 columns of $H_I(x)$ are linearly independent. Suppose this would not hold true. Then those first $c-1$ columns are always linearly dependent and there is a column $k\le c-1$ which gen. is linearly dependent on the foregoing ones, say for $x\in V\supset\{\mathbb{R}^n\backslash p^{-1}(0)\}$ for some polynomial $p:\mathbb{R}^n\to\mathbb{R}$ with $p\neq0$. Let $p_i:=pA^i$, $i\in[0,T-d-k]$, and $\tilde{p}:=\Pi_{i=0}^{T-d-k}p_i$, then $\tilde{p}\neq0$ as A is invertible under condition (C_1), and $\tilde{V}:=\{x;\ A^ix\in V$ for all $i\in[0,T-d-k]\}\supset\mathbb{R}^n\backslash\tilde{p}^{-1}(0)$ is generic in $\mathbb{R}^n$. For $x\in\tilde{V}$ there holds $Ax\in V$ and from the structure of $H_I(x)$ it follows that for such x column k+1 of $H_I(x)$ is linearly dependent on columns 2,...,k, hence also on the first k–1 columns. Analogously it follows by induction that for $x\in\tilde{V}$ all columns $k+i$ of $H_I(x)$ are linearly dependent on the first k–1 columns, $i\in[0,T-d-k]$, and hence gen. $\text{rank}(H_I(x))\le k-1\le c-2$. This contradicts condition (C_2).

So gen. the first c–1 columns of $H_I(x)$ are linearly independent and as $\text{rank}(H_I(x))\le c-1$ for all $x\in\mathbb{R}^n$ it follows that gen. column c of $H_I(x)$ is linearly dependent on the foregoing ones, say on the generic set $V'\subset\mathbb{R}^n$. Then $\tilde{V}':=\{x;\ A^ix\in V'$ for all $0\le i\le\max\{2n-1-c,\ T-d-c\}\}$ is also generic, due to (C_1), and for $x\in\tilde{V}'$ both $\text{rank}(H_I(x))\le c-1$ and $\text{rank}(H_I^e(x))\le c-1$. If $T-d\ge2n-1$, then (C_2) implies that gen. on $\mathbb{R}^n$ the last row of $H_I(w)=H_I(x)$ is linearly dependent on the foregoing ones, and hence the same holds true for $H_I^e(x)$ as it contains less columns than $H_I(x)$. If on the other hand $T-d<2n-1$, then (C_2) implies that gen. the first c–1 rows of $H_I^e(x)$ have rank c–1, and hence again gen. the last row of $H_I^e(x)$ is linearly dependent on the foregoing ones. This proves (L).

(*iii*) Finally we consider the case where condition (C_1) or (C_2) is not satisfied. This step is split in four parts.

(*iii*–1) First suppose that condition (C_2) is not satisfied. Then there exists a $k\le c-1$ such that the matrix consisting of the first c–1 rows of $H_I(w)$ always on $\mathcal{B}_0$ has rank at most k and such that there is a $w_0\in\mathcal{B}_0$ with $\text{rank}(H_I(w_0))=k$, say for w_0 the rows in $J':=\{i_1',...,i_k'\}\subset\{i_1,...,i_{c-1}\}=:J$ are linearly independent, where $0 \le i_1' < \ldots < i_k' < d$. Then gen. on $\mathcal{B}_0$ the rows in J' are linearly independent and the rows in $J \backslash J'$ are linearly dependent on those in J'. As it is supposed that the last row of $H_I(w)$ is not gen. linearly independent from the rows in J it also is not gen. linearly independent from the rows in J'. Now condition (C_2) is satisfied for $c' := k+1$ and $I' := [0, T-1] \backslash (J' \cup \{d\})$.

If condition (C_1) is satisfied then the results of (i) and (ii) imply that there is an R' such that $\mathcal{B}_0 \subset \mathcal{B}_T(R') \in \mathbb{B}^*(d, I') \subset \mathbb{B}^*(d, I)$ as $I' \supset I$, and the desired results follow.

Finally let condition (C_2) be satisfied for c' and I' as defined before and suppose that condition (C_1) is not satisfied. Let $n' := \min\{k; r_k \ne 0\}$. We will consider three cases, i.e., in $(iii\text{-}2)$ that $n' > d$, in $(iii\text{-}3)$ that $n' \le i_1'$, and in $(iii\text{-}4)$ that $i_1' < n' \le d$.

$(iii\text{-}2)$ The case $n' > d$ cannot arise. It would imply that on $\mathcal{B}_0$ the values of $w(t)$, $t \in [1, d+1]$, could be chosen freely. Now for $j \in [1, k]$ and $l \in [1, k-j+1]$ there holds $i_j' + l < i_j' + k - j + 2 \le i_{j+1}' + k - j + 1 \le \ldots \le i_k' + 2 \le d+1$, and hence there is a $w_0 \in \mathcal{B}_0$ with $w_0(i_j' + l) := 0$, $j \in [1, k]$, $l \in [1, k-j+1]$, and $w_0(d+1) = w_0(i_k' + 2) = w_0(i_{k-1}' + 3) = \ldots = w_0(i_1' + k + 1) := 1$. From the structure of $H_I'(w_0)$ it follows that $\text{rank}(H_I'(w_0)) = c'$ and hence that gen. on $\mathcal{B}_0$ $\text{rank}(H_I'(w)) = c'$. This contradicts that the last row in $H_I'(w)$ is not gen. linearly independent from the foregoing ones.

$(iii\text{-}3)$ Next suppose that $n' \le i_1'$. Then consider (R_0'', I'', T'') defined by $R_0'' := s^{-n'} R_0$, $T'' := T - n'$, and $I'' := [0, T''-1] \backslash \{i_1' - n', i_2' - n', \ldots, i_k' - n', d - n'\}$. Note that R_0'' satisfies (C_1) and that $H_I''(w)$ satisfies (C_2) on $\mathcal{B}_T''(R_0'')$, as $H_I'(w)$ satisfies this condition on $\mathcal{B}_0$. As moreover $\bar{d}(R_0'') + c(R_0'') = d - n' + (k+1) \le d - n' + c \le T - n' = T''$ it follows from steps (i) and (ii) that there is an R'' such that $\mathcal{B}_T''(R_0'') \subset \mathcal{B}_T''(R'') \in \mathbb{B}_T^*{}''(d - n', I'')$ and hence $\mathcal{B}_0 = \mathcal{B}_T(R_0) \subset \mathcal{B}_T(\sigma^{n'} R'') \in \mathbb{B}^*(d, I') \subset \mathbb{B}^*(d, I)$, as $c(\sigma^{n'} R'') + \bar{d}(\sigma^{n'} R'') \le k+1+d \le d+c \le T$. Hence the desired results follow.

$(iii\text{-}4)$ Finally suppose that $i_1' < n' \le d$. Let $l \in [2, k+1]$ be such that $i_{l-1}' < n' \le i_l'$, where $i_{k+1}' := d$. Let $H(w) \in \mathbb{R}^{(k+1) \times (k+1)}$ consist of the first $k+1$ columns of $H_I'(w)$ and let $\tilde{H}(w) \in \mathbb{R}^{(k-l+2) \times (k-l+2)}$ be defined by

$$\tilde{H}(w) := \begin{bmatrix} w(i'_l+l) & \cdots & w(i'_l+k+1) \\ w(i'_{l+1}+l) & \cdots & w(i'_{l+1}+k+1) \\ \vdots & & \vdots \\ w(i'_k+l) & \cdots & w(i'_k+k+1) \\ w(d+l) & \cdots & w(d+k+1) \end{bmatrix}.$$

Note that the assumption that $w \notin W(d,I')$ is not gen. satisfied on $\mathcal{B}_0$ implies that $\det(H(w)) \equiv 0$ on $\mathcal{B}_0$. Moreover on $\mathcal{B}_0$ there holds that, for given values of $w(t)$ for $t \in [n'+1,T]$, the values of $w(t)$ for $t \in [1,n']$ can be chosen arbitrarily. Let $w|_{[n'+1,T]} \in \mathcal{B}_0|_{[n'+1,T]}$ be arbitrary and fixed, choose $w(t):=0$ for $t \in [1,n'] \setminus \{i'_{l-j}+j;\ j \in [1,l-1]\}$, and consider $\det(H(w))$ as a polynomial in the variables $\{w(i'_{l-j}+j);\ j \in [1,l-1]\}$. Note that indeed for $j \in [1,l-1]$ $i'_{l-j}+j \leq i'_{l-1}+1 \leq n'$. It is a simple matter of explicitly writing out $H(w)$ to prove that $\det(H(w))$ contains the term $\Pi_{j=1}^{l-1} w(i'_{l-j}+j)$ with coefficient $\det(\tilde{H}(w))$. As $\det(H(w)) \equiv 0$ it follows that also $\det(\tilde{H}(w)) \equiv 0$ on $\mathcal{B}_0$.

Now for $w \in \mathcal{B}_0$ define $\tilde{w} := w|_{[i'+l,d+k+1]} \in \mathbb{R}^{\tilde{T}}$, where $\tilde{T} := d+k-l-i'_l+2$. Further let $\tilde{\mathcal{B}}_0 := \mathcal{B}_0|_{[i'_l+l,d+k+1]} = \mathcal{B}_0|_{[n'+1,n'+\tilde{T}]}$, $\tilde{R}_0 := s^{-n'} R_0 = \tilde{r}_0 + \tilde{r}_1 s + \ldots + \tilde{r}_{\tilde{n}-1} s^{\tilde{n}-1} + s^{\tilde{n}}$ with $\tilde{r}_0 \neq 0$ and $\tilde{n} := n-n'$. Let $\tilde{c} := k-l+2$ and $(\tilde{\imath}_1, \tilde{\imath}_2, \ldots, \tilde{\imath}_{\tilde{c}-1}, \tilde{d}) := (0, i'_{l+1}-i'_l, \ldots, i'_k-i'_l, d-i'_l)$ and let $\tilde{I} := [0,\tilde{T}-1] \setminus \{\tilde{\imath}_1, \ldots, \tilde{\imath}_{\tilde{c}-1}, \tilde{d}\}$. Define $\tilde{A}$ and $\tilde{x}$ in terms of $\tilde{R}_0$ and $\tilde{w}$ analogous to the definition of A and x in terms of R_0 and w respectively as indicated in the notation in the beginning of the proof.

Note that $\tilde{H}(w) = H_{\tilde{I}}(\tilde{w})$ for $\tilde{\mathcal{B}}_0$, that hence $\det(H_{\tilde{I}}(\tilde{w})) \equiv 0$ on $\tilde{\mathcal{B}}_0$, that $\tilde{c}+\tilde{d}=\tilde{T}$, and that $\tilde{R}_0$ satisfies condition (C_1). As $\det(H_{\tilde{I}}(\tilde{w})) \equiv 0$ on $\tilde{\mathcal{B}}_0$ it follows that there is a row in $H_{\tilde{I}}(\tilde{w})$ which gen. is linearly dependent on the foregoing ones. This shows the existence of a $\tilde{d}' \in \{\tilde{\imath}_1, \ldots, \tilde{\imath}_{\tilde{c}-1}, \tilde{d}\}$ such that for gen. $\tilde{w} \in \tilde{\mathcal{B}}_0$ $\tilde{w} \in \tilde{W}(\tilde{d}', \tilde{I}) \subset \mathbb{R}^{\tilde{T}}$. Using arguments similar to those of steps $(iii\text{–}1)$ and (i) it follows that there exist $\{\tilde{\imath}'_1, \ldots, \tilde{\imath}'_{\tilde{k}}, \tilde{d}'\} \subset \{\tilde{\imath}_1, \ldots, \tilde{\imath}_{\tilde{c}-1}, \tilde{d}\}$ and $\tilde{\alpha}_j \in \mathbb{R}$, $j \in [1,\tilde{k}]$, such that $\tilde{A}^{\tilde{d}'} = \Sigma_{j=1}^{\tilde{k}} \tilde{\alpha}_j s^{\tilde{\imath}'_j}$. Let $\tilde{R} := s^{\tilde{d}'} - \Sigma_{j=1}^{\tilde{k}} \tilde{\alpha}_j s^{\tilde{\imath}'_j}$, then $\mathcal{B}_0|_{[n'+1,n'+\tilde{T}]} = \tilde{\mathcal{B}}_0 \subset \mathcal{B}_{\tilde{T}}(\tilde{R}) \in \mathbb{B}^*_{\tilde{T}}$ as $c(\tilde{R}) + \bar{d}(\tilde{R}) \leq \tilde{c} + \tilde{d} = \tilde{T}$. From this and shift invariance of $\mathcal{B}_0$ it follows that $\mathcal{B}_0 \subset \mathcal{B}_T(s^{n'}\tilde{R})$. Let $R := s^{i'_l}\tilde{R}$, then $\mathcal{B}_0 \subset \mathcal{B}_T(R)$, as $i'_l \geq n'$, and $[0,T-1] \setminus I(R) \subset \{\tilde{\imath}_1+i'_l, \ldots, \tilde{\imath}_{\tilde{c}-1}+i'_l, \tilde{d}+i'_l\} = \{i'_l, i'_{l+1}, \ldots, i'_k, d\}$. It was supposed that condition (C_2) is satisfied for c' and I' of step $(iii\text{–}1)$. Hence necessarily $\tilde{d}'=\tilde{d}$ and $\bar{d}(r)=d$. Hence $\mathcal{B}_0 \subset \mathcal{B}_T(R) \in \mathbb{B}^*(d,I') \subset \mathbb{B}^*(d,I)$, which proves the desired results.

This concludes the proof of lemma 3–20–1 and lemma 3–20–2. ■

To prove lemma 3–20–3 we make use of a result stated in lemma 3–20–5 which also plays a role in the proof of proposition 3–25. Let $0 \leq i_1 < i_2 < \ldots < i_{c-1} < d \leq T-c$ and

$$M(w) := \begin{bmatrix} w(i_1+1) & w(i_1+2) & \ldots & w(i_1+T-d) \\ w(i_2+1) & w(i_2+2) & \ldots & w(i_2+T-d) \\ \vdots & \vdots & & \vdots \\ w(i_{c-1}+1) & w(i_{c-1}+2) & \ldots & w(i_{c-1}+T-d) \end{bmatrix}.$$

Lemma 3–20–5 If rank $(M(w)) \leq c-2$ everywhere on $\mathcal{B}_0$, then there exists $R \neq 0$, $\bar{d}(R) \leq i_{c-1}$, $I(R) \supset [0,T-1] \backslash \{i_1,\ldots,i_{c-1}\}$, such that $\mathcal{B}_0 \subset \mathcal{B}_T(R) \in \mathbb{B}_T^*$.

Proof of lemma 3–20–5

There is at least one row of M which is not gen. linearly independent from the foregoing ones, say row k_0. Exactly analogous to the proof of lemma 3–20–1 and lemma 3–20–2 it follows that this row then always is linearly dependent on the foregoing ones and that there exist a_k such that on $\mathcal{B}_0$ $w(i_{k_0}+t) = \Sigma_{k=1}^{k_0-1} a_k w(i_k+t)$, $t \in [1,T-d]$. By shift invariance of $\mathcal{B}_0$ this then also holds true on $[1,T-i_{k_0}]$. Define $R := \sigma^{i_{k_0}} - \Sigma_{k=1}^{k_0-1} a_k \sigma^{i_k}$, then $\mathcal{B}_0 \subset \mathcal{B}_T(R) \in \mathbb{B}_T^*$, while $\bar{d}(R) = i_{k_0} \leq i_{c-1}$ and $I(R) \supset [0,T-1] \backslash \{i_1,\ldots,i_{c-1}\}$. ■

Proof of lemma 3–20–3

Let $I \in J_0$, and define M as before, with $\{i_1,\ldots,i_{c-1},d_0\} := [0,T-1] \backslash I$.

We state that gen. on $\mathcal{B}_0$ rank$(M(w)) = c-1$. For suppose this is not true, then $\det(MM^T) \neq 0$ is not generic, so $\det(MM^T) \equiv 0$ on $\mathcal{B}_0$ and rank$(M(w)) \leq c-2$ on $\mathcal{B}_0$. By lemma 3–20–5 this would imply that there exists $R \neq 0$ such that $\mathcal{B}_0 \subset \mathcal{B}_T(R) \in \mathbb{B}^*(d',I')$, where $d' \leq i_{c-1} < d_0$ and $I' \supset [0,T-1] \backslash \{i,\ldots,i_{c-1}\}$. Hence $\mathcal{B}_0 \subset \mathcal{B}_T(R) \subset W(d',I')$ and $d' < d_0$, which contradicts the definition of d_0.

Now suppose that $\mathcal{B}_T(R_j) \in P_T^0 w \cap \mathbb{B}^*(d_0,I)$, $j=1,2$. Let $R_j(\sigma) = \sigma^{d_0} + \Sigma_{k=1}^{c-1} a_k^j \sigma^{i_k}$, $a^j := (a_1^j,\ldots,a_{c-1}^j)$, $j=1,2$. Using the notation of lemma 3–20–1, this means that $(a^1,1)H_I(w) = (a^2,1)H_I(w) = 0$, so $(a^1-a^2)M(w) = 0$. As gen. on $\mathcal{B}_0$ rank$(M(w)) = c-1$, we get gen. on $\mathcal{B}_0$ $a^1=a^2$; hence $R_1=R_2$, i.e., gen. on $\mathcal{B}_0$ $P_T^0 w \cap \mathbb{B}^*(d_0,I)$ contains at most one model. From lemma 3–20–2 and the discussion following that lemma we know that gen. on $\mathcal{B}_0$ $\mathcal{B}_T(R(d_0,I)) \in P_T^0 w \cap \mathbb{B}^*(d_0,I)$. So gen. on $\mathcal{B}_0$ $P_T^0 w \cap \mathbb{B}^*(d_0,I)$ consists of a singleton, i.e., $\mathcal{B}_T(R(d_0,I))$. ■

This concludes the proof of theorem 3–20. ■

Proof of corollary 3-21

From theorem 3–20 we immediately conclude that if $\mathcal{B}_0 \in \mathbb{B}_T^*$ then gen. on $\mathcal{B}_0$ $\mathcal{B}_0 \in P_T^0 w$. To prove the corollary, due to theorem 3–20 it suffices to show that $\{\mathcal{B}_0 \subset \mathcal{B} \in \mathbb{B}_T^*, c(\mathcal{B})=c(\mathcal{B}_0)\} \Rightarrow \{\mathcal{B}=\mathcal{B}_0\}$. This easily follows from lemma II.3–34 and the fact that for $R \neq 0$ with $\bar{d}(R) \leq T-1$ there holds $c(\mathcal{B}_T(R))=\bar{d}(R)$. ■

Proof of theorem 3-22

(*i*) Obvious from the definition of P^0.

(*ii*) Obvious from theorem 3–20.

(*iii*) Obviously $\mathbb{B}^s_{P_T^0} \subset \mathbb{B}^w_{P_T^0} \subset \mathrm{im}((P_T^0) \subset \mathbb{B}_T^*$, so it suffices to show that $\mathbb{B}_T^* \subset \mathbb{B}^s_{P_T^0}$. This is immediate from corollary 3–21.

(*iv*) Consider P^0 with the action of P_t^0 restricted to models in the set $\mathbb{B}_t^*$.

For monotonicity and shift invariance, consider the inclusion conditions on t which involve modelling $w|_{[1,t-1]}$ or $w|_{[2,t]}$, and $w|_{[1,t]}$. Assume $\mathcal{B}=\mathcal{B}_T(R) \in \mathbb{B}_T$ with $c(r)+\bar{d}(R) \leq t-1$.

For the monotonicity condition, observe that $w|_{[1,t-1]} \in \mathcal{B}_{t-1}(R) \in \mathbb{B}_{t-1}^*$ and $w|_{[1,t]} \in \mathcal{B}_t(R) \in \mathbb{B}_t^*$. From corollary 3–21 it follows that gen. in $w \in \mathcal{B}$ $P_{t-1}^0(w|_{[1,t-1]})=\mathcal{B}_{t-1}(R)$ and $P_t^0(w|_{[1,t]})=\mathcal{B}_t(R)$, and the condition $\mathcal{B}_t(R)|_{[1,t-1]} \subset \mathcal{B}_{t-1}(R)$ is trivial. Note that in fact corollary 3–21 only gives that for example $P_{t-1}^0(w|_{[1,t-1]})=\mathcal{B}_{t-1}(R)$ gen. in $w|_{[1,t-1]} \in \mathcal{B}_{t-1}(R)$. That this also holds true gen. in $w \in \mathcal{B}$ can be derived from the fact that $\bar{d}(R) \leq t-1$ which implies that there is a linear bijection $w|_{[1,t-1]} \rightarrow w$ from $\mathcal{B}_{t-1}(R)$ to $\mathcal{B}$.

For the shift invariance condition, consider two cases for $R=\Sigma_{k=0}^{\bar{d}(R)} a_k \sigma^k$. If $a_0 \neq 0$, then according to propositions II.3–30(*ii*) and II.3–33(*ii*) $w|_{[2,t]} \in \mathcal{B}|_{[2,t]}=\mathcal{B}|_{[1,t-1]}$, while if $a_0=0$, then $w|_{[2,t]} \in \mathcal{B}|_{[2,t]}=\mathcal{B}_{t-1}(\sigma^{-1}R)$. Generically in $w \in \mathcal{B}$ $P_{t-1}^0(w|_{[2,t]})=\mathcal{B}_{t-1}(R)$ in the first case, $P_{t-1}^0(w|_{[2,t]})=\mathcal{B}_{t-1}(\sigma^{-1}R)$ in the second case. The shift invariance condition is trivally satisfied in both cases.

Concerning linearity, let $\mathcal{B}_i \in \mathbb{B}_t^*$, $i=1,2$, then gen. in $(w_1,w_2)\in\mathcal{B}_1\times\mathcal{B}_2$ $P_t^0 w_i=\mathcal{B}_i$, $i=1,2$, while due to theorem 3–20 gen. for $\mathcal{B}\in P_t^0(w_1+w_2)$ $\mathcal{B}_1+\mathcal{B}_2\subset\mathcal{B}$, which proves linearity. Note that theorem 3–20 in fact gives $\mathcal{B}_1+\mathcal{B}_2\subset\mathcal{B}\in P_t^0 w$ gen. in $w\in\mathcal{B}_1+\mathcal{B}_2$, but one easily proves that this then also gen. holds true in $(w_1,w_2)\in\mathcal{B}_1\times\mathcal{B}_2$ with $w:=w_1+w_2$.

Finally we give an example which shows that P^0 is not monotone, not shift invariant, and not linear.

Example. Let $T=20$, $R:=\sigma^{10}+\sigma^9+2\sigma^8+\sigma^7+\sigma^6+2\sigma^5+\sigma^4+\sigma^3+2\sigma^2+\sigma+1$, so $c(R)+\bar{d}(R)=21$, and consider $\mathcal{B}:=\mathcal{B}_{20}(R)$. Further define $R_1:=(\sigma-1)R=\sigma^{11}+\sigma^9-\sigma^8+\sigma^6-\sigma^5+\sigma^3-\sigma^2-1$ with $c(R_1)+\bar{d}(R_1)=19$, and $R_2:=(\sigma-\frac{1}{2})R=\sigma^{11}+\frac{1}{2}\sigma^{10}+\frac{3}{2}\sigma^9+\frac{1}{2}\sigma^7+\frac{3}{2}\sigma^6+\frac{1}{2}\sigma^4+\frac{3}{2}\sigma^3+\frac{1}{2}\sigma-\frac{1}{2}$ with $c(R_2)+\bar{d}(R_2)=20$. Then gen. in $w=\mathcal{B}$ $P_{19}^0(w|_{[1,19]})=P_{19}^0(w|_{[2,20]})=\mathcal{B}_{19}(R_1)$ while $P_{20}^0 w=\{\mathcal{B}_{20}(R_1),\ \mathcal{B}_{20}(R_2)\}$. So P^0 is not monotone as $\mathcal{B}_{20}(R_2)|_{[1,19]}\not\subset\mathcal{B}_{19}(R_1)$, and P^0 is not shift invariant as $\mathcal{B}_{20}(R_2)\not\subset\mathcal{B}_{20}(\sigma.R_1)$. P^0 also is not linear, as for $\mathcal{B}_1=\mathcal{B}_2:=\mathcal{B}$ and for generic $(w_1,w_2)\in\mathcal{B}\times\mathcal{B}$ there holds $P_{20}^0 w_1=P_{20}^0 w_2=P_{20}^0(w_1+w_2)=\{\mathcal{B}_{20}(R_1),\mathcal{B}_{20}(R_2)\}$, while e.g. $\mathcal{B}_{20}(R_1)+\mathcal{B}_{20}(R_1)=\mathcal{B}_{20}(R_1)\not\subset\mathcal{B}_{20}(R_2)$ as would be required for linearity. □

This concludes the proof of theorem 3–22. ■

Proof of proposition 3–25

For $\mathcal{B}_0\in\mathbb{B}_T$ we have to prove that gen. in $w\in\mathcal{B}_0$ the following holds: $\{R\in L(w):=\{R'\neq 0;\ \bar{d}(R')\leq T-1 \text{ and } w\in\mathcal{B}_T(R')\in\mathbb{B}_T^*\}\}\Rightarrow\{\mathcal{B}_0\subset\mathcal{B}_T(R)\}$.

Notation. We will use some of the lemmas and the notation introduced in the proof of theorem 3–20. Further we define $K_0:=\{(d,I);\ \mathcal{B}_0\subset W(d,I)\}$ and $L(w;d,I):=\{R\in L(w);\ I(R)\supset I,\bar{d}(R)=d\}$. Again we can restrict attention to the case $d+(T-\#(I))\leq T$, i.e., $d\leq\#(I)$. □

Remark. According to lemma 3–20–1 gen. on $\mathcal{B}_0$ $L(w;d,I)=\emptyset$ if $(d,I)\notin K_0$. So gen. on $\mathcal{B}_0$ $L(w)=\bigcup\{L(w;d,I);\ (d,I)\in K_0\}$. □

The following lemma is crucial in the proof of proposition 3–25.

Lemma 3–25 Let $(d,I)\in K_0$ be fixed. Then there exist $n\geq 0$ and $R^{(j)}\in\mathbb{R}[s]$, $j\in[0,n]$, such that

(*i*) $d=\bar{d}(R^{(0)})>\bar{d}(R^{(1)})>...>\bar{d}(R^{(n)}), I(R^{(j)})\supset I, j\in[0,n]$;

(*ii*) $\mathcal{B}_0\subset\mathcal{B}_T(R^{(j)})$ for all $j\in[0,n]$;

(*iii*) gen. in $w\in\mathcal{B}_0$ $L(w;d,I)=\text{span}_0\{R^{(j)}, j\in[0,n]\}:=\{R; \exists\alpha_j\in\mathbb{R}, j\in[0,n], \alpha_0\neq 0$, such that $R=\Sigma_{j=0}^{n}\alpha_j R^{(j)}\}$.

We first give an interpretation of this lemma, then give the proof of proposition 3–25 using lemma 3–25, and finally prove lemma 3–25.

Interpretation. Lemma 3–25 has the following interpretation. Let $(d,I)\in K_0$, then by definition of K_0 for every $w\in\mathcal{B}_0$ there is a remarkable law R_w with $w\in\mathcal{B}_T(R_w)\in\mathbb{B}^*(d,I)$. Now the lemma states that gen. in $\mathcal{B}_0$ the class of unfalsified remarkable laws in $\mathbb{B}^*(d,I)$ is independent of $w\in\mathcal{B}_0$ and that this class is spanned by a finite number of "basic laws" $R^{(j)}$ which moreover are true laws for $\mathcal{B}_0$. □

Proposition 3–25 can be proved by using lemma 3–25 in the following way. Generically on $\mathcal{B}_0$ $L(w)=\bigcup\{L(w;d,I); (d,I)\in K_0\}$, and as K_0 is a finite set it suffices to prove that for $(d,I)\in K_0$ gen. in $w\in\mathcal{B}_0$ $\{R\in L(w;d,I)\}\Rightarrow\{\mathcal{B}_0\subset\mathcal{B}_T(R)\}$.

According to lemma 3–25(*ii*), (*iii*), $R\in L(w; d,I)$ gen. is of the form $R=\Sigma_{j=0}^{n}\alpha_j R^{(j)}$, $\alpha_0\neq 0$, with $\mathcal{B}_0\subset\mathcal{B}_T(R^{(j)})$, $j\in[0,n]$. Using lemma 3–25(*i*) this implies that for $w\in\mathcal{B}_0$ $[R^{(j)}(\sigma)w](t)=0$, $t\in[1,T-\bar{d}(R^{(j)})]\supset[1,T-d]$. This implies that $[R(\sigma)w](t)=0$, $t\in[1,T-d]$, and hence $w\in\mathcal{B}_T(R)$ as $\bar{d}(R)=d$ for $\alpha_0\neq 0$. So $\mathcal{B}_0\subset\mathcal{B}_T(R)$, which proves proposition 3–25.

Now finally we prove lemma 3–25.

Remark. First note that if d_0:=min{d; there exists I such that $(d,I)\in K_0$} then it follows from lemma 3–20–3 that for $(d_0,I_0)\in K_0$ $L(w;d_0,I_0)$ gen. is a singleton $\mathcal{B}_T(R(d_0,I_0))$, and according to lemma 3–20–2 $\mathcal{B}_0\subset\mathcal{B}_T(R(d_0,I_0))$, which proves lemma 3–25 for d_0. However, in general for $(d,I)\in K_0$ $L(w; d,I)$ need not gen. be a singleton. As an example, let $T=5$, $\mathcal{B}_0:=\mathcal{B}_5(\sigma-1)$, then $(d,I):=$

$(2,\{3,4\})\in K_0$ and for all $w\in\mathcal{B}_0$ $L(w;\, d,I)\supset\{R_\alpha,\alpha\in\mathbb{R}\}$ where $R_\alpha:=(\sigma-\alpha)(\sigma-1)=\sigma^2-(\alpha+1)\sigma+\alpha$. □

Proof of lemma 3-25

We give the proof by construction. First we define $R^{(j)}$ and then we show that these have the desired properties.

Part (i) and (ii). Let $(d,I)\in K_0$, $[0,T-1]\backslash I=\{i_1,i_2,\ldots,i_{c-1},d\}$, $0\leq i_1<i_2<\ldots<i_{c-1}<d$ $(c+d\leq T)$, and define

$$H_I(w):=\begin{bmatrix} w(i_1+1) & w(i_1+2) & \ldots & w(i_1+T-d) \\ w(i_2+1) & w(i_2+2) & \ldots & w(i_2+T-d) \\ \vdots & \vdots & & \vdots \\ w(i_{c-1}+1) & w(i_{c-1}+2) & \ldots & w(i_{c-1}+T-d) \\ w(d+1) & w(d+2) & \ldots & w(T) \end{bmatrix},$$

and let $M_k(w)$ consist of the first k rows of $H_I(w)$.

As $\mathcal{B}_0\subset W(d,I)$, lemma 3-20-2 implies that there exists $R^{(0)}$ such that $\mathcal{B}_0\subset\mathcal{B}_T(R^{(0)})\in\mathbb{B}^*(d,I)$.

Now note that for $R=\Sigma_{k=1}^{c-1}a_k\sigma^{i_k}+a_c\sigma^d$, $a_c\neq0$, $a:=(a_1,\ldots,a_c)$, there holds $\{R\in L(w;\, d,I)\}\Leftrightarrow\{aH_I(w)=0\}$. If gen. on $\mathcal{B}_0$ $\text{rank}(M_{c-1}(w))=c-1$, then gen. a is unique up to a constant factor and hence gen. $L(w;\, d,I)=\{\alpha R^{(0)};\ 0\neq\alpha\in\mathbb{R}\}$ and lemma 3-25 is shown with $n=0$.

So suppose that not gen. $\text{rank}(M_{c-1}(w))=c-1$; hence not gen. $\det(M_{c-1}(w)M_{c-1}(w)^T)\neq0$, so $\det(M_{c-1}(w)M_{c-1}(w)^T)\equiv0$ and $\text{rank}(M_{c-1}(w))\leq c-2$ on $\mathcal{B}_0$.

Lemma 3-20-5 implies that there exists R' with $\mathcal{B}_0\subset\mathcal{B}_T(R')$ and $\bar{d}(R')\leq i_{c-1}<d=\bar{d}(R^{(0)})$, $I(R')\supset I\cup\{d\}$. Let $R^{(1)}$ be such a law for which $\bar{d}(R^{(1)})$ is maximal. Let $\bar{d}(R^{(1)})=i_{d_1}$ and $I_1:=[0,T-1]\backslash\{i_1,\ldots,i_{d_1}\}$.

Now either gen. on $\mathcal{B}_0$ $\text{rank}(M_{d_1-1}(w))=d_1-1$, in which case gen. $L(w;\, i_{d_1},I_1)=\{\alpha R^{(1)};\ 0\neq\alpha\in\mathbb{R}\}$ and we stop, or $\text{rank}(M_{d_1-1}(w))\leq d_1-2$ on $\mathcal{B}_0$. In the latter case, using lemma 3-20-5, we find a law $R^{(2)}$ of maximal degree in the class of laws with $I(R'')\supset I_1\cup\{i_{d_1}\}$ such that $\mathcal{B}_0\subset\mathcal{B}_T(R'')$. So $\mathcal{B}_0\subset\mathcal{B}_T(R^{(2)})$. Let $\bar{d}(R^{(2)})=i_{d_2}\leq i_{d_1-1}<i_{d_1}=\bar{d}(R^{(1)})$ and $I_2:=[0,T-1]\backslash\{i_1,\ldots,i_{d_2}\}$.

Going on in this way we find a number $n\leq c-1$ such that for $j\in[0,n]$ there exists $R^{(j)}$ with $\mathcal{B}_0\subset\mathcal{B}_T(R^{(j)})$, $I(R^{(j)})\supset I$, $\bar{d}(R^{(j)})<\bar{d}(R^{(j-1)})$, while for $i_{d_n}:=$

$\bar{d}(R^{(n)})$, $I_n:=[0,T-1]\backslash\{i_1,\ldots,i_{d_n}\}$, gen. on $\mathcal{B}_0$ $\text{rank}(M_{d_n-1}(w))=d_n-1$, so gen. on $\mathcal{B}_0$ $L(w;\, i_{d_n},I_n)=\{\alpha R^{(n)};\ 0\neq\alpha\in\mathbb{R}\}$.

In this way we have defined n, $R^{(j)}$, $j\in[0,n]$, with $\bar{d}=\bar{d}(R^{(0)})>\bar{d}(R^{(1)})>\ldots>\bar{d}(R^{(n)})$, $I(R^{(j)})\supset I$, and $\mathcal{B}_0\subset\mathcal{B}_T(R^{(j)})$. This proves (*i*) and (*ii*).

Part (*iii*). If $R=\Sigma_{j=0}^{n}\alpha_jR^{(j)}$, $\alpha_0\neq 0$, then $\bar{d}(R)=d$, $I(R)\supset I$, and $\mathcal{B}_0\subset\mathcal{B}_T(R)\in\mathbb{B}_T^*$, as $c(R)+\bar{d}(R)=T-\#(I(R))+\bar{d}(R)\leq T$, so $R\in L(w;d,I)$. We now have to show that gen. on $\mathcal{B}_0$ if $R\in L(w;\,d,I)$ then there exist α_k, $k\in[0,n]$, with $\alpha_0\neq 0$, such that $R=\Sigma_{j=0}^{n}\alpha_jR^{(j)}$. Without loss of generality we assume R and $R^{(j)}$ to be monic, $j\in[0,n]$.

So let $R\in L(w;d,I)$ be given. If $R=R^{(0)}$ then we are done, else define $R_1:=\{R-R^{(0)}\}.\beta_0$, with β_0 such that R_1 is monic. We state that gen. on $\mathcal{B}_0$ there exists $j\in[1,n]$ such that $\bar{d}(R_1)=\bar{d}(R^{(j)})$. For suppose this does not hold true, then there exists $i_k\in I\backslash\{\bar{d}(R^{(j)}),\ j\in[0,n]\}$ such that in $H_I(w)$ row k is not gen. linearly independent of the foregoing ones; hence by lemma 3–20–1 it is always linearly dependent on them and lemma 3–20–2 implies that there exists R_{i_k}, $\bar{d}(R_{i_k})=i_k$, $I(R_{i_k})\supset[0,T-1]\backslash i_1,\ldots,i_k\}$, with $\mathcal{B}_0\subset\mathcal{B}_T(R_{i_k})$. Now $i_k<\bar{d}(R^{(n)})$ is impossible by definition of n, so $\bar{d}(R^{(n)})<i_k<\bar{d}(R^{(0)})$. Let j be such that $\bar{d}(R^{(j)})<i_k<\bar{d}(R^{(j-1)})=:d_{j-1}$, then this contradicts the construction of $R^{(j)}$ as being of maximal degree in the class of laws $\bar{R}$ such that $I(\bar{R})\supset[0,T-1]\backslash\{i_1,i_2,\ldots,i_{d_{j-1}-1}\}$ and $\mathcal{B}_0\subset\mathcal{B}_T(\bar{R})$.

So indeed gen. on $\mathcal{B}_0$ there exists $j\in[1,n]$, say j_1, such that $\bar{d}(R_1)=\bar{d}(R^{(j_1)})$. If $R_1=R^{(j_1)}$, then stop, else define $R_2:=\{R_1-R^{(j_1)}\}\cdot\beta_1$, where β_1 is such that R_2 is monic. Going on in this way we gen. reduce R to laws of lower degree in the set $\{\bar{d}(R^{(j)}),\ j\in[1,n]\}$. The process gen. will end either if we find a $k\in[1,n]$ such that $R_k=R^{(j_k)}$, or if we get R_k with $\bar{d}(R_k)=\bar{d}(R^{(n)})$. As $\text{rank}(M_{d_n-1}(w))=d_n-1$ gen. on $\mathcal{B}_0$, also gen. $R_k=R^{(n)}$ in the latter case.

In this way we conclude that gen. on $\mathcal{B}_0$ if $R\in L(w;\,d,I)$ then $R=R^{(0)}+\beta_0^{-1}R_1=R^{(0)}+\beta_0^{-1}R^{(j_1)}+\beta_0^{-1}\beta_1^{-1}R_2$ and going on in this way we find α_j, $j\in[0,n]$, with $\alpha_0=1$, such that $R=\Sigma_{j=0}^{n}\alpha_jR^{(j)}$.

This concludes the proof of lemma 3–25 and hence of proposition 3–25. ■

Proof of corollary 3-26

According to proposition 3–25 gen. in $w \in \mathcal{B}_0$ $\mathcal{B}_0 \subset \mathcal{B}_T(R(w))$, so gen. $w \in \mathcal{B}_T(R(w))$ and hence gen. $P_T^* w = \mathcal{B}_T(R(w))$, $R(w) := \mathrm{GCD}\{R;\ R \in L(w)\}$. So gen. on $\mathcal{B}_0$ if $R \in L(w)$ then $\mathcal{B}_0 \subset \mathcal{B}_T(R)$ and $P_T^* w = \mathcal{B}_T(R(w)) \subset \mathcal{B}_T(R)$. ■

Proof of corollary 3-27

Let $R := \mathrm{GCD}\{\tilde{R} \neq 0;\ \bar{d}(\tilde{R}) \leq T-1$ and $\mathcal{B}_0 \subset \mathcal{B}_T(\tilde{R}) \in \mathbb{B}_T^*\}$, $R(w) := \mathrm{GCD}\{\tilde{R};\ \tilde{R} \in L(w)\}$. From corollary 3–26 gen. on $\mathcal{B}_0$ if $\tilde{R} \in L(w)$ then $\mathcal{B}_0 \subset \mathcal{B}_T(\tilde{R}) \in \mathbb{B}_T^*$, hence by using proposition II.3–36(*ii*) gen. on $\mathcal{B}_0$ $\mathcal{B}_T(R) \subset \mathcal{B}_T(R(w)) = P_T^* w$. On the other hand, if $\tilde{R} \neq 0$, $\bar{d}(\tilde{R}) \leq T-1$ and $\mathcal{B}_0 \subset \mathcal{B}_T(\tilde{R}) \in \mathbb{B}_T^*$, then on $\mathcal{B}_0$ $\tilde{R} \in L(w)$, so $\mathcal{B}_T(R(w)) \subset \mathcal{B}_T(\tilde{R})$, which implies $\mathcal{B}_T(R(w)) \subset \mathcal{B}_T(R)$, so gen. on $\mathcal{B}_0$ $P_T^* w \subset \mathcal{B}_T(R)$. ■

Proof of theorem 3-28

(*i*) Obvious from the definition of P^*.

(*ii*) For $\mathcal{B} \in \mathbb{B}_t$ let $R(\mathcal{B}) := \mathrm{GCD}\{R \neq 0;\ \bar{d}(R) \leq t-1$ and $\mathcal{B} \subset \mathcal{B}_t(R) \in \mathbb{B}_t^*\}$.

Monotonicity. Let $\mathcal{B}_0 \in \mathbb{B}_T$, $w \in \mathcal{B}_0$, $P_{t-1}^*(w|_{[1,t-1]}) = \mathcal{B}_{t-1}$, $P_t^*(w|_{[1,t]}) = \mathcal{B}_t$, then we have to prove that gen. $\mathcal{B}_t|_{[1,t-1]} \subset \mathcal{B}_{t-1}$. If $\mathcal{B}_{t-1} = \mathbb{R}^{t-1}$ then this is trivial, so assume $\mathcal{B}_{t-1} \neq \mathbb{R}^{t-1}$. Then from corollary 3–27 we can conclude that, gen. on $\mathcal{B}_0$, $\mathcal{B}_{t-1} = \mathcal{B}_{t-1}(R_{t-1})$ and $\mathcal{B}_t = \mathcal{B}_t(R_t)$, where $R_{t-1} := R(\mathcal{B}_0|_{[1,t-1]})$ and $R_t := R(\mathcal{B}_0|_{[1,t]})$. Now if R is such that $\mathcal{B}_0|_{[1,t-1]} \subset \mathcal{B}_{t-1}(R) \in \mathbb{B}_{t-1}^*$, then $\mathcal{B}_0|_{[2,t]} \subset \mathcal{B}_0|_{[1,t-1]} \subset \mathcal{B}_{t-1}(R)$ and hence $\mathcal{B}_0|_{[1,t]} \subset \mathcal{B}_t(R) \in \mathbb{B}_t^*$ as $c(R) + \bar{d}(R) \leq t-1 \leq t$, hence for these R $\mathcal{B}_t(R_t) \subset \mathcal{B}_t(R)$, cf. proposition II.3–36(*ii*). This implies $\mathcal{B}_t(R_t) \subset \mathcal{B}_t(R_{t-1})$, hence also gen. on $\mathcal{B}_0$ $\mathcal{B}_t|_{[1,t-1]} = \mathcal{B}_t(R_t)|_{[1,t-1]} \subset \mathcal{B}_t(R_{t-1})|_{[1,t-1]} = \mathcal{B}_{t-1}(R_{t-1}) = \mathcal{B}_{t-1}$.

Shift invariance. Let $\mathcal{B}_0 \in \mathbb{B}_T$, $w \in \mathcal{B}_0$, $P_{t-1}^*(w|_{[2,t]}) = \mathcal{B}_{t-1}(R)$, $\mathcal{B}' = P_t^*(w|_{[1,t]})$, then we have to prove that gen. $\mathcal{B}' \subset \mathcal{B}_t(\sigma.R)$. From corollary 3–27 it follows that gen. on $\mathcal{B}_0$ $R = R(\mathcal{B}_0|_{[2,t]})$ and gen. $\mathcal{B}' = \mathcal{B}_t(R')$ where $R' = R(\mathcal{B}_0|_{[1,t]})$. Now if $\tilde{R}$ is such that $\mathcal{B}_0|_{[2,t]} \subset \mathcal{B}_{t-1}(\tilde{R}) \in \mathbb{B}_{t-1}^*$, then $\mathcal{B}_0|_{[1,t]} \subset \mathcal{B}_t(\sigma\tilde{R}) \in \mathbb{B}_t^*$ as $c(\sigma\tilde{R}) + \bar{d}(\sigma\tilde{R}) = c(\tilde{R}) + \bar{d}(\tilde{R}) + 1 \leq t-1+1 = t$. So gen. $\mathcal{B}' = \mathcal{B}_t(R') \subset \mathcal{B}_t(\bar{R})$ where $\bar{R} := \mathrm{GCD}\{\sigma\tilde{R} \neq 0;\ d(\tilde{R}) \leq t-1$ and $\mathcal{B}_0|_{[2,t]} \subset \mathcal{B}_{t-1}(\tilde{R}) \in \mathbb{B}_{t-1}^*\} = \sigma R$.

Linearity. $P_t^*(\alpha w) = P_t^*(w)$ for $\alpha \neq 0$ follows from $L(\alpha w) = L(w)$, $\alpha \neq 0$. Now let $(\mathcal{B}_1, \mathcal{B}_2) \in \mathbb{B}_T^2$, $\mathcal{B}_i = \mathcal{B}_T(R_i)$, $i=1,2$, $(w_1, w_2) \in \mathcal{B}_1 \times \mathcal{B}_2$, $P_T^* w_1 = \mathcal{B}_T(R')$, $P_T^* w_2 = \mathcal{B}_T(R'')$, $P_T^*(w_1 + w_2) = \mathcal{B}_T(R)$, then we have to prove that gen. $\mathcal{B}_T(R') + \mathcal{B}_T(R'') \subset \mathcal{B}_T(R)$.

According to proposition II.3–36(i) $\mathcal{B}_1+\mathcal{B}_2=\mathcal{B}_T(K)$ with K:=LCM(R_1,R_2). According to corollary 3–27 gen. in w_1+w_2, hence also gen. in $(w_1,w_2)\in\mathcal{B}_1\times\mathcal{B}_2$, there holds $R=R(\mathcal{B}_T(K))$. Moreover, gen. $R'=R(\mathcal{B}_1)$ and $R''=R(\mathcal{B}_2)$. Because $\mathcal{B}_i\subset\mathcal{B}_T(K)$, i=1,2, if $\mathcal{B}_T(K)\subset\mathcal{B}_T(\tilde{R})$ then also $\mathcal{B}_i\subset\mathcal{B}_T(\tilde{R})$, i=1,2, so $\mathcal{B}_T(R')\subset\mathcal{B}_T(R)$, $\mathcal{B}_T(R'')\subset\mathcal{B}_T(R)$, and hence also $\mathcal{B}_T(R')+\mathcal{B}_T(R'')\subset\mathcal{B}_T(R)$.

(iii) This is immediate from corollary 3–27. Note that for $\mathcal{B}:=\mathbb{R}^T$ $P_T^*w=\mathbb{R}^T$ gen. in $w\in\mathbb{R}^T$.

(iv) Let $V:=\{\mathcal{B}_T(R);\ R=\text{GCD}\{R_\lambda,\ \lambda\in\Lambda\}$ for some $\{R_\lambda,\ \lambda\in\Lambda\}$ with $\bar{d}(R_\lambda)\le T-1$ and $\mathcal{B}_T(R_\lambda)\in\mathbb{B}_T^*\}$. For every procedure $\mathbb{B}^s_{P_T}\subset\mathbb{B}^w_{P_T}\subset\text{im}(P_T)$, so it suffices to show that $V\subset\mathbb{B}^s_{P_T^*}$ and that $\text{im}(P_T^*)\subset V$. If $\mathcal{B}\in V$, then corollary 3–27 implies that gen. on $\mathcal{B}$ $P_T^*w=\mathcal{B}$, hence $\mathcal{B}\in\mathbb{B}^s_{P_T^*}$. If $\mathcal{B}\in\text{im}(P_T^*)$, then either $\mathcal{B}=\mathbb{R}^T\in V$ or $\exists w\in\mathbb{R}^T$ such that $\mathcal{B}=\mathcal{B}_T(R(w))$, $R(w)=\text{GCD}\{R;\ R\in L(w)\}$. Because for $R\in L(w)$ $\bar{d}(R)\le T-1$ and $\mathcal{B}_T(R)\in\mathbb{B}_T^*$ it follows that $\mathcal{B}\in V$. ■

Proof of proposition 3–32

Let $\mathcal{B}_0\subset\mathbb{B}_T$. As $\tilde{L}(w)\subset L(w)$ we conclude from corollary 3–26 that gen. on $\mathcal{B}_0$ $\{\tilde{R}\in\tilde{L}(w)\}\Rightarrow\{\mathcal{B}_0\subset\mathcal{B}_T(\tilde{R})\}$. So gen. on $\mathcal{B}_0$ $\mathcal{B}_0\subset\mathcal{B}_T(\tilde{R}(w))$ and hence gen. $\tilde{P}_T^*w=\mathcal{B}_T(\tilde{R}(w))$ where $\tilde{R}(w):=\text{GCD}\{\tilde{R};\tilde{R}\in\tilde{L}(w)\}$.

Now define $R:=\text{GCD}\{\tilde{R}\neq0;\ \bar{d}(\tilde{R})\le T-1$ and $\mathcal{B}_0\subset\mathcal{B}_T(\tilde{R})\in\tilde{\mathbb{B}}_T^*\}$. For every $\tilde{R}\neq0$, $\bar{d}(\tilde{R})\le T-1$ with $\mathcal{B}_0\subset\mathcal{B}_T(\tilde{R})\in\tilde{\mathbb{B}}_T^*$ there holds that on $\mathcal{B}_0$ $\tilde{R}\in\tilde{L}(w)$, hence $\mathcal{B}_T(\tilde{R}(w))\subset\mathcal{B}_T(R)$ and gen. $\tilde{P}_T^*w\subset\mathcal{B}_T(R)$. On the other hand gen. on $\mathcal{B}_0$ for $\tilde{R}\in\tilde{L}(w)$ $\mathcal{B}_0\subset\mathcal{B}_T(\tilde{R})\in\tilde{\mathbb{B}}_T^*$, hence gen. $\mathcal{B}_T(R)\subset\mathcal{B}_T(\tilde{R}(w))$ and gen. $\mathcal{B}_T(R)\subset\tilde{P}_T^*w$. ■

Proof of theorem 3–33

It is trivial that $\tilde{P}^*$ is exact. That it is truthful immediately follows from proposition 3–32. Strong prudence is shown by means of proposition 3–32 in a way exactly analogous to the proof of theorem 3–28(iv), along with the characterization of $\text{im}(\tilde{P}_T^*)=\mathbb{B}^w_{\tilde{P}_T^*}=\mathbb{B}^s_{\tilde{P}_T^*}$. Linearity is shown by using proposition 3–32 and the arguments in the proof of theorem 3–28(ii).

Hence it remains to show that $\tilde{P}^*$ is bilaterally monotone. Monotonicity follows as in the proof of theorem 3–28(ii). Finally let $\mathcal{B}_0\in\tilde{\mathbb{B}}_T$, then we have to prove that gen. in $w\in\mathcal{B}_0$ there holds $\sigma\mathcal{B}_{T-t+2}\subset\mathcal{B}_{T-t+1}$, where $\mathcal{B}_{T-t+i}:=\tilde{P}^*_{T-t+i}(w|_{[t-i+1,T]})$, i=1,2. According to proposition 3–32 gen. on $\mathcal{B}_0$ $\mathcal{B}_{T-t+i}=$

$\mathcal{B}_{T-t+i}(R_i)$, with R_i:=GCD$\{\tilde{R}\neq 0;\ \bar{d}(\tilde{R})\leq T-t+i-1$ and $\mathcal{B}_0|_{[t-i+1,T]}\subset\mathcal{B}_{T-t+i}(\tilde{R})\in\tilde{\mathbb{B}}^*_{T-t+i}\}$, i=1,2. Now if $\tilde{R}$ is such that $\bar{d}(\tilde{R})\leq T-t$ and $\mathcal{B}_0|_{[t,T]}\subset\mathcal{B}_{T-t+1}(\tilde{R})\in\tilde{\mathbb{B}}^*_{T-t+1}$, then translation invariance of $\mathcal{B}_0$ implies (see proposition II.3–30(*ii*)) that also $\mathcal{B}_0|_{[t-1,T-1]}=\mathcal{B}_0|_{[t,T]}\subset\mathcal{B}_{T-t+1}(\tilde{R})$, hence $\mathcal{B}_0|_{[t-1,T]}\subset\mathcal{B}_{T-t+2}(\tilde{R})\in\tilde{\mathbb{B}}^*_{T-t+2}$. From this it follows that $\mathcal{B}_{T-t+2}(R_2)\subset\mathcal{B}_{T-t+2}(R_1)$, hence gen. on $\mathcal{B}_0$ $\mathcal{B}_{T-t+2}\subset\mathcal{B}_{T-t+2}(R_1)$, especially $\sigma\mathcal{B}_{T-t+2}\subset\sigma\mathcal{B}_{T-t+2}(R_1)=\mathcal{B}_{T-t+1}(R_1)=\mathcal{B}_{T-t+1}$ gen. on $\mathcal{B}_0$. ■

Proof of proposition 3–34

For $\mathcal{B}_0=\mathbb{R}^T$ the result in (*i*) follows from the truthfulness of P^*_T and $\tilde{P}^*_T$. So let $\mathcal{B}_0\neq\mathbb{R}^T$. For $0\neq R=\Sigma_{k=0}^{T-1}a_k\sigma^k\in\mathbb{R}[\sigma]$ let $l(R)$:=min$\{k;a_k\neq 0\}$. It easily follows that $\{\mathcal{B}_T(R)\in\tilde{\mathbb{B}}_T\}\Leftrightarrow\{l(R)=0\}$ for $R\neq 0$, $\bar{d}(R)\leq T-1$.

Let $\mathcal{B}_0=\mathcal{B}_T(R_0)\in\mathbb{B}_T$. From corollary 3–27 and proposition 3–32 it follows that gen. on $\mathcal{B}_0$ $P^*_Tw=\mathcal{B}_T(R)$ and $\tilde{P}^*_Tw=\mathcal{B}_T(\tilde{R})$, where R:=GCD$\{R'\neq 0;\ \bar{d}(R')\leq T-1$ and $\mathcal{B}_0\subset\mathcal{B}_T(R')\in\mathbb{B}^*_T\}$ and $\tilde{R}$:=GCD$\{\tilde{R}'\neq 0;\ \bar{d}(\tilde{R}')\leq T-1$ and $\mathcal{B}_0\subset\mathcal{B}_T(\tilde{R}')\in\tilde{\mathbb{B}}^*_T\}$.

(*i*) Let $\mathcal{B}_0\in\tilde{\mathbb{B}}_T$, so $l(R_0)=0$, and let $R'\neq 0$, $\bar{d}(R')\leq T-1$, such that $\mathcal{B}_0\subset\mathcal{B}_T(R')\in\mathbb{B}^*_T$. Then according to lemma II.3–34 there exists F' such that $R'=F'R_0$. If l:=$l(R')=l(F')\neq 0$ then define $\tilde{F}'$ by $\tilde{F}'$:=$\sigma^{-l}F'$ and $\tilde{R}'$:=$\tilde{F}'R_0$. So $\mathcal{B}_0\subset\mathcal{B}(\tilde{R}')\in\tilde{\mathbb{B}}^*_T$ as $l(\tilde{R}')=0$ and $c(\tilde{R}')+\bar{d}(\tilde{R}')=c(R')+\bar{d}(R')-l\leq T-1\leq T$. Now $R'=\sigma^l\tilde{R}'$ and it follows that R=GCD$\{R'\neq 0,\ \bar{d}(R')\leq T-1$ and $\mathcal{B}_0\subset\mathcal{B}_T(R')\in\mathbb{B}^*_T\}$=GCD$\{\tilde{R}'\neq 0;\ \bar{d}(\tilde{R}')\leq T-1$ and $\mathcal{B}_0\subset\mathcal{B}_T(\tilde{R}')\in\tilde{\mathbb{B}}^*_T\}=\tilde{R}$ and hence gen. $P^*_Tw=\tilde{P}^*_Tw$.

(*ii*) Let $\mathcal{B}_0\in\mathbb{B}_T$, $\mathcal{B}_0\notin\tilde{\mathbb{B}}_T$, so $l(R_0)\geq 1$. If $\bar{d}(R')\leq T-1$ and $\mathcal{B}_0\subset\mathcal{B}_T(R')\in\mathbb{B}^*_T$ then there exists F' such that $R'=F'R_0$ and hence $l(R')\geq 1$, so $\mathcal{B}_T(R')\notin\tilde{\mathbb{B}}^*_T$. This implies that gen. on $\mathcal{B}_0$ $\tilde{L}(w)=\emptyset$, as according to corollary 3–26 gen. on $\mathcal{B}_0$ $\{R\in\tilde{L}(w)\subset L(w)\}\Rightarrow\{\mathcal{B}_0\subset\mathcal{B}_T(R)\in\tilde{\mathbb{B}}^*_T\}$. So gen. on $\mathcal{B}_0$ $\tilde{R}(w)=0$ and $\tilde{P}^*_Tw=\mathbb{R}^T$. ■

Proof of lemma 3–35

First suppose that $g\in\mathbb{R}^{T+1}$ has a realization of dimension n, i.e., there exist $g^e\in\mathbb{R}^{\mathbb{Z}_+}$ and $(A,B,C)\in\mathbb{R}^{n\times n}\times\mathbb{R}^{n\times 1}\times\mathbb{R}^{1\times n}$ such that $g^e|_{[0,T]}=g$ and $S(g^e)=S(A,B,C,g_0)$. Consider the input u defined by $u(0)$:=1 and $u(t)$:=0 for $t\neq 0$. The corresponding output y in $S(g^e)$ is given by $y(t)=0$ for $t<0$ and $y(t)=g^e_t$ for $t\geq 0$. As $(u,y)\in S(A,B,C,g_0)$ it is easily seen that $g_t=CA^{t-1}B$, $t\in[1,T]$. Define $R\in\mathbb{R}[s]$ as the characteristic polynomial of A. Then $\bar{d}(R)=n$, $R(A)=0$, and hence $\tilde{g}\in\mathcal{B}_T(R)$.

Next suppose that $\tilde{g}\in\mathcal{B}_T(R)$ with $\bar{d}(R)=n$. Let $R=\sum_{k=0}r_ks^k$, then without loss of generality we may assume that $r_n=1$. Define $(A,B,C)\in\mathbb{R}^{n\times n}\times\mathbb{R}^{n\times 1}\times\mathbb{R}^{1\times n}$ by $B:=(g_1,\dots,g_n)^T$, $C:=(1,0,\dots,0)$ and $A:=\begin{bmatrix} 0 & I_{n-1} \\ & \rho \end{bmatrix}$, where $0\in\mathbb{R}^{n-1}$, I_{n-1} is the identity matrix in $\mathbb{R}^{(n-1)\times(n-1)}$, and $\rho:=(-r_0,-r_1,\dots,-r_{n-1})$. A direct calculation gives that $g_t=CA^{t-1}B$, $t\in[1,T]$. Define $g^e\in\mathbb{R}^{\mathbb{Z}_+}$ by $g_0^e:=g_0$ and $g_t^e:=CA^{t-1}B$ for $t\in\mathbb{N}$. It is easily seen that $S(g^e)=S(A,B,C,g_0)$. Hence g has a realization of dimension n. ■

CHAPTER IV

Proof of proposition 3-2

In this proof and the next one we use the following result.

Lemma 3-2 If $H=\mathcal{B}\cap l_2^q$, $\mathcal{B}\in\mathbb{B}$, and $\mathcal{B}$ has a minimal realization $\mathcal{B}_s(A,B,C,D):= \{(v,x,w)\in(\mathbb{R}^m\times\mathbb{R}^n\times\mathbb{R}^q)^{\mathbb{Z}};\ \begin{bmatrix}\sigma x\\ w\end{bmatrix}=\begin{bmatrix}A & B\\ C & D\end{bmatrix}\begin{bmatrix}x\\ v\end{bmatrix}\}$, then $H=\{w;\ \exists(v,x)$ such that $(v,x,w)\in\mathcal{B}_s(A,B,C,D)\cap l_2^{m+n+q}\}$.

Proof of lemma 3-2

If $(v,x,w)\in\mathcal{B}_s(A,B,C,D)\cap l_2^{m+n+q}$, then $w\in\mathcal{B}\cap l_2^q=H$. On the other hand, if $w\in H$ and $(v,x,w)\in\mathcal{B}_s(A,B,C,D)$, then it suffices to show that $(v,x)\in l_2^{m+n}$. According to proposition II.3–22 (A,B,C,D) is perfectly observable. From this and the linearity and shift invariance of $\mathcal{B}$ it follows that there exists a linear map $L:(\mathbb{R}^q)^n\to\mathbb{R}^n$ such that for $(v,x,w)\in\mathcal{B}_s$ there holds $x(t)=L(w|_{[t,t+n-1]})$, $t\in\mathbb{Z}$. Hence $x\in l_2^n$ if $w\in l_2^q$. Moreover, $Dv=w-Cx$ and D is injective, see proposition II.3–22, hence $v\in l_2^m$. ∎

Next we prove the proposition. The result for $\mathcal{B}_s^2$ is contained in the lemma. To show that $H\in\mathbb{B}_2$ has a $\mathcal{B}_{i/s/o}^2$ realization, let $H=\mathcal{B}\cap l_2^q$, where $\mathcal{B}\in\mathbb{B}$ has a minimal input/state/output realization $\mathcal{B}_{i/s/o}$, i.e., $\Pi\mathcal{B}=\{(u,y)\in(\mathbb{R}^m\times\mathbb{R}^{q-m})^{\mathbb{Z}};\ \exists x\in(\mathbb{R}^n)^{\mathbb{Z}}$ such that $\begin{bmatrix}\sigma x\\ y\end{bmatrix}=\begin{bmatrix}\tilde{A} & \tilde{B}\\ \tilde{C} & \tilde{D}\end{bmatrix}\begin{bmatrix}x\\ u\end{bmatrix}\}$, cf. corollary II.3–23. Then $\mathcal{B}_{i/s/o}^2:=\mathcal{B}_{i/s/o}\cap l_2^{m+n+(q-m)}$ is an l_2–input/state/output realization of H. Indeed, if $(u,x,y)\in\mathcal{B}_{i/s/o}\cap l_2^{m+n+(q-m)}$, then $\Pi\begin{bmatrix}u\\ y\end{bmatrix}\in\mathcal{B}\cap l_2^q=H$, while for $w\in H$ there holds, according to lemma 3–2, that $x\in l_2^n$, hence with $\Pi w=\begin{bmatrix}u\\ y\end{bmatrix}$ also $(u,x,y)\in\mathcal{B}_{i/s/o}\cap l_2^{m+n+(q-m)}$. The results for ${}^R\mathcal{B}_s^2$ and ${}^R\mathcal{B}_{i/s/o}^2$ follow by considering $\mathcal{R}H$, where $\mathcal{R}$ denotes the time reverse operator. ∎

Proof of proposition 3-3

First we state two results which will be useful in the sequel. Then we use these results to prove the proposition. Finally we prove the lemmas.

Lemma 3-3-1 If $A \in \mathbb{R}^{n \times n}$, then $\{x \in l_2^n;\ \sigma x = Ax\} = \{0\}$.

Lemma 3-3-2 Let $H \in \mathbb{B}_2$ and let $\mathcal{B}^*$ be the closure of H in $(\mathbb{R}^q)^{\mathbb{Z}}$ with respect to the topology of pointwise convergence. If H has a realization $\mathcal{B}_s^2(A,B,C,D)$ with (A,B) controllable, then $\mathcal{B}_s(A,B,C,D)$ is a realization of $\mathcal{B}^*$.

To prove the proposition we now first show that in a realization $\mathcal{B}_s^2(A,B,C,D)$ of $H \in \mathbb{B}_2$ for which n is minimal (for fixed m) the pair (A,B) is controllable. Let $R := \mathrm{im}[B\ AB\ \ldots\ A^{n-1}B] \subset \mathbb{R}^n$ and $d := \dim(R)$. As $AR \subset R$ and $\mathrm{im}(B) \subset R$ it follows that there is a choice of basis in $\mathbb{R}^n$ such that in this basis $A = \begin{bmatrix} A_1 & A_3 \\ 0 & A_2 \end{bmatrix}$ and $B = \begin{bmatrix} B_1 \\ 0 \end{bmatrix}$. So in a corresponding partition $\begin{bmatrix} x_1 \\ x_2 \end{bmatrix}$ of x in this basis there holds for $(v,x,w) \in \mathcal{B}_s^2$ that $\sigma x_2 = A_2 x_2$. From lemma 3-3-1 we conclude $x_2 = 0$, hence $\mathcal{B}_s^2(A_1,B_1,C,D)$ also is a realization of H. As n is minimal it follows that $\dim(R) = d = n$.

Hence for any m there exist controllable realizations of H with n minimal, for given number m of driving variables. According to lemma 3-3-2 these induce realizations of $\mathcal{B}^*$. According to proposition II.3-22 $\mathcal{B}^*$ has a minimal realization, which induces a realization of H, see lemma 3-2 in the proof of proposition 3-2. So a minimal realization of H exists, and $\mathcal{B}_s^2(A,B,C,D)$ is a minimal realization of H if and only if $\mathcal{B}_s(A,B,C,D)$ is a minimal realization of $\mathcal{B}^*$.

Now if a realization $\mathcal{B}_s^2(A,B,C,D)$ is minimal for H, then (A,B) is controllable, and $\mathcal{B}_s(A,B,C,D)$ is a minimal realization of $\mathcal{B}^*$, hence (A,B,C,D) is perfectly observable and D injective. On the other hand, if in a realization $\mathcal{B}_s^2(A,B,C,D)$ (A,B,C,D) is perfectly observable, D injective and (A,B) controllable, then it easily follows that $(A\ B)$ is surjective, hence $\mathcal{B}_s(A,B,C,D)$ is a minimal realization of $\mathcal{B}^*$. This implies that $\mathcal{B}_s^2(A,B,C,D)$ is a minimal realization of H.

Finally we prove lemma 3–3–1 and lemma 3–3–2.

Proof of lemma 3–3–1

For $A=0$ the result is trivial. Hence assume $A\neq 0$. Let $p(s):=\det(sI-A)$ be the characteristic polynomial of A. As $p(A)=0$ it follows that for solutions of $\sigma x=Ax$ there holds $p(\sigma)x_i=0$, $i=1,\ldots,n$. To prove the lemma it suffices to show that with $V:=\{w\in\mathbb{C}^{\mathbb{Z}};\ p(\sigma)w=0\}$ there holds $\{w\in V, \sum_{t\in\mathbb{Z}}|w(t)|^2<\infty\}\Rightarrow\{w=0\}$.

Let $p(s)=s^k\cdot\Pi_{i=1}^{M}(s-\lambda_i)^{m_i}$, $|\lambda_1|\geq\ldots\geq|\lambda_M|>0$, $k\in\mathbb{Z}_+$, $\lambda_i\neq\lambda_j$ for $i\neq j$, $\Sigma_{i=1}^{M}m_i=n-k$. Then $\dim(V)=n-k$. Moreover one easily shows by induction on j that the trajectories $w_{ij}(t):=t^j\lambda_i^t$, $j=0,\ldots,m_i-1$, $i=1,\ldots,M$, give $n-k$ independent solutions in V. Hence $w\in V$ if and only if there exist $\alpha_{ij}\in\mathbb{C}$, $j=1,\ldots,m_i$, $i=1,\ldots,M$, such that $w(t)=\Sigma_{i=1}^{M}\Sigma_{j=1}^{m_i}\alpha_{ij}t^{j-1}\lambda_i^t$. Now consider such a w with $\sum_{t\in\mathbb{Z}}|w(t)|^2<\infty$. If $w\neq 0$ then let $i_+:=\max\{i;\ \exists j$ such that $\alpha_{ij}\neq 0\}$ and $i_-:=\min\{i;\ \exists j$ such that $\alpha_{ij}\neq 0\}$. Taking $t\to+\infty$ it can be shown that $|\lambda_{i_-}|<1$ and taking $t\to-\infty$ it can be shown that $|\lambda_{i_+}|^{-1}<1$, hence $|\lambda_{i_-}|<|\lambda_{i_+}|$ while $i_+\geq i_-$. This contradicts the ordering of the λ's. Hence $w=0$. ■

Proof of lemma 3–3–2

If in $\mathcal{B}_s:=\mathcal{B}_s(A,B,C,D)$ the pair (A,B) is controllable, then for any $t_0,t_1\in\mathbb{Z}$ with $t_0\leq t_1$ there holds $\mathcal{B}_s|_{[t_0,t_1]}=\mathcal{B}_s^2|_{[t_0,t_1]}$, where $\mathcal{B}_s^2:=\mathcal{B}_s^2(A,B,C,D)$. Indeed, let $(v,x,w)\in\mathcal{B}_s$ and define (v',x',w') as follows. Let $(v',x',w')(t):=0$ for $t\in\mathbb{Z}\backslash[t_0-n,t_1+n]$, choose v' on $[t_0-n,t_0-1]$ such that $[B\ldots A^{n-1}B]\cdot\mathrm{col}(v'(t_0-1),\ldots,v'(t_0-n))=x(t_0)$ and on $[t_1+1,t_1+n]$ such that $[B\ldots A^{n-1}B]\cdot\mathrm{col}(v'(t_1+n),\ldots v'(t_1+1))=-A^nx(t_1+1)$, while $v'|_{[t_0,t_1]}:=v|_{[t_0,t_1]}$. Compute $(x',w')|_{t_0-n,t_1+n]}$ according to $\begin{pmatrix}\sigma x'\\ w'\end{pmatrix}=\begin{pmatrix}A & B\\ C & D\end{pmatrix}\begin{pmatrix}x'\\ v'\end{pmatrix}$. It easily follows that $(v',x',w')\in\mathcal{B}_s$, and as it has compact support it is in $\mathcal{B}_s^2$, while $(v',x',w')|_{[t_0,t_1]}=(v,x,w)|_{[t_0,t_1]}$.

Now let H have a realization $\mathcal{B}_s^2$ with (A,B) controllable, then we have to show that $\mathcal{B}^*=\mathcal{B}':=\{w;\ \exists(v,x)$ such that $(v,x,w)\in\mathcal{B}_s\}$.

First let $w\in\mathcal{B}'$, $(v,x,w)\in\mathcal{B}_s$. We conclude from the foregoing that for all $-\infty<t_0\leq t_1<+\infty$ $(v,x,w)|_{[t_0,t_1]}\in\mathcal{B}_s^2|_{[t_0,t_1]}$, so especially $w|_{[t_0,t_1]}\in H|_{[t_0,t_1]}\subset\mathcal{B}^*|_{[t_0,t_1]}$. Completeness of $\mathcal{B}^*$ implies that $w\in\mathcal{B}^*$, so $\mathcal{B}'\subset\mathcal{B}^*$. That $\mathcal{B}^*\subset\mathcal{B}'$ is seen as follows. According to theorem II.3–21 $\mathcal{B}'\in\mathbb{B}$. Moreover, as $\mathcal{B}_s^2\subset\mathcal{B}_s$, it follows that $\mathcal{B}'\subset H$. As $\mathcal{B}^*$ is the closure of H with respect to the topology of pointwise convergence it follows from proposition II.3–3 that $\mathcal{B}^*\subset\mathcal{B}'$. ■

This concludes the proof of proposition 3–3. ■

Proof of corollary 3-4

(*i*) Let $H\in\mathbb{B}_2$ have a minimal realization $\mathcal{B}_s^2(A,B,C,D)$, so according to proposition 3–3 the pair (A,B) is controllable. Then $\mathcal{B}_s(A,B,C,D)$ is a minimal realization of $\mathcal{B}^*$ as defined in lemma 3–3–2 in the proof of proposition 3–3. That $\mathcal{B}^*$ is controllable follows along the lines of the first part of the proof of lemma 3–3–2. Finally $H=\mathcal{B}^*\cap l_2^q$, as for $H=\mathcal{B}\cap l_2^q$, $\mathcal{B}\in\mathbb{B}$, there holds $H\subset\mathcal{B}^*\subset\mathcal{B}$, so $H\subset\mathcal{B}^*\cap l_2^q\subset\mathcal{B}\cap l_2^q=H$.

(*ii*) This follows along the lines of the first part of the proof of lemma 3–3–2, using the fact that for minimal realizations $\mathcal{B}_s(A,B,C,D)$ of $\mathcal{B}\in\mathbb{B}_c$ there holds that (A,B) is controllable, see Willems [74, section 4.8.2].

(*iii*) In the proof of proposition 3–3 we derived that $\mathcal{B}_s^2(A,B,C,D)$ is a minimal realization of H if and only if $\mathcal{B}_s(A,B,C,D)$ is a minimal realization of $\mathcal{B}^*$ as defined in lemma 3–3–2. Hence the result follows from proposition II.3–25. ■

Proof of proposition 3-5

Let $\mathcal{B}_{i/s/o}^2:=\{(u,x,y)\in l_2^{m+n+(q-m)};\ \begin{pmatrix}\sigma x\\ y\end{pmatrix}=\begin{pmatrix}\tilde{A} & \tilde{B}\\ \tilde{C} & \tilde{D}\end{pmatrix}\begin{pmatrix}x\\ u\end{pmatrix}\}$ be a minimal input/state/output realization of $H\in\mathbb{B}_2$ such that $\Pi H=\{\begin{pmatrix}u\\ y\end{pmatrix};\ \exists x\in l_2^n$ such that $(u,x,y)\in\mathcal{B}_{i/s/o}^2\}$. It follows from proposition 3–3 that then $(\tilde{A},\tilde{B})$ is controllable and that $(\tilde{A},\tilde{C})$ is observable, i.e., $\mathrm{col}(\tilde{C},\tilde{C}\tilde{A},\ldots,\tilde{C}\tilde{A}^{n-1})$ is injective. The last statement follows from proposition 3–3 and the definition of perfect observability in section II.3.3, which implies that $\{(u,y)|_{[0,n-1]}=0\}\Rightarrow\{x(0)=0\}$.

In $\mathcal{B}_{i/s/o}^2$ there exists a linear map $L:u\to x$, as for $u=0$ it follows from lemma 3–3–1 in the proof of proposition 3–3 that $x=0$. As $y=\tilde{C}x+\tilde{D}u$ there exists a linear map $F:u\to y$ in H. So the remaining questions are in which case $u\in l_2^m$ can be chosen arbitrarily and in which cases F is causal or anticausal.

First we prove the implication ($\Leftarrow$) for (*i*), (*ii*) and (*iii*). Note that there is a choice of basis for $\mathbb{R}^n$ such that in this basis $\tilde{A}=\begin{bmatrix} A_+ & 0 & 0 \\ 0 & A_- & 0 \\ 0 & 0 & A_0 \end{bmatrix}$ with $\sigma(A_+)\subset\mathbb{C}_+$, $\sigma(A_-)\subset\mathbb{C}_-$, $\sigma(A_0)\subset\mathbb{C}_0$, e.g., for $\tilde{A}$ in real Jordan form. Let $x=\begin{bmatrix} x_+ \\ x_- \\ x_0 \end{bmatrix}$, $B=\begin{bmatrix} B_+ \\ B_- \\ B_0 \end{bmatrix}$ and $C=(C_+\ C_-\ C_0)$ be corresponding partitions.

If $\sigma(\tilde{A})\cap\mathbb{C}_0=\emptyset$, then for $u\in l_2^m$ let x_+,x_- be defined by $x_+(t):=\Sigma_{k=1}^{\infty} A_+^{k-1}B_+u(t-k)$ and $x_-(t):=-\Sigma_{k=0}^{\infty}(A_-^{-1})^{k+1}B_-u(t+k)$, and let $y:=C_+x_++C_-x_-+\tilde{D}u$. It easily follows that $(x_+,x_-)\in l_2^n$ and that $\begin{bmatrix} \sigma x_+ \\ \sigma x_- \\ y \end{bmatrix}=\begin{bmatrix} A_+ & 0 & B_+ \\ 0 & A_- & B_- \\ C_+ & C_- & \tilde{D} \end{bmatrix}\begin{bmatrix} x_+ \\ x_- \\ u \end{bmatrix}$, so $(u,\begin{bmatrix} x_+ \\ x_- \end{bmatrix},y)\in \mathcal{B}_{i/s/o}^2$. Hence $u\in l_2^m$ is free. Moreover, if $\sigma(A)\subset\mathbb{C}_+$ then clearly the map $L:u\to x$ where $x=x_+$ is causal, hence $F:u\to y$ is causal. Similarly, if $\sigma(A)\subset\mathbb{C}_-$ then $L:u\to x$ is anticausal, hence $F:u\to y$ also.

Next we prove (*i*) ($\Rightarrow$). Suppose $\lambda\in\sigma(\tilde{A})$ with $|\lambda|=1$, say $\tilde{A}v=\lambda v$ where $\|v\|=1$. It suffices to show that there is an $u\in l_2^m$ such that there exists no $x\in l_2^n$ with $\sigma x=\tilde{A}x+\tilde{B}u$. Then there is no $y\in l_2^{q-m}$ with $(u,y)\in\Pi H$, as due to observability $\{(u,y)\in l_2^{m+(q-m)}\}\Rightarrow\{x\in l_2^n\}$. We then conclude that u is not free in l_2^n and hence that there is no $F:l_2^m\to l_2^{q-m}$ with $\Pi H=\mathrm{gr}(F)$.

To construct u such that $\{x\in l_2^n;\ \sigma x=\tilde{A}x+\tilde{B}u\}=\emptyset$, let $a_1,a_2\in(\mathbb{R}^m)^n$ be such that $[A^{n-1}B...B]a_1=\mathrm{Re}(v)$ and $[A^{n-1}B...B]a_2=\mathrm{Im}(v)$. This is possible as (A,B) is controllable. Define $u_i\in l_2^m$ by $u_i|_{[-n,-1]}:=a_i$ and $u_i(t):=0$ for $t\notin[-n,-1]$, $i=1,2$. Define $X_i:=\{x\in(\mathbb{R}^n)^{\mathbb{Z}};\ \sigma x=\tilde{A}x+\tilde{B}u_i\}$, $i=1,2$, and $X_0:=\{x\in(\mathbb{R}^n)^{\mathbb{Z}};\ \sigma x=\tilde{A}x\}$. Then $X_i=x_i+X_0$, where x_i is defined as that element $x_i\in X_i$ for which $x_i|_{(-\infty,-n]}=0$, $i=1,2$. It is easily seen that $x_1(t)=\mathrm{Re}(\lambda^t v)$ and $x_2(t)=\mathrm{Im}(\lambda^t v)$ for $t\geq0$. As $|\lambda|=1$ it follows that at least one of the series $\sum_{t\geq0}\|x_1(t)\|^2$ and $\sum_{t\geq0}\|x_2(t)\|^2$ diverges, say $\sum_{t\geq0}\|x_1(t)\|^2=\infty$. We now prove that then $X_1\cap l_2^m=\emptyset$. Let $x\in X_1$, i.e., $x=\bar{x}+x_1$ with $\sigma\bar{x}=\tilde{A}\bar{x}$. In an appropriate basis $\tilde{A}=\mathrm{diag}(A_{+0},A_{+1},A_-,A_0)$ with $\sigma(A_{+0})=\{0\}$, $\sigma(A_{+1})\subset\mathbb{C}_+\backslash\{0\}$, $\sigma(A_-)\subset\mathbb{C}_-$, and $\sigma(A_0)\subset\mathbb{C}_0$. Let $\bar{x}=\mathrm{col}(\bar{x}_{+0},\bar{x}_{+1},\bar{x}_-,\bar{x}_0)$ be a corresponding partition of $\bar{x}$. It easily follows that $\bar{x}_{+0}=0$. Further, in order that $\sum_{t\geq0}\|x(t)\|^2<\infty$ it is clearly necessary that $\bar{x}_-=0$, while $\sum_{t\leq0}\|x(t)\|^2<\infty$ implies that $\bar{x}_{+1}=0$ and $\bar{x}_0=0$ (note that $x_1(t)=0$ for $t\leq-n$). Hence if $x\in l_2^n$, then $\bar{x}=0$. However, $x_1\notin l_2^n$, which proves the desired result

Finally we prove (*ii*) ($\Rightarrow$) and (*iii*) ($\Rightarrow$). Suppose $\sigma(\tilde{A})\cap\mathbb{C}_-\neq\emptyset$, then we show that there is $(u,y)\in\Pi H$ with $u|_{(-\infty,0]}=0$ but $y|_{(-\infty,0]}\neq0$. Note that controllability implies that (A_-,B_-) is controllable and that perfect observability implies that $C_-\neq0$. Let $a\in\mathbb{R}^n$ and $a\notin\ker(C_-)$, and let b be such that $[A_-^{n-1}B_-...B_-]b=A_-^{n+1}a$. Define $u\in l_2^m$ by $u|_{[1,n]}=b$ and $u(t)=0$ for $t\notin[1,n]$.

Let $y(t):=\tilde{D}u(t)-C_{-}\Sigma_{k=0}^{\infty}(A_{-}^{-1})^{k+1}B_{-}u(t+k)$. Then one easily verifies that $(u,y)\in\Pi H$ and $y(0)\neq 0$. This proves (*ii*) ($\Rightarrow$). The proof of (*iii*) ($\Rightarrow$) is completely analogous. ■

Proof of proposition 4-1

We only prove $(iii)_{+}$, as $(iii)_{-}$ follows in an analogous way and $(i)_{\pm}$, $(ii)_{\pm}$ are trivial. For $(iii)_{+}$ we need to prove that $\bar{w}\in H$ can be obtained as l_2-limit of trajectories in H with left compact support, i.e., for any $\varepsilon>0$ there should be a $k\in\mathbb{Z}$ and $\tilde{w}\in(\sigma^{*})^{k}H_{+}$ with $\|\bar{w}-\tilde{w}\|^{2}<\varepsilon$.

Let $\mathcal{B}_{i/s/o}^{2}$ $(\tilde{A},\tilde{B},\tilde{C},\tilde{D})$ be a minimal input/state/output realization of H and let $\begin{bmatrix}\bar{u}\\ \bar{y}\end{bmatrix}:=\Pi\bar{w}$, $(\bar{u},\bar{x},\bar{y})\in\mathcal{B}_{i/s/o}^{2}$. Define a linear map $L:\mathbb{R}^{n}\to(\mathbb{R}^{m})^{n}$ as follows. Let $\{e_1,\ldots,e_n\}$ be a basis of $\mathbb{R}^{n}$ and let $a_i\in(\mathbb{R}^{m})^{n}$ be such that $[A^{n-1}B\ldots B]a_i=e_i$, $i=1,\ldots,n$. Such a_i exist as (A,B) is controllable. Define $Le_i:=a_i$ and for $x\in\mathbb{R}^{n}$, $x=\Sigma_{i=1}^{n}x_ie_i$, let $Lx:=\Sigma_{i=1}^{n}x_ia_i$. Let $M:=\|L^{T}(I+F^{T}F)L\|^{2}$ with $F\in\mathbb{R}^{n(q-m)\times nm}$ defined by

$$F:=\begin{bmatrix} D & 0 & \cdots & 0\\ CB & D & & \vdots\\ \vdots & & & 0\\ CA^{n-2}B & \cdots & CB & D\end{bmatrix}$$

Let $T\in\mathbb{Z}$ be such that $\sum_{t<T}\|\bar{w}(t)\|^{2}<\frac{1}{4}\varepsilon$ and $\|\bar{x}(T)\|^{2}<\varepsilon/(4M)$. Define $(\tilde{u},\tilde{x},\tilde{y})\in\mathcal{B}_{i/s/o}^{2}$ by $(\tilde{u},\tilde{x},\tilde{y})|_{(-\infty,T-n-1]}:=0$, $\tilde{u}|_{[T-n,T-1]}:=L\bar{x}(T)$, and $\tilde{u}|_{[T,\infty)}:=\bar{u}|_{[T,\infty)}$. Then clearly $\tilde{x}(T)=\bar{x}(T)$ and hence $(\tilde{u},\tilde{x},\tilde{y})|_{[T,\infty)}=(\bar{u},\bar{x},\bar{y})|_{[T,\infty)}$. Define $\tilde{w}:=\Pi\begin{bmatrix}\tilde{u}\\ \tilde{y}\end{bmatrix}$, then $\tilde{w}\in(\sigma^{*})^{T-n}H_{+}$ and $\|\bar{w}-\tilde{w}\|^{2}=\sum_{t<T}\|\bar{w}(t)-\tilde{w}(t)\|^{2}\leq 2(\sum_{t<T}\|\bar{w}(t)\|^{2}+\sum_{t<T}\|\tilde{w}(t)\|^{2})<\frac{1}{2}\varepsilon+2(\|L\bar{x}(T)\|^{2}+\|FL\bar{x}(T)\|^{2})\leq\frac{1}{2}\varepsilon+2M\|\bar{x}(T)\|^{2}<\varepsilon$, as desired. ■

Proof of proposition 4-4

Consider $H':=\{(v,w)\in l_2^{m}\times l_2^{q};\ \exists x\in l_2^{n}$ such that $\begin{bmatrix}\sigma x\\ w\end{bmatrix}=\begin{bmatrix}A & B\\ C & D\end{bmatrix}\begin{bmatrix}x\\ v\end{bmatrix}\}$. Then $\mathcal{B}_{s}^{2}$ is a minimal input/state/output realization of H'. If $\sigma(A)\cap\mathbb{C}_0=\emptyset$ then according to proposition 3-5(*i*) there exists $L:v\to w$ such that $H'=\mathrm{gr}(L)$, hence $H=\mathrm{im}(L)$, which proves (*i*). The results in (*ii*) follow from proposition 3-5(*ii*) and (*iii*). Finally, (*iii*) follows from corollary 3-4(*iii*) and the fact that (A,B) is controllable, see proposition 3-3. Indeed, it is a well-known result from linear control theory that in this case $\det(sI-(A+BF))$ may be any monic (real) polynomial of degree n, by appropriate choice of F. See e.g. Kailath [33, section 7.1]. ■

Proof of proposition 4-6

In the proof of (*ii*) we use some results on the solutions of the discrete time Lyapunov equation. For similar and more general results for the continuous time case we refer to Glover [17, theorem 3.3], and Kailath [33, section 2.6.2]. We will first state and prove a lemma which subsequently is used to prove the proposition.

Lemma 4-6 Let $(A,C)\in\mathbb{R}^{n\times n}\times\mathbb{R}^{q\times n}$ be observable, i.e., $\mathrm{col}(C,\ldots,CA^{n-1})$ is injective. Consider the discrete time Lyapunov equation $A^TKA+C^TC=K$.

(*i*) If there exists a nonsingular solution K, then $\sigma(A)\cap\mathbb{C}_0=\varnothing$.

(*ii*) If $\sigma(A)\subset\mathbb{C}_+$, then there exists a unique solution K and moreover $K>0$; if $\sigma(A)\subset\mathbb{C}_-$, then there exists a unique solution K and moreover $K<0$.

Proof of lemma 4-6

(*i*) If $A^TKA+C^TC=K$, then by repeatedly applying this identity we conclude that $K=(A^T)^nKA^n+\Sigma_{t=0}^{n-1}(A^T)^t\,C^TCA^t$, so $K-(A^T)^nKA^n>0$ as (A,C) is observable. Now suppose $\sigma(A)\cap\mathbb{C}_0\neq\varnothing$, $Ax=\lambda x$ with $|\lambda|=1$ and $x=x_1+ix_2\neq 0$, $x_1,x_2\in\mathbb{R}^n$. Then $x^*(K-(A^T)^nKA^n)x=(1-|\lambda|^{2n})x^*Kx=0$, hence $(K-(A^T)^nKA^n)x_i=0$, $i=1,2$, which is in contradiction with $K-(A^T)^nKA^n>0$.

(*ii*) Let $\sigma(A)\subset\mathbb{C}_+$ and suppose $A^TK_iA+C^TC=K_i$, $i=1,2$. Then with $K_0:=K_1-K_2$ it follows that $K_0=A^TK_0A=(A^T)^nK_0A^n\to 0$ for $n\to\infty$, hence $K_0=0$. Moreover, $K:=\Sigma_{t=0}^{\infty}(A^T)^t\,C^TCA^t>0$ clearly is a solution. Analogously, if $\sigma(A)\subset\mathbb{C}_-$ then $K:=-\Sigma_{t=0}^{\infty}[(A^{-1})^T]^{t+1}\,C^TC(A^{-1})^{t+1}<0$ clearly is a solution, and it is the only one, as for $A^TK_iA+C^TC=K_i$, $i=1,2$, it follows that for $K_0:=K_1-K_2$ $K_0=(A^{-1})^TK_0A^{-1}=[(A^{-1})^T]^n\,K_0(A^{-1})^n\to 0$ for $n\to\infty$, hence $K_0=0$. ■

(*i*) We now first prove proposition 4-6(*i*). As in the proof of proposition II.3-22 we use for $z\in(\mathbb{R}^d)^{\mathbb{Z}}$ the notation $z^{--}:=z|_{(-\infty,-1]}$, $z^+:=z|_{[0,\infty)}$. To prove ($\Leftarrow$), let $\begin{bmatrix}A&B\\C&D\end{bmatrix}$ be a $(\begin{bmatrix}K&0\\0&I_m\end{bmatrix},\begin{bmatrix}K&0\\0&I_q\end{bmatrix})$ Pontryagin isometry, then for $(v,x,w)\in\mathcal{B}_s^2$ there holds $\|\sigma x\|_K^2+\|w\|^2=\|x\|_K^2+\|v\|^2$, hence $\|w\|^2=\|v\|^2$. To prove ($\Rightarrow$), let $x_0\in\mathbb{R}^n$ and $\mathcal{B}_s^2(x_0):=\{(v,x,w)\in\mathcal{B}_s^2;\ x(0)=x_0\}$. Controllability implies that there is $(v^0,x^0,w^0)\in\mathcal{B}_s^2(x_0)$ with $(v^0,x^0,w^0)|_{[n,\infty)}=0$. Analogous to part of the proof of proposition 4–1 it is easily seen that we can take $v^0|_{[0,n-1]}=Lx_0$ for a linear map $L:\mathbb{R}^n\to(\mathbb{R}^m)^n$, hence $w^0|_{[0,n-1]}=\tilde{L}x_0$ for a linear map $\tilde{L}:\mathbb{R}^n\to(\mathbb{R}^q)^n$. Define $K:=\tilde{L}^T\tilde{L}-L^TL$. Because $\|w^0\|^2=\|v^0\|^2$ we conclude that $\|(v^0)^{--}\|^2-\|(w^0)^{--}\|^2=\Sigma_{t=0}^{n-1}\{\|w^0(t)\|^2-\|v^0(t)\|^2\}=x_0^TKx_0$. As $x_0\in\mathbb{R}^n$ is arbitrary due to controllability, if follows that $K=K^T$ is uniquely defined by $\mathcal{B}_s^2$. Moreover, for any $(v,x,w)\in\mathcal{B}_s^2(x_0)$ there exists $(v^0,x^0,w^0)\in\mathcal{B}_s^2$ with $(v,x,w)^{--}=(v^0,x^0,w^0)^{--}$ and $(v^0,x^0,w^0)|_{[n,\infty)}=0$, due to controllability. Hence $\|v^{--}\|^2-\|w^{--}\|^2=x_0^TKx_0$.

K is nonsingular, as for $Kx_0=0$ it follows that for $(v,x,w)\in\mathcal{B}_s^2(x_0)$ with $v^+=0$ there holds $\|v^{--}\|=\|w^{--}\|$, hence $w^+=0$ and $x_0=0$ due to perfect observability. We finally show that $\begin{bmatrix} A & B\\ C & D\end{bmatrix}$ is a $(\begin{bmatrix} K & 0\\ 0 & I_m\end{bmatrix},\begin{bmatrix} K & 0\\ 0 & I_q\end{bmatrix})$ Pontryagin isometry. Let $(a,b)\in\mathbb{R}^n\times\mathbb{R}^m$ and $\begin{bmatrix} c\\ d\end{bmatrix}:=\begin{bmatrix} A & B\\ C & D\end{bmatrix}\begin{bmatrix} a\\ b\end{bmatrix}$. Due to controllability there exists a $(v,x,w)\in\mathcal{B}_s^2$ with $(x(0),v(0),x(1),w(0))=(a,b,c,d)$. As $(\sigma v,\sigma x,\sigma w)\in\mathcal{B}_s^2(c)$ we conclude from the foregoing that $\|c\|_K^2=\|(\sigma v)^{--}\|^2-\|(\sigma w)^{--}\|^2=\|v^{--}\|^2-\|w^{--}\|^2+\|b\|^2-\|d\|^2=\|a\|_K^2+\|b\|^2-\|d\|^2$, hence $\|c\|_K^2+\|d\|^2=\|a\|_K^2+\|b\|^2$ which is the desired result.

(*ii*–1) Suppose (*i*) is satisfied, then K satisfies $A^TKA+C^TC=K$. Indeed, let x_0 be arbitrary, then take $(v,x,w)\in\mathcal{B}_s^2(x_0)$ with $v(0)=0$, so $\begin{bmatrix} x(1)\\ w(0)\end{bmatrix}=\begin{bmatrix} A & B\\ C & D\end{bmatrix}\begin{bmatrix} x_0\\ 0\end{bmatrix}$, and $\begin{bmatrix} A & B\\ C & D\end{bmatrix}$ being an isometry implies $\|Ax_0\|_K^2+\|Cx_0\|^2=\|x_0\|_K^2$, i.e., $x_0^T(A^TKA+C^TC-K)x_0=0$. Now perfect observability of (A,B,C,D) implies observability of (A,C), which is seen by taking $(v,x,w)\in\mathcal{B}_s^2$ with $(v,w)|_{[0,n-1]}=0$. As K is nonsingular, it follows from lemma 4–6(*i*) that $\sigma(A)\cap\mathbb{C}_0=\emptyset$ and from proposition 3–5(*i*) that $L(A,B,C,D)$ exists. That it is an isometry follows from (*i*). This proves (*ii*–1).

(*ii*–2) To prove (*ii*–2), note that in(K) does not depend upon a choice of coordinates in $\mathbb{R}^n$. Choose these in such a way that $A=\begin{bmatrix} A_+ & 0\\ 0 & A_-\end{bmatrix}$ with $\sigma(A_+)\subset\mathbb{C}_+$ and $\sigma(A_-)\subset\mathbb{C}_-$. Let $C=(C_+,C_-)$ and $K=\begin{bmatrix} K_+ & K_{+-}\\ K_{+-}^T & K_-\end{bmatrix}$ be correspoding partitions of C and K. As $A^TKA+C^TC=K$ it follows that $A_+^TK_+A_++C_+^TC_+=K_+$ and $A_-^TK_-A_-+C_-^TC_-=K_-$. Moreover, (A_+,C_+) and (A_-,C_-) are both observable. From lemma 4–6(*ii*) we conclude that $K_+>0$ and $K_-<0$. From this we easily get (*ii*–2).

(*ii*–3) This result is an immediate consequence of (*ii*–2) and proposition 3–5(*ii*) and (*iii*). ■

Proof of proposition 4-7

Let $M:=\begin{bmatrix} A+BF & BR \\ C+DF & DR \end{bmatrix} \in \mathbb{R}^{(n+q)\times(n+m)}$, $S\in\mathbb{R}^{n\times n}$ invertible and $\bar{K}=\bar{K}^T\in\mathbb{R}^{n\times n}$ invertible. It is easily shown that $\begin{bmatrix} S & 0 \\ 0 & I_q \end{bmatrix} M \begin{bmatrix} S^{-1} & 0 \\ 0 & I_m \end{bmatrix}$ is a $(\begin{bmatrix} \bar{K} & 0 \\ 0 & I_m \end{bmatrix}, \begin{bmatrix} \bar{K} & 0 \\ 0 & I_q \end{bmatrix})$ Pontryagin isometry if and only if M is a $(\begin{bmatrix} K & 0 \\ 0 & I_m \end{bmatrix}, \begin{bmatrix} K & 0 \\ 0 & I_q \end{bmatrix})$ Pontryagin isometry, where $K:=S^T\bar{K}S$. Hence it suffices to prove the proposition for the case $S=I$.

Now M is a $(\begin{bmatrix} K & 0 \\ 0 & I_m \end{bmatrix}, \begin{bmatrix} K & 0 \\ 0 & I_q \end{bmatrix})$ Pontryagin isometry if and only if for all $(a,b)\in\mathbb{R}^n\times\mathbb{R}^m$ and $\begin{bmatrix} c \\ d \end{bmatrix} := M \begin{bmatrix} a \\ b \end{bmatrix}$ there holds

$$\begin{bmatrix} a \\ b \end{bmatrix}^T \begin{bmatrix} K & 0 \\ 0 & I_m \end{bmatrix} \begin{bmatrix} a \\ b \end{bmatrix} = \begin{bmatrix} c \\ d \end{bmatrix}^T \begin{bmatrix} K & 0 \\ 0 & I_q \end{bmatrix} \begin{bmatrix} c \\ d \end{bmatrix} = \begin{bmatrix} a \\ b \end{bmatrix}^T \begin{bmatrix} X_{11} & X_{12} \\ X_{21} & X_{22} \end{bmatrix} \begin{bmatrix} a \\ b \end{bmatrix},$$

where a direct calculation shows that $X_{11}=(A+BF)^TK(A+BF)+(C+DF)^T(C+DF)$, $X_{22}=R^T(B^TKB+D^TD)R$, and $X_{21}=X_{12}^T=R^T[B^TK(A+BF)+D^T(C+DF)]$. Hence M is such an isometry if and only if $X_{11}=K$, $X_{22}=I_m$ and $X_{12}=0$. Now $X_{22}=I_m$ is equivalent with (R) as R is invertible, and then $X_{12}=0$ is equivalent with (F) as R is invertible and B^TKB+D^TD also, according to (R). Given (F) and (R), $X_{11}=K$ is equivalent with (ARE) by using the expression (F) for F. ∎

Proof of lemma 4-8

According to proposition 3-3 (A,B,C,D) is perfectly observable and (A,B) controllable, hence stabilizable, see Kailath [33, section 7.1]. The result follows immediately from Payne and Silverman [57, theorem 2.1 and lemma 4.5]. ∎

Proof of theorem 4-9

According to propositions 4-6 and 4-7 $L:=L(A_+,B_+,C_+,D_+)$ is a causal, time invariant isometry with $\text{im}(L)=H$. According to corollary 4-2 it suffices to show that $H_+=Ll_2(\mathbb{Z}_+,\mathbb{R}^m)$, as L_+ is unique (up to isomorphisms of $\mathbb{R}^m$). As L is causal $Ll_2(\mathbb{Z}_+,\mathbb{R}^m)\subset H_+$. It remains to show that for $(v,x,w)\in\mathbb{B}_s^2(A_+,B_+,C_+,D_+)$ with $w^{--}=0$ there holds $v^{--}=0$. From corollary 3-4(*iii*) and proposition 3-3 it follows that (A_+,B_+,C_+,D_+) is perfectly observable and that D_+ is injective. Hence $w^{--}=0$ implies $x|_{(-\infty,-n]}=0$, and as $D_+v=w-C_+x$ also $v|_{(-\infty,-n]}=0$. From $\begin{bmatrix} \sigma x \\ w \end{bmatrix} = \begin{bmatrix} A_+ & B_+ \\ C_+ & D_+ \end{bmatrix} \begin{bmatrix} x \\ v \end{bmatrix}$ it follws by induction that $(v,x)(t)=0$ for $t=-n+1,\dots,-1$, hence

$v^{--}=0$. ■

Proof of proposition 4-11

Let $\tilde{L}_+$ be the forward scattering representation of $\tilde{H}$ and let $L:=\mathcal{R}\tilde{L}_+\mathcal{R}$. Then L clearly is isometric and time invariant and $\text{im}(L)=\mathcal{R}\text{im}(\tilde{L}_+)=\mathcal{R}\tilde{H}=H$. According to corollary 4-2 it now suffices to show that $Ll_2(\mathbb{Z}_-,\mathbb{R}^m)=H_-$ in order to conclude that $L=L_-$. Define $l_2(\mathbb{Z}_{++},\mathbb{R}^m):=\{v\in l_2(\mathbb{Z},\mathbb{R}^m);\ v(t)=0 \text{ for } t\leq 0\}=\sigma^* l_2(\mathbb{Z}_+,\mathbb{R}^m)$, then $Ll_2(\mathbb{Z}_-,\mathbb{R}^m)=\mathcal{R}\tilde{L}_+l_2(\mathbb{Z}_{++},\mathbb{R}^m)=\mathcal{R}\tilde{L}_+\sigma^*l_2(\mathbb{Z}_+,\mathbb{R}^m)=\mathcal{R}\sigma^*\tilde{L}_+l_2(\mathbb{Z}_+,\mathbb{R}^m)=\mathcal{R}\sigma^*\tilde{H}_+$ $=\sigma\mathcal{R}\tilde{H}_+=\sigma\{w\in H;\ w(t)=0 \text{ for } t>0\}=\{w\in H;\ w(t)=0 \text{ for } t\geq 0\}=H_-$. ■

Proof of corollary 5-3

It suffices to prove that $P=L_+L_+^*=L_-L_-^*$. We will prove $P=L_+L_+^*$, as the other result follows analogously. $P:l_2^q\to l_2^q$ is uniquely determined by the conditions (i) $P^*=P$; (ii) $P^2=P$; (iii) $\text{im}(P)=H$, see e.g. Akhiezer and Glazman [2, section 31]. Now $L_+L_+^*$ clearly satisfies (i), while (ii) follows from the fact that L_+ is isometric, so $L_+^*L_+=I$. As $\text{im}(L_+)=H$ it only remains to show that $H\subset\text{im}(L_+L_+^*)$. Let $w\in H$, then $w=L_+L_+^*w$ which is seen as follows. Let v be such that $L_+v=w$, then for any v' $\langle w,L_+v'\rangle=\langle L_+v,L_+v'\rangle=\langle v,v'\rangle$ as L_+ is isometric, so by definition $L_+^*w=v$ and $L_+L_+^*w=L_+v=w$. Hence $H\subset\text{im}(L_+L_+^*)$. ■

Proof of theorem 5-4

In this proof we use some standard results on the discrete Fourier transform and its inverse, the Z-transform. We will state these results in a lemma, give an outline of the proof of the lemma and finally prove the theorem by means of this lemma.

Notation. Let $\mathbb{C}_0:=\{z\in\mathbb{C};\ |z|=1\}$ and for $f:\mathbb{C}_0\to\mathbb{C}^d$ let $\|f\|^2:=(2\pi)^{-1}\int_{-\pi}^{\pi}[f(e^{i\theta})]^*f(e^{i\theta})d\theta$, where $*$ denotes complex conjugate transpose. We define $L_2^d:=\{f:\mathbb{C}_0\to\mathbb{C}^d;\ \|f\|^2<\infty\}$ and on L_2^d $\langle f,g\rangle:=(2\pi)^{-1}\int_{-\pi}^{\pi}[g(e^{i\theta})]^*f(e^{i\theta})d\theta$. With this inner product L_2^d is a Hilbert space. Let $l_2^d(\mathbb{C}):=\{x\in(\mathbb{C}^d)^{\mathbb{Z}};\ \sum_{t\in\mathbb{Z}}[x(t)]^*x(t)<\infty\}$. We define the discrete Fouriertransform

$F:L_2^d \to l_2^d(\mathbb{C})$ by $(Ff)(k):=\hat{f}(k):= (2\pi)^{-1}\int_{-\pi}^{\pi} e^{ik\theta}f(e^{i\theta})d\theta$ and the Z–transform $Z:l_2^d(\mathbb{C})\to L_2^d$ by $(Zx)(e^{i\theta}):=\Sigma_{k=-\infty}^{\infty} e^{-ik\theta}x(k)$. These transformations are well-defined in a limit-in-mean sense. □

Lemma 5-4 (*i*) F is unitary from L_2^d to $l_2^d(\mathbb{C})$, i.e., it is isometric and surjective; Z is the inverse of F.

(*ii*) Let $H\in\mathbb{B}_2$ have minimal realization $\mathcal{B}_s^2(A,B,C,D)$ with $\sigma(A)\cap\mathbb{C}_0=\emptyset$ and let $L:=L(A,B,C,D)$ be the corresponding driving operator, then $Z(Lv)= G\cdot Z(v)$ with $G(z):=D+C(zI-A)^{-1}B$, $z\in\mathbb{C}_0$; if L is isometric then $Z(v)= G^*\cdot Z(Lv)$ where $G^*(z):=[G(z)]^*$.

(*iii*) For L as in (*ii*) $\|L\|_2:=\sup\{\|Lv\|;\ \|v\|=1\}=\|G\|_\infty$.

Proof of lemma 5-4 (outline)

The result in (*i*) is the well-known Fourier–Plancherel theorem, cf. Kato [40, section V.2.2].

To prove (*ii*), let a basis in $\mathbb{R}^n$ be chosen such that $A=\begin{pmatrix}A_+ & 0\\ 0 & A_-\end{pmatrix}$ with $\sigma(A_+)\subset\mathbb{C}_+$, $\sigma(A_-)\subset\mathbb{C}_-$, and let $B=\begin{pmatrix}B_+\\ B_-\end{pmatrix}$ and $C=(C_+\ C_-)$ be corresponding partitions. Then for $v\in l_2^m$ $Lv=L_1v+L_2v+Dv$ where $(L_1v)(t):=C_+\Sigma_{k=1}^{\infty}A_+^{k-1}B_+v(t-k)$ and $(L_2v)(t):= -C_-\Sigma_{k=0}^{\infty}(A_-^{-1})^{k+1}B_-v(t+k)$, cf. the proof of proposition 3–5. By the definition of Z–transform we obtain $Z(Lv)=[C_+(\Sigma_{k=1}^{\infty}z^{-k}A_+^{k-1})B_+-C_-(\Sigma_{k=0}^{\infty}z^k(A_-^{-1})^{k+1})B_-+D]Z(v)$ and as $\sigma(A_+)$ and $\sigma(A_-^{-1})$ are contained in $\mathbb{C}_+$ there holds for $z\in\mathbb{C}_0$ that $\Sigma_{k=1}^{\infty}z^{-k}A_+^{k-1}= z^{-1}(I-z^{-1}A_+)^{-1}=(zI-A_+)^{-1}$ and $\Sigma_{k=0}^{\infty}z^k(A_-^{-1})^{k+1}=A_-^{-1}(I-zA_-^{-1})^{-1}= -(zI-A_-)^{-1}$. From this we obtain G. If L is in addition isometric, then for $v\in l_2^m$ with $Lv=w$ there holds for arbitrary v' with $Lv'=:w'$ that $\langle G^*Zw-Zv,Zv'\rangle= \langle Zw,GZv'\rangle-\langle Zv,Zv'\rangle=\langle Zw,Zw'\rangle-\langle Zv,Zv'\rangle=\langle w,w'\rangle-\langle v,v'\rangle=\langle v,L^*Lv'\rangle-\langle v,v'\rangle=0$ as $L^*L=I$. Hence $G^*Zw= Zv$ as desired.

To prove (*iii*) note that for $\|v\|=1$ $\|Lv\|=\|GZ(v)\|\le\|G\|_\infty$, hence it suffices to prove that for every $\varepsilon>0$ there is a $v\in l_2^m$ with $\|v\|=1$ and $\|Lv\|>\|G\|_\infty-\varepsilon$.

As $\|G(z)\|$ is continuous on $\mathbb{C}_0$ which is compact it follows that there is a $\theta_0\in[-\pi,\pi]$ such that $\|G(e^{i\theta_0})\|=\sup\{\|G(z)\|;\ z\in\mathbb{C}_0\}=:\|G\|_\infty$. Let $a\in\mathbb{C}^m$ with $\|a\|=1$ be such that $\|G(e^{i\theta_0})a\|=\|G\|_\infty$. Continuity of $G(z)$ implies that there exists a $\delta>0$ such that for $|\theta-\theta_0|<\delta$ $\|G(e^{i\theta})a\|>\|G\|_\infty-\varepsilon$. Then for $|\theta+\theta_0|<\delta$ there holds $\|G(e^{i\theta})\bar{a}\|=\|G(e^{-i\theta})a\|>\|G\|_\infty-\varepsilon$. Now define $u\in l_2^m$ as follows. If $\theta_0=0$, then take $a\in\mathbb{R}^m$ (this is possible), and let $u(e^{i\theta}):=(\pi/\delta)^{1/2}\cdot a\cdot I_{(-\delta,\delta)}(\theta)$, where $I_V(\theta)=1$

for $\theta\in V$ and 0 for $\theta\notin V$. If $\theta_0\neq 0$, then take $\delta<|\theta_0|$ and let $u(e^{i\theta}):=(\pi/2\delta)^{1/2}\cdot\{a\cdot I_{(\theta_0-\delta,\theta_0+\delta)}(\theta)+\bar{a}\cdot I_{(-\theta_0-\delta,-\theta_0+\delta)}(\theta)\}$. Let $v:=Fu$. Then $\|v\|=\|u\|=1$ and $v\in l_2^m$, i.e., v is real, as $\overline{v(k)}=(2\pi)^{-1}\int_{-\pi}^{\pi}e^{-ik\theta}\overline{u(e^{i\theta})}d\theta=(2\pi)^{-1}\int_{-\pi}^{\pi}e^{-ik\theta}u(e^{-i\theta})d\theta=v(k)$, $k\in\mathbb{Z}$. Now $\|Lv\|_2=\|Gu\|$ which is easily seen to be larger than $\|G\|_\infty-\varepsilon$. ■

Theorem 5–4 is easily proved by means of this lemma. Let $L:=L_+L_+^*-L_+'L_+'^*$ and $G:=G_+G_+^*-G_+'G_+'^*$, then for $w\in l_2^q$ there holds $Z(Lw)=Z(L_+L_+^*w)-Z(L_+'L_+'^*w)=G_+Z(L_+^*w)-G_+'Z(L_+'^*w)=G_+G_+^*Zw-G_+'G_+'^*Zw=GZw$, and hence $\|L\|_2=\sup\{\|GZw\|;\ \|w\|=1\}=\|G\|_\infty$ which follows from the proof of part (*iii*) of the lemma. The result for G_- follows in an analogous way. ■

Proof of proposition 5-5

We first state a preliminary result. Let $L_1,L_2\subset\mathbb{R}^n$ be linear subspaces with $\dim(L_1)>\dim(L_2)$, then $L_1\cap L_2^\perp\neq\{0\}$, which easily follows from $\dim(L_1)+\dim(L_2^\perp)=\dim(L_1)+n-\dim(L_2)>n$.

Now let $(m,n,m',n'):=(m(H),n(H),m(H'),n(H'))$ and suppose $m>m'$. It suffices to prove that $H\cap(H')^\perp\neq\{0\}$, as from $0\neq w\in H\cap(H')^\perp$ we conclude $g(H,H')=\|P-P'\|\geq\frac{1}{\|w\|}\|Pw-P'w\|=1$, while $g(H,H')\leq 1$ always holds true.

According to corollary 3–4 and proposition 2–2 there exist controllable systems $\mathcal{B},\mathcal{B}'\in\mathbb{B}$ such that $H=\mathcal{B}\cap l_2^q$ and $H'=\mathcal{B}'\cap l_2^q$, where $(m(\mathcal{B}),n(\mathcal{B}),m(\mathcal{B}'),n(\mathcal{B}'))=(m,n,m',n')$. For $T\in\mathbb{Z}_+$ let $\mathcal{B}(T):=\{w\in\mathcal{B};\ w(t)=0$ for $t<0$ and for $t\geq T\}\subset H$. We will show that there is a T such that there exists $0\neq w\in\mathcal{B}(T)$ with $w|_{[0,T-1]}\perp\mathcal{B}'|_{[0,T-1]}$. As $\mathcal{B}'|_{[0,T-1]}=H'|_{[0,T-1]}$ this implies that $0\neq w\in H\cap(H')^\perp$, as desired.

Now take T such that $(T-2n)m>n'+Tm'$, i.e., $T>(n'+2nm)/(m-m')$. As $\mathcal{B}$ is controllable we conclude that in $\mathcal{B}(T)$ the m inputs can be chosen freely on $[n,T-n-1]$, hence $\dim(\mathcal{B}(T)|_{[0,T-1]})\geq(T-2n)m$. From a minimal input/state/output realization of $\mathcal{B}'$ it is evident that $\dim(\mathcal{B}'|_{[0,T-1]})\leq n'+Tm'$. From the result stated in the first lines of this proof it follows that there exists $0\neq x\in[\mathcal{B}(T)|_{[0,T-1]}]\cap[\mathcal{B}'|_{[0,T-1]}]^\perp$. Define $w\in\mathcal{B}(T)$ by $w|_{[0,T-1]}:=x$ and $w(t):=0$ for $t\notin[0,T-1]$. Then $0\neq w\in H\cap(H')^\perp$. ■

Proof of proposition 5-7

From propositions 4–6 and 4–7 characterizing isometries we conclude that $\|x_+(t+1)\|_{K_+}^2+\|w(t)\|^2=\|x_+(t)\|_{K_+}^2+\|v_+(t)\|^2$. As $x_+\in l_2^n$ this implies $\|x_+(0)\|_{K_+}^2=\|v_+^{--}\|^2-\|w^{--}\|^2=\|w^+\|^2-\|v_+^+\|^2$. From these propositions and the fact that $R\mathcal{B}_s^-=\mathcal{B}_s^2(A_-,B_-,C_-,D_-)$ we conclude that $\|x_-(t-1)\|_{\tilde{K}_+}^2+\|w(t)\|^2=\|x_-(t)\|_{\tilde{K}_+}^2+\|v_-(t)\|^2$, so $\|x_-(0)\|_{\tilde{K}_+}^2=\|v_-^{++}\|^2-\|w^{++}\|^2$.

Next we prove the expressions for $\|x_0\|_{++}$ and $\|x_0\|_{--}$. We use the facts that in $\mathcal{B}_s^+$ $x_+(0)=\Sigma_{k=1}^{\infty}A_+^{k-1}B_+v_+(-k)$ and in $\mathcal{B}_s^-$ $x_-(0)=\Sigma_{k=1}^{\infty}A_-^{k-1}B_-v_-(-k)$, that L_+ and L_- are isometries, that $\sigma x_+=x_-$ and $\sigma H=H$, to obtain that $\|x_0\|_{++}^2 = \inf\{\ \|v_+^{--}\|^2;\ (v_+,x_+,w)\in\mathcal{B}_s^+,\ x_+(0)=x_0\} = \inf\{\ \|v_+\|^2;\ (v_+,x_+,w)\in\mathcal{B}_s^+,\ x_+(0)=x_0\} = \inf\{\ \|w\|^2;\ (v_+,x_+,w)\in\mathcal{B}_s^+,\ x_+(0)=x_0\} = \inf\{\ \|w\|^2;\ (v_-,x_-,w)\in\mathcal{B}_s^-,\ x_-(-1)=x_0\} = \inf\{\ \|w\|^2;\ (v_-,x_-,w)\in\mathcal{B}_s^-,\ x_-(0)=x_0\} = \inf\{\ \|v_-\|^2;\ (v_-,x_-,w)\in\mathcal{B}_s^-,\ x_-(0)=x_0\} = \inf\{\ \|v_-^{++}\|^2;\ (v_-,x_-,w)\in\mathcal{B}_s^-,\ x_-(0)=x_0\}=\|x_0\|_{--}^2 = \inf\{\ \|x_0\|_-^2+\|w^{++}\|^2;\ (v_-,x_-,w)\in\mathcal{B}_s^-,\ x_-(0)=x_0\} = \|x_0\|_-^2 + \inf\{\ \|w^+\|^2;\ (v_-,x_-,w)\in\mathcal{B}_s^-,\ x_-(-1)=x_0\} = \|x_0\|_-^2+\inf\{\|w^+\|^2;\ (v_+,x_+,w)\in\mathcal{B}_s^+,\ x_+(0)=x_0\} = \|x_0\|_-^2+\|x_0\|_+^2+\inf\{\ \|v_+^+\|^2;\ (v_+,x_+,w)\in\mathcal{B}_s^+,\ x_+(0)=x_0\}=\|x_0\|_+^2+\|x_0\|_-^2$. It remains to prove that $\|x_0\|_{++}^2=x_0^TQ_+^{-1}x_0$. This is a well-known result from linear quadratic control. For $v\in l_2^m$ define $Tv^{--}:=\Sigma_{k=1}^{\infty}A_+^{k-1}B_+v(-k)$, then for $(v_+,x_+,w)\in\mathcal{B}_s^+$ there holds $Tv_+^{--}=x_+(0)$. Let $T^*:\mathbb{R}^n\to l_2(\mathbb{N},\mathbb{R}^m)$ denote the adjoint operator of T, then $\|x_0\|_{++}^2=\inf\{\|v_+^{--}\|^2;\ Tv_+^{--}=x_0\}=\|T^*(TT^*)^{-1}x_0\|^2=x_0^T(TT^*)^{-1}x_0$, and a direct calculation shows that $TT^*=Q_+$. ■

Proof of lemma 5-8

Let $\mathcal{B}_s^2(A,B,C,D)$ and ${}^R\mathcal{B}_s^2(\tilde{A},\tilde{B},\tilde{C},\tilde{D})$ be minimal realizations of $H\in\mathbb{B}_2$. A change of corrdinates $x\to x':=Sx$ leads to a transformation of parameters $(A,B,C,D)\to(SAS^{-1},SB,CS^{-1},D)$ and $(\tilde{A},\tilde{B},\tilde{C},\tilde{D})\to(S\tilde{A}S^{-1},S\tilde{B},\tilde{C}S^{-1},\tilde{D})$. From (ARE) it follows that $(K_+,\tilde{K}_+)\to((S^{-1})^TK_+S^{-1},(S^{-1})^T\tilde{K}_+S^{-1})$. Now let $K_+^{1/2}\tilde{K}_+^{-1}K_+^{1/2}=U\Lambda^2U^T$ with U orthogonal and $\Lambda=\mathrm{diag}(\lambda_1,\ldots,\lambda_n)$, $\lambda_1\geq\ldots\geq\lambda_n>0$. A direct calculation shows that $S:=\Lambda^{-1/2}U^TK_+^{1/2}$ gives the desired result. ■

Proof of proposition 5-9

We will show that $\sigma(\hat{A})\subset\mathbb{C}_+$. According to propostion 3–5(i) there then exists a map $v \rightarrow \hat{w}$ with domain l_2^m for $\mathcal{B}_s^2(\hat{A},\hat{B},\hat{C},\hat{D})$. As $\hat{D}=D_+$ is injective we can conclude that $m(\hat{H})=c_\infty(\hat{H})=m$. As $\mathcal{B}_s^2(\hat{A},\hat{B},\hat{C},\hat{D})$ is a realization of $\hat{H}$ we conclude that $n(\hat{H})\leq\bar{n}$.

To show that $\sigma(\hat{A})\subset\mathbb{C}_+$ we use the fact that with $D_b:=D_+$ (A_b,B_b,C_b,D_b) are the parameters of the forward scattering representation of H with the corresponding solution K_b of (ARE) given by $\Lambda=\text{diag}(\lambda_1,\ldots,\lambda_n)$. Define $\hat{\Lambda}:=\text{diag}(\lambda_1,\ldots,\lambda_{\bar{n}})$ and $\Lambda_2:=\text{diag}(\lambda_{\bar{n}+1},\ldots,\lambda_n)$. From propositions 4-6 and 4-7 we conclude that $I=B_b^T\Lambda B_b+D_b^TD_b$ from (R_b), that $0=B_b^T\Lambda A_b+D_b^TC_b$ from (F_b), and hence from (ARE_b) that $\Lambda=A_b^T\Lambda A_b+C_b^TC_b$.

Take partitions as in step 3 of the algorithm for balancing. Let $\alpha\in\sigma(\hat{A})$ and $\hat{A}\hat{x}=\alpha\hat{x}$ for some $0\neq\hat{x}\in\mathbb{C}^{\bar{n}}$. Let $x:=\begin{bmatrix}\hat{x}\\0\end{bmatrix}\in\mathbb{C}^n$. A direct computation of $x^*(\text{ARE}_b)x$ shows that then $\hat{x}^*[A_{21}^T\Lambda_2A_{21}+\hat{C}^T\hat{C}]\hat{x}=(1-|\alpha|^2)\hat{x}^*\hat{\Lambda}\hat{x}$. As $\Lambda>0$ it follows that $|\alpha|\leq1$. It suffices to show that $|\alpha|\neq1$. As $\Lambda_2>0$, $|\alpha|=1$ would imply that $A_{21}\hat{x}=0$ and hence that $A_bx=\alpha x$, which contradicts the fact that $\sigma(A_b)\subset\mathbb{C}_+$, cf. proposition 4-6. ■

CHAPTER V

Proof of proposition 2-8

The minimax property is well-known and shown e.g. in Stewart [67, theorem 6.5 of section 6.6]. As the property is of crucial importance in the sequel we give its proof explicitly.

Let A have (SVD) $A=U\Sigma V^T$. For $x\in\mathbb{R}^{n_2}$ and $L\subset\mathbb{R}^{n_2}$ let $y:=V^Tx$ and $L':=V^TL$. Then $\min\{\max_{0\neq x\in L}\frac{\|Ax\|}{\|x\|};\ \dim(L)\geq n_2-k\}=\min\{\max_{0\neq y\in L'}\{\Sigma_{i=1}^{n_2}\sigma_i^2y_i^2\}^{1/2};\ \Sigma_{i=1}^{n_2}y_i^2=1,\ \dim(L')\geq n_2-k\}=:\alpha_k$. Taking $L':=\text{span}\{e_{k+1},...,e_{n_2}\}$ shows that $\alpha_k\leq\sigma_{k+1}$. On the other hand, if $\dim(L')\geq n_2-k$, then $V:=L'\cap\text{span}\{e_1,...e_{k+1}\}\neq\{0\}$, hence V contains a y' of norm 1. This implies that $\alpha_k\geq\sigma_{k+1}$.

Finally note that α_k is achieved for $L':=\text{span}\{e_{k+1},...,e_{n_2}\}$, i.e., for $L:=VL'=\text{span}\{v_{k+1},...,v_{n_2}\}=L^*_{n_2-k}$. ■

Proof of lemma 2-9

If $n_1=n_2$, $A=\text{diag}(d_1,...,d_{n_2})$ with $d_1\geq...\geq d_{n_2}\geq0$ and $L=\mathbb{R}^{n_2}$, then it easily follows that $\varepsilon_A(\mathbb{R}^{n_2})$ is well-defined and equal to $(d_1,...,d_{n_2})$. To prove the general case, let $L\subset\mathbb{R}^{n_2}$ with dim $(L)=d$ and let the columns of $B\in\mathbb{R}^{n_2\times d}$ form an orthonormal basis of L. Further let B^TA^TAB have (SVD) $U\Sigma U^T$ with $\Sigma=\text{diag}(\sigma_1,...,\sigma_d)$, $\sigma_1\geq...\geq\sigma_d\geq0$. Then it follows that for $0\neq x\in L$, say $x=BUy$, there holds $\frac{\|Ax\|}{\|x\|}=\frac{\|\Sigma^{1/2}y\|}{\|y\|}$. From this and the fact that $U^TB^TBU=I_d$ it is easily seen that $\varepsilon_A(L)|_{[1,d]}=\varepsilon_{\Sigma^{1/2}}(\mathbb{R}^d)$ which was already seen to be well-defined. Hence $\varepsilon_A(L)$ is also well-defined and is equal to $(\sigma_1^{1/2},...,\sigma_d^{1/2},0,...0)$. ■

Proof of proposition 2-10

If $k\leq n_2-r$ then $\varepsilon_A(L_k^*)=0$ and the results easily follow. We hence will assume that $k>n_2-r$.

(*i*) As $\dim(L\cap(\text{span}\{x_1,...,x_{j-1}\})^\perp)\geq k-j+1$ it follows from proposition 2-8 that for $j=1,...,\dim(L)$ $\varepsilon_j(L)\geq\sigma_{n_2-k+j}$. For L_k^* it follows by induction that for

$j=1,\ldots,k \quad \varepsilon_j(L_k^*)=\sigma_{n_2-k+j}$.

(*ii*) First suppose that $n_1=n_2$ and that $A=\text{diag}(d_1,\ldots,d_{n_2})$ with $d_1\geq\ldots\geq d_{n_2-k}>d_{n_2-k+1}\geq\ldots\geq d_r>d_{r+1}=\ldots=d_{n_2}=0$. We then have to prove that $\varepsilon_A(L)=(d_{n_2-k+1},\ldots,d_{n_2},0,\ldots,0)$ implies that $L=L_k^*$, where $L_k^*=\text{span}\{e_{n_2-k+1},\ldots,e_{n_2}\}$.

Let $0<\delta_1<\ldots<\delta_s$ denote the distinct values in $\{d_{n_2-k+1},\ldots,d_r\}$ and let δ_j have multiplicity m_j, $\Sigma_{j=1}^s m_j=\max\{0,r-n_2+k\}$. Define $M_j:=n_2-r+\Sigma_{i=1}^j m_i$, $j\in[1,s]$, and $M_0:=n_2-r$. We will show by induction that for $j=0,\ldots,s$ $\text{span}\{e_{n_2-M_j+1},\ldots,e_{n_2}\}\subset L$. For $j=s$ this gives the desired result, as $n_2-M_s+1=n_2-k+1$. First consider $j=0$. Proposition 2–8 implies that for $\dim(L')\geq k+1$ $\varepsilon_1(L')\geq d_{n_2-k}$, hence $\dim(L)=k$. As $\varepsilon_i(L)=0$ for $i\in[r-n_2+k+1,k]$ there is a subspace $V_0\subset L$ with $\dim(V_0)\geq n_2-r$ and $Ax=0$ for $x\in V_0$. Hence $V_0=\text{span}\{e_{r+1},\ldots,e_{n_2}\}$ which proves the result for $j=0$. Next suppose that $\text{span}\{e_{n_2-M_{j-1}+1},\ldots,e_{n_2}\}\subset L$ for some $j\in[1,s]$. As $\varepsilon_A(L)\big|_{[(\Sigma_{i=j+1}^s m_i)+1,k]}=(\delta_j,\ldots,\delta_j,\delta_{j-1},\ldots,\delta_{j-1},\ldots,\delta_1,\ldots,\delta_1,0,\ldots,0)$, where 0 appears n_2-r and δ_i appears m_i times, $i=1,\ldots,j$, it easily follows that there is a subspace $V_j\subset L$ with $\dim(V_j)=m_j$, $V_j\perp\text{span}\{e_{n_2-M_{j-1}+1},\ldots,e_{n_2}\}$ and $\max_{0\neq x\in V_j}\frac{\|Ax\|}{\|x\|}=\delta_j$. Hence $V_j=\text{span}\{e_{n_2-M_j+1},\ldots,e_{n_2-M_{j-1}}\}$ which concludes the inductive proof.

We finally prove the result for general A. Let $A\in\mathbb{R}^{n_1\times n_2}$ have (SVD) $A=U\Sigma V^T$ and define $D:=\text{diag}(\sigma_1,\ldots,\sigma_{n_2})$ where $\sigma_1\geq\ldots\geq\sigma_{n_2-k}>\sigma_{n_2-k+1}\geq\ldots\geq\sigma_r>\sigma_{r+1}=\ldots=\sigma_{n_2}=0$. Suppose $\varepsilon_A(L)=(\sigma_{n_2-k+1},\ldots,\sigma_{n_2},0,\ldots,0)$, and let $L':=V^TL$. With $y:=V^Tx$ there holds $\frac{\|Ax\|}{\|x\|}=\frac{\|Dy\|}{\|y\|}$. Using the orthogonality of V we conclude that $\varepsilon_A(L)=\varepsilon_D(L')$. We have shown that hence $L'=\text{span}\{e_{n_2-k+1},\ldots,e_{n_2}\}$, so $L=VL'=L_k^*$.

(*iii*) The proof of ($\Leftarrow$) is direct. To prove ($\Rightarrow$), let $\varepsilon_A(L)=(\sigma_{n_2-k+1},\ldots,\sigma_r,0,\ldots,0)$. Then $\dim(L)=k$. Let the singular values of A satisfy $\sigma_1\geq\ldots\geq\sigma_{c_1}>\sigma_{c_1+1}=\ldots=\sigma_{n_2-k}=\sigma_{n_2-k+1}=\ldots=\sigma_{c_2}>\sigma_{c_2+1}\geq\ldots\geq\sigma_{n_2}\geq0$. From the proof of (*ii*) it easily follows that $L':=\text{span}\{v_{c_2+1},\ldots,v_{n_2}\}\subset L$. Further there is a subspace $L''\subset L$ with $L''\perp L'$, $\dim(L'')=k-\dim(L')$, such that $\max_{0\neq x\in L''}\frac{\|Ax\|}{\|x\|}=\sigma_{n_2-k}$. Hence $L''\subset\text{span}\{v_{c_1+1},\ldots,v_{c_2}\}$. ■

Proof of proposition 2–11

It follows from definition 2–2 that $e^D(d,a)=\frac{\|Sa\|}{\|a\|}$ and from definition 2–4 that $\varepsilon^D(d,M)=\varepsilon_S(M^\perp)$, which is well-defined according to lemma 2–9. ■

Proof of proposition 2-12

The ordering of the misfits is lexicographical, cf. definition 2–5, and according to proposition 2–11 $\varepsilon^D(d,M)=\varepsilon_S(M^\perp)$ with S the empirical covariance matrix corresponding to d. This enables us to use proposition 2–10.

(*i*) This is trivial from the definition of $P^D_{c_{tol}}$.

(*ii*) Let $L^*:=\text{span}\{\tilde{x}_1,\ldots,\tilde{x}_N\}$. As $\text{rank}(S)=r$ $\dim(L^*)=r\leq c_{tol}$, and $\varepsilon^D(d,L^*)=\varepsilon_S((L^*)^\perp)=0$, hence minimal. To show optimality of L^* it remains to prove that $\{\dim(M)\leq r,\ \varepsilon^D(d,M)=0\}\Rightarrow\{M=L^*\}$. If $\varepsilon^D(d,M)=\varepsilon_S(M^\perp)=0$ then $\|Sa\|=0$ for all $a\in M^\perp$, so $M^\perp\subset\ker(S)$ and hence $M\supset\text{im}(S)=L^*$. If in addition $\dim(M)\leq r=\dim(L^*)$ then $M=L^*$.

(*iii*) First note that $(M^*_{c_{tol}})^\perp=L^*_{n-c_{tol}}$ as defined in section 2.1.2. Let $M\in\mathbb{M}$ with $c^D(M)\leq c_{tol}$, then $\dim(M^\perp)\geq n-c_{tol}$. According to proposition 2–10(*i*) then $\varepsilon^D(d,M)=\varepsilon_S(M^\perp)\geq\varepsilon_S(L^*_{n-c_{tol}})=\varepsilon^D(d,M^*_{c_{tol}})$, while according to proposition 2–10(*ii*) $\sigma_{c_{tol}}>\sigma_{c_{tol}+1}$ implies that $\varepsilon_S(M^\perp)=\varepsilon_S(L^*_{n-c_{tol}})$ if and only if $M^\perp=L^*_{n-c_{tol}}$, i.e., $M=M^*_{c_{tol}}$.

(*iv*) As for $c^D(M)\leq c_{tol}$ there holds $\dim(M^\perp)\geq n-c_{tol}$ and $\sigma_{c_{tol}}\geq\sigma_r>0$ we conclude from proposition 2–10(*i*) and (*iii*) that M has minimal misfit if and only if $M^\perp=L'+L''$, where $L':=\text{span}\{u_j;\ \sigma_j<\sigma_{c_{tol}}\}$ with $\dim(L')=n-c_1-\dim(M(\sigma_{c_{tol}}))$ and $L''\subset M(\sigma_{c_{tol}})$ with $\dim(L'')=n-c_{tol}-\dim(L')=c_1-c_{tol}-\dim(M(\sigma_{c_{tol}}))$. Then $M=(L')^\perp\cap(L'')^\perp=M^*_{c_1}+L$, where $L:=M(\sigma_{c_{tol}})\cap(L'')^\perp$ has dimension $\dim(M(\sigma_{c_{tol}}))-\dim(L'')=c_{tol}-c_1$. ■

Proof of proposition 2-13

(*i*) This is evident as $\varepsilon^D_1(d,\{0\})=\sigma_1<\varepsilon^{tol}_1$.

(*ii*) Let $L^*:=\text{span}\{\tilde{x}_1,\ldots,\tilde{x}_N\}$, then $\varepsilon^D_1(d,L^*)=0<\varepsilon^{tol}_1$. If $c^D(M)<r$ then $\dim(M^\perp)\geq n-r+1$ and according to proposition 2–10(*i*) $\varepsilon^D_1(d,M)\geq\sigma_r\geq\varepsilon^{tol}_1$. So the minimal achievable complexity is r. To show optimality of L^* it remains to prove that $\{c^D(M)=r,\ \varepsilon^D(d,M)=0\}\Rightarrow\{M=L^*\}$, for which we refer to the last part of the proof of proposition 2–12(*ii*).

(*iii*) First note that $(M^*_k)^\perp=L^*_{n-k}$ as defined in section 2.1.2, hence $\varepsilon^D(d,M^*_k)=\varepsilon_S(L^*_{n-k})=(\sigma_{k+1},\ldots,\sigma_r,0,\ldots,0)$, especially $\varepsilon^D_i(d,M^*_k)\leq\sigma_{k+1}<\varepsilon^{tol}_1$ for all $i\in[1,n]$. If $c^D(M)<k$ then $\dim(M^\perp)\geq n-k+1$ and according to proposition 2–10(*i*) $\varepsilon^D_1(d,M)\geq\sigma_k\geq\varepsilon^{tol}_1$. Hence the minimal achievable complexity is k. As $\sigma_k>\sigma_{k+1}$ we conclude from proposition 2–10(i) and (ii) that among models with $c^D(M)=k$ $\varepsilon^D(d,M)$ is uniquely minimized by taking $M:=M_k^*$. ■

Proof of proposition 2–18

Let the data be generic, so S_{xx} and S_{yy} are invertible. Let $M\in\mathbb{M}$ and $M_2^\perp:=\{a_2\in\mathbb{R}^{n_2};\ \exists a_1\in\mathbb{R}^{n_1}$ such that $(a_1,a_2)\in M^\perp\}$. Suppose that $(a_1,a_2)\in M^\perp$ and $(a_1',a_2)\in M^\perp$, then $(a_1-a_1',0)\in M^\perp$ and as the projection in M on the first n_1 coordinates is surjective it follows that $a_1=a_1'$. Hence there exists a linear map $A:\mathbb{R}^{n_2}\to\mathbb{R}^{n_1}$ such that $M^\perp=\{(a_1,a_2);\ a_2\in M_2^\perp,\ a_1=Aa_2\}$. For $a=(a_1,a_2)\in M^\perp$ let $\alpha:=S_{yy}^{1/2}a_2$, then it follows that $e^P(d,a)=\{a^T\begin{bmatrix}S_{xx} & S_{xy}\\ S_{yx} & S_{yy}\end{bmatrix}a\}^{1/2}/\{a_2^TS_{yy}a_2\}^{1/2}=\frac{\|Q^{1/2}\alpha\|}{\|\alpha\|}$ where $Q:=S_{yy}^{-1/2}(A^TS_{xx}A+S_{yy}+A^TS_{xy}+S_{yx}A)S_{yy}^{-1/2}$. As $\alpha\perp\alpha'$ if and only if $a_2\perp_{(2)}a_2'$ it follows from definition 2–17 that for generic data $\varepsilon^P(d,M)=\varepsilon_{Q^{1/2}}(S_{yy}^{1/2}M_2^\perp)$ which is well-defined according to lemma 2–9. ■

Proof of lemma 2–19

Let $L_2\subset\mathbb{R}^{n_2}$ be given and $M\in\mathbb{M}(L_2)$. In the proof of proposition 2–18 it was shown that there exists an $A:\mathbb{R}^{n_2}\to\mathbb{R}^{n_1}$ such that $M^\perp=\{(a_1,a_2);\ a_2\in L_2,\ a_1=Aa_2\}$ and that for generic data $\varepsilon^P(d,a)=\frac{\|Q^{1/2}\alpha\|}{\|\alpha\|}$ with $\alpha:=S_{yy}^{1/2}a_2$ and $Q:=S_{yy}^{-1/2}(A^TS_{xx}A+S_{yy}+A^TS_{xy}+S_{yx}A)S_{yy}^{-1/2}$. Define $\Delta:=S_{xx}^{-1/2}(A+S_{xx}^{-1}S_{xy})S_{yy}^{-1/2}$, then a direct calculation shows that $Q=I-V\Sigma^T\Sigma V^T+\Delta^T\Delta$. So $\varepsilon^P(d,M)$ clearly is minimal on $\mathbb{M}(L_2)$ if and only if $\Delta\alpha=0$ for all $\alpha\in S_{yy}^{1/2}L_2$, i.e., $(\Delta S_{yy}^{1/2})|_{L_2}=0$. As $S_{xx}>0$ it follows that $A|_{L_2}=(-S_{xx}^{-1}S_{xy})|_{L_2}$ which corresponds to $M=M^*(L_2)$. From the last line of the proof of proposition 2–18 it follows that $\varepsilon^P(d,M^*(L_2))=\varepsilon_{(I-V\Sigma^T\Sigma V^T)^{1/2}}(S_{yy}^{1/2}L_2)=\varepsilon_{(I-\Sigma^T\Sigma)^{1/2}}(V^TS_{yy}^{1/2}L_2)$. ■

Proof of corollary 2–20

Let $M\in\mathbb{M}$ with $c^P(M)\leq n_2-k$, hence $\dim(M_2^\perp)\geq k$, and let $L_2:=V^TS_{yy}^{1/2}M_2^\perp$. It follows from lemma 2–19 and proposition 2–10(i) that for generic data $\varepsilon^P(d,M)\geq\varepsilon^P(d,M^*(M_2^\perp))=\varepsilon_{(I-\Sigma^T\Sigma)^{1/2}}(L_2)\geq\varepsilon_{(I-\Sigma^T\Sigma)^{1/2}}(L_k^*)$, where $L_k^*=\text{span}\{e_1,\ldots,e_k\}$ as $(I-\Sigma^T\Sigma)^{1/2}$ has eigenvalues $\lambda_1\geq\ldots\geq\lambda_{n_2}\geq0$ with $\lambda_i=(1-\sigma_{n_2-i+1}^2)^{1/2}$, $i\in[1,n]$, and as e_{n_2-i+1} is an eigenvector corresponding to λ_i.

Hence minimal misfit is achieved by the model $M^*(M_2^\perp)$ with $V^TS_{yy}^{1/2}M_2^\perp=L_k^*$. A direct calculation then shows that $M_2^{\perp}=\text{span}\{a_2^{(i)};\ i\in[1,k]\}$ and that $-S_{xx}^{-1}S_{xy}a_2^{(i)}=-\sigma_i a_1^{(i)}$, so $M^*(M_2^{\perp})=\{(x,y)\in\mathbb{R}^{n_1}\times\mathbb{R}^{n_2};\ <-S_{xx}^{-1}S_{xy}a_2,x>+<a_2,y>=0$ for all $a_2\in M_2^{\perp}\}=\{(x,y)\in\mathbb{R}^{n_1}\times\mathbb{R}^{n_2};\ <a_2^{(i)},y>=\sigma_i\cdot<a_1^{(i)},x>,\ \ i\in[1,k]\}=M_k^*$. Moreover lemma 2–19 implies that $\varepsilon^P(d,M_k^*)=\varepsilon_{(I-\Sigma^T\Sigma)^{1/2}}(L_k^*)=((1-\sigma_k^2)^{1/2},\ldots,(1-\sigma_1^2)^{1/2},\ 0,\ldots,0)$. ■

Proof of proposition 2–21

According to lemma 2–19 it suffices to determine those subspaces $M_2^{\perp}\subset\mathbb{R}^{n_2}$ for which $\varepsilon_{(I-\Sigma^T\Sigma)^{1/2}}(L_2)$ is minimal, where $L_2:=V^TS_{yy}^{1/2}M_2^{\perp}$, and to accept the models $M^*(M_2^{\perp})$. The requirement $c^P(M)\le c_{tol}$ is equivalent to $\dim(L_2)=n_2-c^P(M)\ge n_2-c_{tol}=:k$. Let $\lambda_1\ge\ldots\ge\lambda_{n_2}\ge 0$ denote the singular values of $(I-\Sigma^T\Sigma)^{1/2}$, i.e., $\lambda_i=(1-\sigma_{n_2-i+1}^2)^{1/2}$, so especially for $r<n_2$ $\lambda_1=\ldots=\lambda_{n_2-r}=1$ and for $r^*>0$ $\lambda_{n_2-r^*+1}=\ldots=\lambda_{n_2}=0$. In the notation of section 2.1.2 $L_k^*=\text{span}\{e_1,\ldots e_k\}$, and in the proof of corollary 2–20 it was shown that $M^*(S_{yy}^{-1/2}VL_k^*)=M_k^*$, $\varepsilon_1^P(d,M_k^*)=(1-\sigma_k^2)^{1/2}$.

(*i*) If $c_{tol}<n_2-r$ then $k>r$ and hence $\lambda_{n_2-k}=\lambda_{n_2-k+1}=1$. According to proposition 2–10(*i*) and (*iii*) the optimal models are obtained by taking $L_2=L'+L''$ with $L'=L_r^*$ and $L''\subset\text{span}\{e_i;\ i\in[r+1,n_2]\}$ with $\dim(L'')=n_2-c_{tol}-r$, i.e., $L_2\supset L_r^*$ and $\dim(L_2)=n_2-c_{tol}$. This is equivalent to $\dim(M_2^{\perp})=n_2-c_{tol}$ and $M\subset M^*(S_{yy}^{-1/2}VL_r^*)=M_r^*$.

(*ii*) Clearly $c^P(M_{r^*}^*)=n_2-r^*\le c_{tol}$ and $\varepsilon^P(d,M_{r^*}^*)=(1-\sigma_{r^*}^2)^{1/2}=0$, hence $\varepsilon^P(d,M_{r^*}^*)=0$. So it suffices to prove that $\{c^P(M)\le n_2-r^*,\ \varepsilon^P(d,M)=0\}\Rightarrow\{M=M_{r^*}^*\}$. If $\varepsilon^P(d,M)=0$, then there holds $M^{\perp}\subset\ker(\begin{pmatrix}S_{xx} & S_{xy}\\ S_{yx} & S_{yy}\end{pmatrix}) = \begin{pmatrix}S_{xx}^{-1/2}U & 0\\ 0 & S_{yy}^{-1/2}V\end{pmatrix}\cdot\ker(\begin{pmatrix}I & \Sigma\\ \Sigma^T & I\end{pmatrix}) = \begin{pmatrix}S_{xx}^{-1/2}U & 0\\ 0 & S_{yy}^{-1/2}V\end{pmatrix}\cdot\text{span}\{\begin{pmatrix}e_i'\\ -e_i''\end{pmatrix};\ i\in[1,r^*]\}=\text{span}\{(a_1^{(i)},-a_2^{(i)});\ i\in[1,r^*]\}$, where e_i' and e_i'' denote the i–th unit vectors in $\mathbb{R}^{n_1}$ and $\mathbb{R}^{n_2}$ respectively. If in addition $c^P(M)\le n_2-r^*$, then $\dim(M^{\perp})=\dim(M_2^{\perp})\ge r^*$, hence $M^{\perp}=\text{span}\{(a_1^{(i)},-a_2^{(i)});\ i\in[1,r^*]\}$ and $M=M_{r^*}^*$.

(*iii*) If $\sigma_{n_2-c_{tol}}>\sigma_{n_2-c_{tol}+1}$ then $\lambda_{n_2-k}>\lambda_{n_2-k+1}$ and according to proposition 2–10(*ii*) we get $L_2=L_k^*$ with corresponding model $M_k^*=M_{n_2-c_{tol}}^*$.

(*iv*) If $\sigma_{n_2-c_{tol}}=\sigma_{n_2-c_{tol}+1}$ then $\lambda_{n_2-k}=\lambda_{n_2-k+1}$, so according to proposition 2–10(*i*) and (*iii*) the optimal models are obtained by taking $L_2=L'+L''$ where $L'=\text{span}\{e_1,\ldots,e_{c_1}\}$ and $L''\subset\text{span}\{e_{c_1+1},\ldots,e_{c_2}\}$ with $\dim(L'')=k-\dim(L')=n_2-c_{tol}-c_1$. The corresponding models are $M^*(L_2)=M_{c_1}^*\cap L$ where $L^{\perp}\subset M(\sigma_{n_2-c_{tol}})^{\perp}$ with $\dim(L^{\perp})=\dim(L'')$, so $M(\sigma_{n_2-c_{tol}})\subset L$ and $c^P(L)=n_2-\dim(L'')=c_{tol}+c_1$. ■

Proof of proposition 2-22

(i) Clearly $\varepsilon_1^P(d,M_{n_2}^*)=(1-\sigma_{n_2}^2)^{1/2}<\varepsilon_1^{tol}$ and $c^P(M_{n_2}^*)=0$, hence it suffices to show that $\{c^P(M)=0,\ \varepsilon^P(d,M)\leq\varepsilon^P(d,M_{n_2}^*)\}\Rightarrow\{M=M_{n_2}^*\}$. This follows from lemma 2-19 as for $c^P(M)=0$ $M_2^\perp=\mathbb{R}^{n_2}$ and $M^*(\mathbb{R}^{n_2})=M_{n_2}^*$.

(ii) If $c^P(M)<n_2$ then $\dim(M_2^\perp)\geq1$ and according to corollary 2-20 $\varepsilon_1^P(d,M)\geq(1-\sigma_1^2)^{1/2}\geq\varepsilon_1^{tol}$. Hence $c^P(M)=n_2$, so $M_2^\perp=\{0\}$ and hence $M^\perp=\{0\}$, i.e., $M=\mathbb{R}^{n_1+n_2}$.

(iii) Let $k:=r$ and note that $(1-\sigma_k^2)^{1/2}<\varepsilon_1^{tol}\leq1=(1-\sigma_{k+1}^2)^{1/2}$. The result then follows from the proof of (iv).

(iv) If $c^P(M)<n_2-k$ then according to corollary 2-20 $\varepsilon_1^P(d,M)\geq(1-\sigma_{k+1}^2)^{1/2}\geq\varepsilon_1^{tol}$. As $c^P(M_k^*)=n_2-k$ and $\varepsilon_1^P(d,M_k^*)=(1-\sigma_k^2)^{1/2}<\varepsilon_1^{tol}$ it follows that the minimal achievable complexity is n_2-k. As $\sigma_k>\sigma_{k+1}$ it follows from proposition 2-21(iii) with $c_{tol}:=n_2-k$ that the optimal model is M_k^*. ■

Proof of theorem 4-4

Let $\mathcal{B}$ be as defined in step 5 of the algorithm. Due to assumption 4-3(ii) $\mathcal{B}$ is uniquely defined. It follows from steps 3.0 and 3.2 that $v_t(L_t)\perp v_t(\mathcal{B}_{t-1}^\perp+s\mathcal{B}_{t-1}^\perp)$ and it inductively follows from assumption 4-3(iii) that $\mathcal{B}_t^\perp$ as defined in steps 2.2 and 3.2 indeed exactly consists of the t-th order laws claimed by $\mathcal{B}$ as defined in section II.3.2.4. Then (iv) follows from the definition of L_t^D in section II.3.2.5 and (ii) is implied by proposition II.3-10.

We prove (i) and (iii) by induction. For $t=0$ the restriction $c(\mathcal{B})\leq c_{tol}$ implies that at least e_0^{tol} zero order laws should be accepted, cf. definition 4-1. As for such laws $e^D(\tilde{w},r)=\|S^{1/2}(\tilde{w},0)v_0(r)^T\|/\|v_0(r)\|$ it follows from assumption 4-3(ii) and proposition 2-10(ii) that the unique optimal solution is given by V_0 as defined in step 2.2, with corresponding misfit $(\sigma_{q-e_0^{tol}+1},\ldots,\sigma_q,0,\ldots,0)$ where σ_k is the k-th singular value of $S^{1/2}(\tilde{w},0)$. This shows (iii) for $t=0$. Note that due to assumption 4-3(ii) and the lexicographic ordering of misfits, cf. definitions 3-4 and 3-9, it is suboptimal to accept more than e_0^{tol} laws.

Next suppose that optimality of step 3.2 is shown for steps τ with $\tau\leq t-1$ for some $t\geq1$. According to definition 4-1 it follows from the requirement $c(\mathcal{B})\leq c_{tol}$ and the fact that e_τ^{tol} equations of order τ have been accepted,

$\tau \le t-1$, that at least e_t^{tol} equations of order t have to be accepted. Moreover, according to definition 3–3 we have to minimize the misfit of the newly accepted laws in $[v_t(\mathcal{B}_{t-1}^{\perp}+s\mathcal{B}_{t-1}^{\perp})]^{\perp}$. Now for $R\in\mathbb{R}_t^{1\times q}[s]$ there holds $\{v_t(r)\perp v_t(\mathcal{B}_{t-1}^{\perp}+s\mathcal{B}_{t-1}^{\perp})\} \Leftrightarrow \{\exists a\in\mathbb{R}^{q_t}$ such that $v_t(r)=a^T P_t\}$. Hence for such laws t we get $e^D(\tilde{w},r)=\|r\tilde{w}\|/\|r\|=(v_t(r)S(\tilde{w},t)v_t(r)^T)^{1/2}/\|v_t(r)\|=\|S_t^{1/2}a\|/\|a\|$, where $S_t:=P_tS(\tilde{w},t)P_t^T$. Here we used the fact that $P_tP_t^T=I_{q_t}$. Assumption 4–3(ii) and proposition 2–10(ii) imply that the unique optimal solution for step t is given by V_t of step 3.2. The corresponding misfit is $(\sigma_{q_t-e_t^{tol}+1},\ldots,\sigma_{q_t},0,\ldots,0)$ where σ_k is the k–th singular value of $S_t^{1/2}$, which shows (iii) for step t. This concludes the inductive proof of (i) and (iii). ■

Proof of theorem 4–6

Due to assumption 4–5(ii) $\mathcal{B}$ in step 5 of the algorithm is uniquely defined. Now (ii) and (iv) follow from assumption 4–5(iii) as in the proof of theorem 4–4(ii) and (iv).

We prove (i) and (iii) by induction. For $t=0$ it follows from the ordering of definition 3–12 that c_0 should be minimized, i.e., the number of zero order laws should be maximized. As for such laws $e^D(\tilde{w},r)=\|S^{1/2}(\tilde{w},0)v_0(r)^T\|/\|v_0(r)\|$, it follows from proposition 2–10(i) and from step 2.1 of the algorithm that the requirement $\varepsilon_{0,1}^D(\tilde{w},\mathcal{B})<\bar{\varepsilon}_0^{tol}$ implies that at most e_0 laws can be accepted. Given the ordering of definition 3–12 the misfit of these e_0 laws should be minimized. Proposition 2–10(ii) implies that the unique optimal solution is given by V_0 as defined in step 2.2, with corresponding misfit $(\sigma_{q-e_0+1},\ldots,\sigma_q,$ $0,\ldots,0)$ where σ_k is the k–th singular value of $S^{1/2}(\tilde{w},0)$. This shows (iii) for $t=0$.

Next suppose that optimality of step 3.2 is shown for steps τ with $\tau\le t-1$ for some $t\ge 1$. According to definition 3–3 we have to minimize the misfit of newly accepted laws $r\in v_t^{-1}\{[v_t(\mathcal{B}_{t-1}^{\perp}+s\mathcal{B}_{t-1}^{\perp})]^{\perp}\}$. Now r is in this set if and only if there exists an $a\in\mathbb{R}^{q_t}$ such that $v_t(r)=a^TP_t$ and then $e^D(\tilde{w},r)=\|S_t^{1/2}a\|/\|a\|$ with $S_t:=P_tS(\tilde{w},t)P_t^T$.

First assume that $e_t''\le e_t'$. The ordering of definition 3–12 implies that a maximal number of t–th order laws should be accepted. Due to the requirement $\varepsilon_{t,1}^D(\tilde{w},\mathcal{B})<\bar{\varepsilon}_t^{tol}$ it follows from proposition 2–10(i) and step 3.1 of the algorithm that at most e_t'' t–th order laws can be accepted. The ordering of definition 3–12 and proposition 2–10(*ii*) then imply that the unique optimal solution is given by V_t as defined in step 3.2, with corresponding misfit $(\sigma_{q_t-e''_t+1},\ldots,\sigma_{q_t},0,\ldots,0)$, where σ_k is the k–th singular value of $S_t^{1/2}$, which shows (*iii*) for step t.

Finally assume that $e''_t>e'_t$. Assumption 4–5(*iii*) implies that $q-e'_t$ independent laws or orders $\tau\leq t-1$ have been accepted. It then follows that at most e'_t independent laws of order t may be accepted. Indeed, otherwise the resulting set of laws cannot be bilaterally row proper and from proposition II.3–8 it follows that then there would exist an equivalent set of laws with more than $q-e'_t$ laws of order at most t–1. According to steps 3.1 of the algorithm for $\tau\leq t-1$ this would necessarily lead to unacceptable misfits. Assumption 4–5(*ii*), the ordering of definition 3–12 and proposition 2–10(*ii*) imply that the unique optimal solution in this case is given by V_t as defined in step 3.2 with misfit as given in (*iii*).

This concludes the inductive proof of (*i*) and (*iii*). ■

Proof of theorem 4–8

Assumption 4–7(*ii*) implies that step 3.1 is well–defined, assumption 4–7(*iii*) that step 3.2 is well–defined and that $\mathcal{B}$ in step 5 is uniquely defined. It follows from steps 3.0 and 3.2 that $v_t(L_t)\perp v_t(F_{t-1}s^t)+v_t(\mathcal{B}_{t-1}^{\perp})$ and it inductively follows from assumption 4–7(*iv*) that $\mathcal{B}_t^{\perp}$ as defined in steps 2.2 and 3.2 indeed exactly consists of the t–th order laws claimed by $\mathcal{B}$ as defined in section II.3.2.4. Then (*iv*) follows from the definition of L_t^P in section II.3.2.6 and (*ii*) is implied by proposition II.3–10.

We prove (*i*) by induction. For t=0 the optimality of step 2.2 follows from theorem 4–4, cf. definition 3–7. Next suppose that optimality of step 3.2 is shown for steps τ with $\tau\leq t-1$ for some $t\geq1$. According to definition 4–1 it follows from the requirement $c(\mathcal{B})\leq c_{tol}$ and the fact that e_τ^{tol} equations of order τ have been accepted, $\tau\leq t-1$, that at least e_t^{tol} equations of order t have to be accepted. Moreover, according to definition 3–7 we have to minimize the misfit of the newly accepted laws in $[v_t(F_{t-1}s^t)+v_t(\mathcal{B}_{t-1}^{\perp})]^{\perp}$. Now for $r\in\mathbb{R}_t^{1\times q}[s]$ there holds $\{v_t(r)\perp[v_t(F_{t-1}s^t)+v_t(\mathcal{B}_{t-1}^{\perp})]\}\Leftrightarrow\{\exists a_-\in\mathbb{R}^{q_t},\ a_+\in\mathbb{R}^{q-n_{t-1}}$ such that $v_t(r)=(a_-^T,a_+^T)P_t\}$. Let $a:=\begin{bmatrix}a_-\\a_+\end{bmatrix}$, $\tilde{z}_i^{(t)}:=P_t\cdot\mathrm{col}(\tilde{w}(t_0+i),\ldots,\tilde{w}(t_0+i+t))$, $i\in[0,t_1-t_0-t]$, then for such laws r we get $e^P(\tilde{w},r)=\|r\tilde{w}\|/\|r^*\tilde{w}\|_t=$

$(a^T P_t S(\tilde{w},t) P_t^T a)^{1/2} / \|a_+^T P_{2t}\tilde{w}\|_t = e^P(\tilde{z}^{(t)},a)$, where the last, static, predictive misfit is defined as in definition 2–15 with $n_1:=q_t$, $n_2:=q-n_{t-1}$, $\tilde{x}_i:=P_{1t}\cdot \mathrm{col}(\tilde{w}(t_0+i),...,\tilde{w}(t_0+i+t-1))$ and $\tilde{y}_i:=P_{2t}\cdot\tilde{w}(t_0+i+t)$, $i\in[0,t_1-t_0-t]$. Due to the requirement $v_t(r)\perp v_t(\mathcal{B}_{t-1}^{\perp})$ at least e_t^{tol} independent a_+ functionals need to be accepted. Defining $\tilde{c}_{tol}:=q-n_{t-1}-e_t^{tol}$, cf. section 2.2.1, we obtain the optimal relations by applying proposition 2–21(*iii*), due to assumption 4–7(*iii*). Note that assumption 4–7(*ii*) implies that the genericity conditions in proposition 2–21 are satisfied, cf. section 2.2.3. So proposition 2–21(*iii*) implies the optimality of step 3.2 in step t of the recursion, cf. the definition of $M^*_{e_t^{tol}}$ in section 2.2.3. This concludes the inductive proof of (*i*).

Finally the expression for the misfit in (*iii*) follows from (*i*), (*iv*), and corollary 2–20. ■

Proof of theorem 4–10

Assumption 4–9(*ii*) implies that step 3.1 is well-defined, assumption 4–9(*iii*) that step 3.2 is well-defined and that $\mathcal{B}$ in step 5 is uniquely defined. Now (*ii*) and (*iv*) follow from assumption 4–9(*iv*) as in the proof of theorem 4–8(*ii*) and (*iv*).

We prove (*i*) by induction. For $t=0$ the optimality of step 2.2 follows from theorem 4–6, cf. definition 3–7. Next suppose that optimality of step 3.2 is shown for steps τ with $\tau\leq t-1$ for some $t\geq 1$. According to definitions 3–7 and 3–12 we have to minimize the misfit of a maximal number of newly accepted laws $r\in v_t^{-1}\{[v_t(F_{t-1}s^t)+v_t(\mathcal{B}_{t-1}^{\perp})]^{\perp}\}$. For such laws $e^P(\tilde{w},r)=e^P(\tilde{z}^{(t)},a)$ with $\tilde{z}^{(t)}$ and a as definied in the proof of theorem 4–8(*i*). First suppose that in step 3.1 $e_t''\leq e_t'$. Due to the requirement $\varepsilon_{t,1}^P(\tilde{w},\mathcal{B})<\bar{\varepsilon}_t^{tol}$ if follows from proposition 2–22(*iv*) and the ordering of definition 3–12 that the unique optimal solution is given by step 3.2, cf. the proof of theorem 4–8(*i*). Note that assumption 4–9(*ii*) implies that the genericity conditions in proposition 2–22 are satisfied, cf. section 2.2.3. Next suppose that in step 3.1 $e_t''>e_t'$. Assumption 4–9(*iv*) implies that at step t at most e_t' independent laws of order t may be accepted for the reason given in the proof of theorem 4–6. Assumption 4–9(*iii*), the ordering of definition 3–12 and proposition 2–22(*iv*) imply that the unique optimal solution in this case is given by step 3.2.

Finally the expression for the misfit in (*iii*) follows from (*i*), (*iv*), and corollary 2–20. ■

Proof of proposition 5-2

As a simple example take $\mathcal{B}=(\mathbb{R}^q)^{\mathbb{Z}}$. For any $w\in\mathcal{B}$ and any $\mathcal{T}$ of the finite length there exist $\mathcal{B}'\in\mathbb{B}$ such that $w|_{\mathcal{T}}\in\mathcal{B}|_{\mathcal{T}}$ and $\dim(\mathcal{B}')\leq q\cdot\#(\mathcal{T})$, e.g., $\mathcal{B}':=\{w\in(\mathbb{R}^q)^{\mathbb{Z}};\ (\sigma^{\#(\mathcal{T})}-1)w=0\}$. Hence $\mathcal{B}\notin P^u(w|_{\mathcal{T}})$ for any $w|_{\mathcal{T}}\in(\mathbb{R}^q)^{\mathcal{T}}$. ■

Proof of proposition 5-3

We give the proof for $P^D_{c_{tol}}$, $P^D_{\varepsilon_{tol}}$ and $\bar{\bar{P}}^D_{\varepsilon_{tol}}$, as similar arguments hold true for the other procedures.

First suppose that c_{tol} is given. Let $e_{tol}:=e(c_{tol})$ be the equation structure corresponding to c_{tol}, cf. definition 4–1. If $e_{tol}=0$, then it follows from definition 3–9 that $P^D_{c_{tol}}$ is not consistent for the same reasons as given for P^u in the proof of proposition 5–2. If there is $t\in\mathbb{Z}_+$ with $e_t^{tol}\geq 1$, then $\mathcal{B}\in\mathbb{B}$ with $e_t^*(\mathcal{B})=0$ cannot be exactly identified, hence $P^D_{c_{tol}}$ is not consistent.

Next suppose that ε_{tol} is given and consider $P^D_{\varepsilon_{tol}}$. If $\varepsilon_{t,1}^{tol}\leq 0$ for some $t\in\mathbb{Z}_+$, then for every $\mathcal{B}\in\mathbb{B}$ (exact) identification is impossible for $\mathcal{T}$ sufficiently large, cf. the interpretation following assumption 4–5. If on the other hand $\varepsilon_{t,1}^{tol}>0$ for all $t\in\mathbb{Z}_+$, then especially $\varepsilon_{0,1}^{tol}>0$ so $P^D_{\varepsilon_{tol}}$ will accept laws of order zero for $w|_{\mathcal{T}}\in(\mathbb{R}^q)^{\mathcal{T}}$ of sufficiently small norm. Not having this sufficiently small norm clearly is not a generic property for any $\mathcal{B}\in\mathbb{B}$, as $\mathcal{B}$ is linear. Especially if $\mathcal{B}\in\mathbb{B}$ with $e_0^*(\mathcal{B})=0$, then $P^D_{\varepsilon_{tol}}$ cannot exactly identify $\mathcal{B}$ for generic time series, hence $P^D_{\varepsilon_{tol}}$ is not consistent.

Finally for $\bar{\bar{P}}^D_{\varepsilon_{tol}}$ note that $\varepsilon_{t,1}^{tol}<0$ for some $t\in\mathbb{Z}_+$ implies that for every $\mathcal{B}\in\mathbb{B}$ identification is impossible for $\mathcal{T}$ sufficiently large, that $\varepsilon_{t,1}^{tol}>0$ for some $t\in\mathbb{Z}_+$ implies that, e.g., $(\mathbb{R}^q)^{\mathbb{Z}}$ cannot be exactly identified for generic time series, and that $\varepsilon_{t,1}^{tol}=0$ for all $t\in\mathbb{Z}_+$ implies that $\bar{\bar{P}}^D_{\varepsilon_{tol}}$ is not consistent for the reasons given in the proof of proposition 5–2. ■

Proof of theorem 5-4

For given $\mathcal{B}\in\mathbb{B}_c$ we have to prove that generically in $w\in\mathcal{B}$ there holds $P_e^D(\tilde{w})=P_e^P(\tilde{w})=\{\mathcal{B}\}$ for $\tilde{w}=w|_{\mathcal{T}}$ with $\#(\mathcal{T})$ sufficiently large.

In the proof we use the following lemma. Let $\bar{d}(\mathcal{T}):=(\#(\mathcal{T})-q)/(q+1)$ and let $S(\tilde{w},t)$ denote the empirical covariance matrix of order t as defined in sections 3.3.1 and 4.2.1. Further for $V\subset(\mathbb{R}^{1\times q})^{t+1}$ let $V^T:=\{v\in(\mathbb{R}^q)^{t+1};\ v^T\in V\}$.

Lemma 5-4 For $t\leq\bar{d}(\mathcal{T})$ generically in $w\in\mathcal{B}$ $\ker(S(w|_{\mathcal{T}},t))=[v_t(\mathcal{B}_t^{\perp})]^T$.

Proof of lemma 5-4

Evidently $[v_t(\mathcal{B}_t^{\perp})]^T\subset\ker(S(w|_{\mathcal{T}},t))$ and hence it suffices to show that gen. in $w\in\mathcal{B}$ $\dim(\ker(S(w|_{\mathcal{T}},t)))\leq\dim(v_t(\mathcal{B}_t^{\perp}))$, i.e., $\mathrm{rank}(S(w|_{\mathcal{T}},t))\geq q(t+1)-\dim(v_t(\mathcal{B}_t^{\perp}))=q(t+1)-\Sigma_{k=0}^{t}(t+1-k)e_k^*$, where $\{e_t^*;\ t\in\mathbb{Z}_+\}$ is the tightest equation structure of $\mathcal{B}$, cf. proposition IV.2–2(i) and the remark following definition II.3–9. Let $T:=\#(\mathcal{T})$ and relabel the time instants such that $\mathcal{T}=[1,T]$. Then $S(w|_{\mathcal{T}},t)=\frac{1}{T-t}H(w)H(w)^T$, where $H(w)\in\mathbb{R}^{q(t+1)\times(T-t)}$ is defined by

$$H(w):=\begin{bmatrix} w(1) & w(2) & \dots & w(T-t) \\ w(2) & w(3) & \dots & w(T-t+1) \\ \vdots & \vdots & & \vdots \\ w(t+1) & w(t+2) & \dots & w(T) \end{bmatrix}.$$

It hence suffices to prove that there exists a $w\in\mathcal{B}$ with $\mathrm{rank}(H(w))\geq q(t+1)-\Sigma_{k=0}^{t}(t+1-k)e_k^*=:r_w$, as this then also holds true generically on $\mathcal{B}$.

Notation. We use the following notation and concepts.

Let $\mathcal{B}$ have a minimal input/state/output realization $\mathcal{B}_{i/s/o}(\tilde{A},\tilde{B},\tilde{C},\tilde{D})$, i.e., there exists a permutation matrix Π such that $\Pi\mathcal{B}=\{(u,y)\in(\mathbb{R}^m)^{\mathbb{Z}}\times(\mathbb{R}^p)^{\mathbb{Z}};$ $\exists x\in(\mathbb{R}^n)^{\mathbb{Z}}$ such that $\begin{bmatrix}\sigma x\\ y\end{bmatrix}=\begin{bmatrix}\tilde{A} & \tilde{B}\\ \tilde{C} & \tilde{D}\end{bmatrix}\begin{bmatrix}x\\ u\end{bmatrix}\}$, where n is the number of states, m the number of inputs, and $p:=q-m$ the number of outputs. We refer to definition II.3–24. As $\mathcal{B}\in\mathbb{B}_c$ it follows that $(\tilde{A},\tilde{B},\tilde{C})$ is a minimal triple, by which we mean that $(\tilde{A},\tilde{B})$ is controllable, cf. section IV.3.2, and that $(\tilde{A},\tilde{C})$ is observable, cf. the proof of proposition IV.3–5, i.e., $\mathrm{rank}([\tilde{B}\ \tilde{A}\tilde{B}\dots\tilde{A}^{n-1}\tilde{B}])=\mathrm{rank}(\mathrm{col}(\tilde{C},\tilde{C}\tilde{A},\dots,\tilde{C}\tilde{A}^{n-1}))=n$.

For given $F\in\mathbb{R}^{m\times p}$ let $(A,C):=(\tilde{A}+\tilde{B}F\tilde{C},\tilde{C}+\tilde{D}F\tilde{C})$ and for $(u,x,y)\in\mathcal{B}_{i/s/o}(\tilde{A},\tilde{B},\tilde{C},\tilde{D})$

let $v:=u-F\tilde{C}x$. Then $\Pi\mathcal{B}=\{(u,y);\ \exists(v,x)$ such that $\begin{pmatrix}\sigma x\\ u\\ y\end{pmatrix}=\begin{pmatrix}A & \tilde{B}\\ F\tilde{C} & I\\ C & \tilde{D}\end{pmatrix}\begin{pmatrix}x\\ v\end{pmatrix}\}$. Hence $\tilde{w}\in\mathcal{B}|_{[1,T]}$ if and only if there exist $x_0\in\mathbb{R}^n$ and $v|_{[1,T]}\in(\mathbb{R}^m)^T$ such that for all $\tau\in[1,T]$ $\Pi\tilde{w}(\tau)=\begin{pmatrix}\tilde{u}(\tau)\\ \tilde{y}(\tau)\end{pmatrix}$ with $\tilde{u}(\tau)=F\tilde{C}x(\tau)+v(\tau)$ and $\tilde{y}(\tau)=Cx(\tau)+\tilde{D}v(\tau)$, where $x(\tau):=A^{\tau-1}x_0+\Sigma_{k=1}^{\tau-1}A^{k-1}\tilde{B}v(\tau-k)$, $\tau\in[1,T]$.

We call $A\in\mathbb{R}^{n\times n}$ cyclic if there exists an $x_0\in\mathbb{R}^n$ such that $\det([x_0\ Ax_0 \ldots A^{n-1}x_0])\neq 0$.

For $z\in(\mathbb{R}^d)^{\mathbb{Z}}$ let $H(z)\in\mathbb{R}^{d(t+1)\times(T-t)}$ be defined by

$$H(z):=\begin{bmatrix} z(1) & z(2) & \ldots & z(T-t)\\ z(2) & z(3) & \ldots & z(T-t+1)\\ \vdots & \vdots & & \vdots\\ z(t+1) & z(t+2) & \ldots & z(T)\end{bmatrix}.$$

Finally, for $M_i\in\mathbb{R}^{n_1\times n_2}$, $i=1,2$, let $M_1\sim M_2$ denote that M_1 can be transformed into M_2 by means of elementary row operations, i.e., $\{M_1\sim M_2\} :\Leftrightarrow \{\exists S\in\mathbb{R}^{n_1\times n_1}$ nonsingular such that $SM_1=M_2\}$. □

We now show that there exists a $w\in\mathcal{B}$ with $\mathrm{rank}(H(w))\geq r_w$, by choosing appropriate F, x_0, and $v|_{[1,T]}$. The proof is split in the following three parts.

(*i*) $\exists F\in\mathbb{R}^{m\times p}$ such that $A:=\tilde{A}+\tilde{B}F\tilde{C}$ is invertible and cyclic;

(*ii*) for F as in (*i*) and for $w\in\mathcal{B}$ with $\Pi w=\begin{pmatrix}u\\ y\end{pmatrix}$ and $\begin{pmatrix}\sigma x\\ u\\ y\end{pmatrix}=\begin{pmatrix}A & \tilde{B}\\ F\tilde{C} & I\\ C & \tilde{D}\end{pmatrix}\begin{pmatrix}x\\ v\end{pmatrix}$ there holds $\mathrm{rank}(H(w))=\mathrm{rank}(\begin{bmatrix}H(v)\\ \tilde{M}\end{bmatrix})$, where $\tilde{M}:=\mathrm{col}(\tilde{C},\tilde{C}\tilde{A},\ldots,\tilde{C}\tilde{A}^t).[x(1)\ x(2)\ldots x(T-t)]$;

(*iii*) there exist $x_0\in\mathbb{R}^n$ and $v|_{[1,T]}\in(\mathbb{R}^m)^T$ such that for $x(1):=x_0$ and $x(\tau+1):=Ax(\tau)+\tilde{B}v(\tau)$, $\tau\in[1,T-t-1]$, there holds $\mathrm{rank}(\begin{bmatrix}H(v)\\ \tilde{M}\end{bmatrix})\geq r_w$.

The desired result then follows from (*ii*) and (*iii*).

(*i*) We have to prove that there is an $F\in\mathbb{R}^{m\times p}$ such that $A:=\tilde{A}+\tilde{B}F\tilde{C}$ is (1) cyclic and (2) invertible. As these are algebraic conditions in F, cf Kailath [33, lemma 7.1–2], it suffices to prove that (1) and (2) can be satisfied individually, as this even implies that conditions (1) and (2) are simultaneously satisfied for generic F.

For (1) we refer to Kailath [33, lemma 7.1–2 and section 7.2.3].

Next we consider (2). As $(\tilde{A},\tilde{B},\tilde{C})$ is a minimal triple it follows from

Kailath [33, lemma 7.1–1] that there exist $\beta\in\mathbb{R}^m$ and $\gamma\in\mathbb{R}^p$ such that for $b:=\tilde{B}\beta$ and $c:=\gamma^T\tilde{C}$ also $(\tilde{A},b,c)$ is a minimal triple. Let a basis in $\mathbb{R}^n$ be chosen such that $(\tilde{A},b)$ is in control canonical form, cf. Kalman, Falb and Arbib [39, section 2.4], i.e., if $\det(sI-\tilde{A})=s^n+\Sigma_{k=0}^{n-1}a_ks^k$ and $a:=-(a_0,a_1,\dots,a_{n-1})$, then in this basis $\tilde{A}=\begin{bmatrix}0 & I_{n-1}\\ & a\end{bmatrix}$, $b=(0,\dots,0,1)^T$, and $c=(c_1,\dots,c_n)$ for some $c_i\in\mathbb{R}$, $i\in[1,n]$. If $a_0\neq 0$ then $\tilde{A}$ is invertible, hence $F=0$ satisfies (2). If $a_0=0$ then observability of $(\tilde{A},c)$ implies that $c_1\neq 0$, hence for $F:=\beta\gamma^T$ there holds $A=\begin{bmatrix}0 & I_{n-1}\\ & a'\end{bmatrix}$ with $a':=(c_1,c_2-a_1,\dots,c_n-a_{n-1})$, and as $c_1\neq 0$ A is invertible.

(*ii*) Let F be as in (*i*), $(A,C):=(\tilde{A}+\tilde{B}F\tilde{C},\tilde{C}+\tilde{D}F\tilde{C})$, $\begin{bmatrix}\sigma x\\ u\\ y\end{bmatrix}=\begin{bmatrix}A & \tilde{B}\\ F\tilde{C} & I\\ C & \tilde{D}\end{bmatrix}\begin{bmatrix}x\\ v\end{bmatrix}$ and $w:=\Pi\begin{bmatrix}u\\ y\end{bmatrix}$. Then by means of elementary row operations it follows that $H(w)\sim\begin{bmatrix}H(u)\\ H(y)\end{bmatrix}=$ $\begin{bmatrix}H(u)\\ H(\tilde{C}x)+H(\tilde{D}u)\end{bmatrix}\sim\begin{bmatrix}H(F\tilde{C}x)+H(v)\\ H(\tilde{C}x)\end{bmatrix}\sim\begin{bmatrix}H(v)\\ H(\tilde{C}x)\end{bmatrix}$.

Noting that for $1\leq\tau_1\leq\tau_2\leq T$ there holds $x(\tau_2)=A^{\tau_2-\tau_1}x(\tau_1)+\Sigma_{k=1}^{\tau_2-\tau_1}A^{k-1}\tilde{B}v(\tau_2-k)$, it follows from the Hankel structure of $H(v)$ and $H(\tilde{C}x)$ that $\begin{bmatrix}H(v)\\ H(\tilde{C}x)\end{bmatrix}\sim\begin{bmatrix}H(v)\\ M\end{bmatrix}$ where $M:=\mathrm{col}(\tilde{C},\tilde{C}A,\dots,\tilde{C}A^t).[x(1)\ x(2)\dots x(T-t)]$. For example, $x(\tau+1)=Ax(\tau)+\tilde{B}v(\tau)$, $\tau\in[1,T-t]$, hence by subtracting $\tilde{C}\tilde{B}$ times the matrix consisting of the first m rows of $H(v)$ from the matrix consisting of rows $p+1,p+2,\dots,2p$ of $H(\tilde{C}x)$ this latter matrix is transformed into $\tilde{C}A[x(1)\ x(2)\dots x(T-t)]$, and similarly for the other rows.

As $\mathrm{col}(\tilde{C},\tilde{C}A,\dots,\tilde{C}A^t)\sim\mathrm{col}(\tilde{C},\tilde{C}\tilde{A},\dots,\tilde{C}\tilde{A}^t)$ it follows that $H(w)\sim\begin{bmatrix}H(v)\\ \tilde{M}\end{bmatrix}$ with $\tilde{M}:=\mathrm{col}(\tilde{C},\tilde{C}\tilde{A},\dots,\tilde{C}\tilde{A}^t).[x(1)\ x(2)\dots x(T-t)]$, hence $\mathrm{rank}(H(w))=\mathrm{rank}(\begin{bmatrix}H(v)\\ \tilde{M}\end{bmatrix})$.

(*iii*) Finally we prove that there exist $x_0\in\mathbb{R}^n$ and $v|_{[1,T]}\in(\mathbb{R}^m)^T$ such that for $x(1):=x_0$ and $x(\tau+1):=Ax(\tau)+\tilde{B}v(\tau)$, $\tau\in[1,T-t-1]$, there holds $\mathrm{rank}(\begin{bmatrix}H(v)\\ \tilde{M}\end{bmatrix})\geq r_w$.

Let $r_x:=p(t+1)-\Sigma_{k=0}^{t}(t+1-k)e_k^*=r_w-m(t+1)$.

Denoting the i–th component of v by v_i, $i\in[1,m]$, we define $v|_{[1,T]}\in(\mathbb{R}^m)^T$ by $v_{m-k}(t+r_x+k(t+1)+1):=1$, $k\in[0,m-1]$, and 0 elsewhere. Note that $t+r_x+(m-1)(t+1)+1=r_w\leq q(t+1)\leq T-t$, as $t\leq\bar{d}(\mathcal{T})=(T-q)/(q+1)$. It is a simple matter of explicitly writing out $H(v)$ to conclude that $\mathrm{rank}(H(v))=m(t+1)$ and that in $H(v)$ the first r_x columns are zero. As $v|_{[1,r_x]}=0$ there holds that $x(\tau)=A^{\tau-1}x_0$, $\tau\in[1,r_x]$. Now suppose that there is an $x_0\in\mathbb{R}^n$ such that $\mathrm{rank}(\mathrm{col}(\tilde{C},\tilde{C}\tilde{A},\dots,\tilde{C}\tilde{A}^t).[x_0\ Ax_0\dots A^{r_x-1}x_0])=r_x$. Then the matrix consisting of the first r_x columns of $\tilde{M}$ has rank r_x, and it easily follows that the first $r_x+m(t+1)=r_w$ columns of $\begin{bmatrix} H(v) \\ \tilde{M} \end{bmatrix}$ are linearly independent, hence rank($\begin{bmatrix} H(v) \\ \tilde{M} \end{bmatrix}$)$\geq r_w$, as desired.

It remains to prove that there is an $x_0 \in \mathbb{R}^n$ such that $\tilde{M}_0 := \mathrm{col}(\tilde{C},\tilde{C}\tilde{A},\ldots,\tilde{C}\tilde{A}^t).[x_0\ Ax_0 \ldots A^{r_x-1}x_0]$ has rank r_x. It follows from Willems [73, table 2 and theorem 6(vi)] that $r_x \leq n$ and that $\mathrm{rank}(\mathrm{col}(\tilde{C},\tilde{C}\tilde{A},\ldots,\tilde{C}\tilde{A}^t)) = \Sigma_{k=0}^{t}(p-\Sigma_{j=0}^{k}e_j^*) = p(t+1) - \Sigma_{k=0}^{t}(t+1-k)e_k^* = r_x$. Now suppose that $\mathrm{rank}(\tilde{M}_0) < r_x$ for all $x_0 \in \mathbb{R}^n$. Then there would exist a $k < r_x$ such that for generic $x_0 \in \mathbb{R}^n$ the $(k+1)$-th column of $\tilde{M}_0$ is linearly dependent on the foregoing ones, say for $x_0 \in V \subset \mathbb{R}^n$ where V is generic. Then $\tilde{V} := \bigcap_{i=0}^{n-1-k} A^{-i}V := \{x \in \mathbb{R}^n;\ A^i x \in V$ for all $i \in [0,n-1-k]\}$ is also generic, as A is invertible. For every $x_0 \in \tilde{V}$ $\mathrm{rank}(\mathrm{col}(\tilde{C},\tilde{C}\tilde{A},\ldots,\tilde{C}\tilde{A}^t).[x_0\ Ax_0 \ldots A^{n-1}x_0]) \leq k < r_x$. However, as A is cyclic, for generic x_0 $\det([x_0\ Ax_0 \ldots A^{n-1}x_0]) \neq 0$ and hence generically on $\mathbb{R}^n$ $\mathrm{rank}(\mathrm{col}(\tilde{C},\tilde{C}\tilde{A},\ldots,\tilde{C}\tilde{A}^t).[x_0\ Ax_0 \ldots A^{n-1}x_0]) = \mathrm{rank}(\mathrm{col}(\tilde{C},\tilde{C}\tilde{A},\ldots,\tilde{C}\tilde{A}^t)) = r_x$. This contradiction shows that there is an $x_0 \in \mathbb{R}^n$ with $\mathrm{rank}(\tilde{M}_0) = r_x$, as desired.

This concludes the proof of lemma 5–4. ■

Using this lemma we now first prove the theorem for P_e^D. Let $V_t^D := v_t(L_t^D)$ be the descriptive complementary spaces of $\mathcal{B}$ as defined in section II.3.2.5 and let $n_t := \dim(V_t^D) = e_t^*(\mathcal{B})$, see proposition II.3–10. Moreover let $t^* := \max\{t;\ n_t \neq 0\}$ and let $\{t;\ n_t \neq 0\} = \{t_1,\ldots,t_c\}$ with $0 \leq t < t_2 < \ldots < t_{c-1} < t_c := t^*$. We now show that gen. in $w \in \mathcal{B}$ $P_e^D(w|_{\mathcal{T}}) = \{\mathcal{B}\}$ for every $\mathcal{T}$ with $\bar{d}(\mathcal{T}) \geq t^*$, i.e., $\#(\mathcal{T}) \geq t^*(q+1)+q$.

Let $\mathcal{T}$ be such that $\bar{d}(\mathcal{T}) \geq t^*$. Now P_e^D only accepts laws with descriptive misfit zero. Note that for $t \leq \bar{d}(\mathcal{T})$ and $r \in \mathbb{R}_t^{1\times q}[s]$ there holds $\{e^D(\tilde{w},r)=0\} \Leftrightarrow \{v_t(r) \in \ker(S(\tilde{w},t))\}$. It hence follows from lemma 5–4 that gen. on $\mathcal{B}$ P_e^D accepts no law of order $t < t_1$, as then $v_t(\mathcal{B}_t^\perp) = \{0\}$. For step $t = t_1$ gen. on $\mathcal{B}$ $\ker(S(w|_{\mathcal{T}},t)) = V_{t_1}^D$ and hence gen. P_e^D exactly identifies these laws on step t_1. Now suppose that for some $k \in [2,c]$ P_e^D gen. on steps t_i identifies $V_{t_i}^D$, $i \in [1,k-1]$, and gen. no laws are accepted on steps $t \leq t_{k-1}$ with $t \notin \{t_1,\ldots,t_{k-1}\}$. For steps $t \in [t_{k-1}+1, t_k-1]$ there holds that $\mathcal{B}_t^\perp = \mathcal{B}_{t-1}^\perp + s\mathcal{B}_{t-1}^\perp$, as $L_t^D = \{0\}$, and it follows by induction from lemma 5–4 that gen. on $\mathcal{B}$ the rows of P_t as defined in step 3.0 of the algorithm of section 4.2.2 span $[v_t(\mathcal{B}_t^\perp)]^\perp$ and that gen. on $\mathcal{B}$ $\ker(P_t S(w|_{\mathcal{T}},t)P_t^T) = \{0\}$. So gen. no laws are accepted for steps $t \in [t_{k-1}+1, t_k-1]$, and gen. the rows of P_{t_k} span $[v_{t_k}(\mathcal{B}_{t_k-1}^\perp + s\mathcal{B}_{t_k-1}^\perp)]^\perp$. Lemma 5–4 and the algorithm of section 4.2.2 imply that hence gen. on $\mathcal{B}$ for step t_k P_e^D exactly identifies $v_{t_k}(\mathcal{B}_{t_k}^{\perp})\cap[v_{t_k}(\mathcal{B}_{t_{k-1}}^{\perp}+s\mathcal{B}_{t_{k-1}}^{\perp})]^{\perp}=V_{t_k}^D$. It follows by induction that on step $\bar{d}(\mathcal{T})$ P_e^D gen. has identified exactly $V_{t_i}^D$, $t_i\leq t^*$, i.e., $P_e^D(w|_{\mathcal{T}})=\{\mathcal{B}\}$. Note that the number of steps is finite and that a finite intersection of generic sets is generic.

This proves consistency of P_e^D. The consistency of P_e^P follows from this result. Note that for order zero the procedures coincide, while for steps $t\geq 1$ $\{e^P(\tilde{w},r)=0\}\Leftrightarrow\{e^D(\tilde{w},r)=0\}$, provided that $\|r^*\tilde{w}\|_t\neq 0$ where r^* is the leading coefficient vector of r, cf. definition 3–5. As always on step 0 the set of zero order laws $V_0^P=V_0^D$ is identified it follows that gen. on $\mathcal{B}$ in all steps $1\leq t\leq\bar{d}(\mathcal{T})$ $\|r^*\tilde{w}\|_t\neq 0$, as it is required that $r^*\perp F_{t-1}$, hence $r^*\perp V_0^P$. The consistency of P_e^D then implies the consistency of P_e^P. ■

Proof of theorem 5–5

Lemma 5–4 and the proof of theorem 5–4 show the following. We recall that for given interval of observation $\mathcal{T}$ the procedures P_e^D and P_e^P only consider the model class $\mathbb{B}(\mathcal{T})$ for identification, cf. definitions 3–8, 3–13 and 3–14.

Lemma 5–5 Generically on $\mathcal{B}\in\mathbb{B}_c$, $P_e^D(w|_{\mathcal{T}})=P_e^P(w|_{\mathcal{T}})=\mathcal{B}(\mathcal{T}):=\{w'\in(\mathbb{R}^q)^{\mathbb{Z}};\ r(\sigma)w'=0 \text{ for all } r\in\mathcal{B}_t^{\perp} \text{ where } t:=\bar{d}(\mathcal{T})\}$.

P_e^D and P_e^P evidently are exact. Moreover, as $\mathcal{B}_t^{\perp}\subset\mathcal{B}_{t+1}^{\perp}$ it follows that $\mathcal{B}([t_0,t_1+1])=\mathcal{B}([t_0-1,t_1])\subset\mathcal{B}([t_0,t_1])$ and lemma 5–5 implies that P_e^D and P_e^P are bilaterally monotone, i.e., the identified model becomes more strict if more observations are available. To prove linearity, let $\mathcal{B}_1,\mathcal{B}_2\in\mathbb{B}_c$, then $\mathcal{B}_1+\mathcal{B}_2\in\mathbb{B}_c$, $(\mathcal{B}_i)_t^{\perp}\supset(\mathcal{B}_1+\mathcal{B}_2)_t^{\perp}$, $i=1,2$, hence $(\mathcal{B}_1+\mathcal{B}_2)(\mathcal{T})\supset\mathcal{B}_1(\mathcal{T})+\mathcal{B}_2(\mathcal{T})$ and lemma 5–5 implies that P_e^D and P_e^P are linear. That these procedures are truthful is evident from lemma 5–5. Finally, for given $\mathcal{T}$ the procedures P_e^D and P_e^P only identify models in the class $\mathbb{B}(\mathcal{T})$, cf. definition 3–8, and according to lemma 5–5 models in $\mathbb{B}(\mathcal{T})$ are strongly corroborable by P_e^D and P_e^P, which proves that these procedures are strongly prudential. ■

Proof of theorem 5-10

We first state and prove two lemmas which will be used to prove the theorem.

Notation. For $A=A^T\in\mathbb{R}^{n\times n}$ let $\sigma(A):=(\sigma_1,\ldots,\sigma_n)$, with $\sigma_1\geq\ldots\geq\sigma_n$, denote the ordered set of eigenvalues of A. Let $A=U\Sigma U^T$, $\Sigma=\text{diag}(\sigma_1,\ldots,\sigma_n)$, $UU^T=U^TU=I_n$, and for $k\leq l$ with $\sigma_{k-1}\neq\sigma_k$, $\sigma_l\neq\sigma_{l+1}$, let $\mathcal{U}_{[k,l]}\subset\mathbb{R}^n$ denote the space spanned by the columns $k,k+1,\ldots,l$ of U. Consider the collection of finite dimensional linear spaces $\mathcal{L}:=\{L;\ \exists n\in\mathbb{N}$ with $L\subset\mathbb{R}^n$, and L linear$\}$. A sequence $(L_k;\ k\in\mathbb{N})$ is defined to converge in the *Grassmannian* topology if there is an $L\in\mathcal{L}$, $L\subset\mathbb{R}^n$, such that for k sufficiently large $L_k\subset\mathbb{R}^n$, $\dim(L_k)=\dim(L)$, and such that there exist choices of bases in L_k which converge in Euclidean sense to a basis of L. We denote this by $L_k\xrightarrow{(G)}L$. The sequence is defined to converge in the *gap* topology if there is an $L\in\mathcal{L}$, $L\subset\mathbb{R}^n$, such that $L_k\subset\mathbb{R}^n$ for k sufficiently large and such that $g(L_k,L)\to0$ for $k\to\infty$, where g denotes the gap between L and L_k, cf. definition IV.5-1 and Kato [40, section IV.2.1], or Stewart [66, section 2.1]. □

Lemma 5-10-1 The Grassmannian and gap topologies are equivalent.

Lemma 5-10-2 (*i*) The mapping $\sigma:A\to\sigma(A)$ is continuous; (*ii*) if in $\sigma(A_0)$ for some $k\leq l$ $\sigma_{k-1}\neq\sigma_k$ and $\sigma_l\neq\sigma_{l+1}$, then the mapping $A\to\mathcal{U}_{[k,l]}$ is continuous in A_0, in the gap topology; (*iii*) if $A_0=A_0^T>0$ then the mapping $A\to A^{-1/2}$ is continuous in A_0.

Proof of lemma 5-10-1

Let $\{L_k;\ k\in\mathbb{N}\}$ be given. Note that convergence in the Grassmannian or gap topology both imply that there is an $n\in\mathbb{N}$ such that $L\subset\mathbb{R}^n$ and $L_k\subset\mathbb{R}^n$ with $\dim(L_k)=\dim(L)=:d$ for k sufficiently large, cf. Stewart [66, section 2.1].

First suppose that $L_k\xrightarrow{(G)}L$. Then there are matrices B and B_k of full column rank with $L_k=\text{im}(B_k)$, $L=\text{im}(B)$, and $\|B_k-B\|\to0$ if $k\to\infty$. The orthogonal projection operators P and P_k on L and L_k respectively are given by $P=B(B^TB)^{-1}B^T$ and $P_k=B_k(B_k^TB_k)^{-1}B_k^T$, hence $\|P_k-P\|\to0$, and Stewart [66, theorem 2.2] implies that $g(L_k,L)\to0$ for $k\to\infty$, cf. lemma IV.5-2.

Next suppose that $g(L_k,L)\to 0$. Let $L=\mathrm{im}(B)$ where B has full column rank d. For all $\varepsilon>0$ there is a K_ε such that for $k\geq K_\varepsilon$ there exist $\{b_1(k,\varepsilon),\dots,b_d(k,\varepsilon)\}\subset L_k$ such that $B(k,\varepsilon):=(b_1(k,\varepsilon),\dots,b_d(k,\varepsilon))$ $\|B-B(k,\varepsilon)\|<\varepsilon$, cf. definition IV.5–1. For ε sufficiently small $B(k,\varepsilon)$ has full column rank and hence for k sufficiently large the columns of $B(k,\varepsilon)$ form a basis of L_k, i.e., $L_k\xrightarrow{(G)}L$. ■

Proof of lemma 5-10-2

(i) We refer to Stewart [67, corollary 6.5.11].

(ii) Let $A=U\Sigma U^T$ with $UU^T=U^TU=I_n$, $\Sigma=\mathrm{diag}(\sigma_1,\dots,\sigma_n)$, $\sigma_1\geq\dots\geq\sigma_n$, and let $k\leq l$ be such that $\sigma_{k-1}\neq\sigma_k$ and $\sigma_l\neq\sigma_{l+1}$. Let $U_1\in\mathbb{R}^{n\times(l-k+1)}$ consist of columns $k,\dots,l$ of U and $U_2\in\mathbb{R}^{n\times(n-l+k-1)}$ of the remaining columns of U, so $\mathcal{U}_{[k,l]}=\mathrm{im}(U_1)$. Let $\Sigma_1:=\mathrm{diag}(\sigma_k,\dots,\sigma_l)$, $\Sigma_2:=\mathrm{diag}(\sigma_1,\dots,\sigma_{k-1},\sigma_{l+1},\dots,\sigma_n)$, so $A=(U_1\ \ U_2)\begin{bmatrix}\Sigma_1 & 0\\ 0 & \Sigma_2\end{bmatrix}(U_1\ \ U_2)^T$. Then the so-called separation of Σ_1 and Σ_2 is $\min\{|\sigma_{j_1}-\sigma_{j_2}|;\ j_1\in[k,l],\ j_2\in[1,k-1]\cup[l+1,n]\}>0$. Now suppose that $A_i=A_i^T\to A$. Let A_i have eigenvalues $\sigma_1^{(i)}\geq\dots\geq\sigma_n^{(i)}$, then according to ($i$) $\sigma_j^{(i)}>\sigma_k^{(i)}$ for $j<k$ and $\sigma_j^{(i)}<\sigma_l^{(i)}$ for $j>l$ if i is sufficiently large. Let $A_i=(U_1^{(i)}\ \ U_2^{(i)})\begin{bmatrix}\Sigma_1^{(i)} & 0\\ 0 & \Sigma_2^{(i)}\end{bmatrix}(U_1^{(i)}\ \ U_2^{(i)})^T$ be a decomposition of A_i analogous to the one of A, so especially $\mathrm{im}(U_1^{(i)})=\mathcal{U}_{[k,l]}^{(i)}$. It follows from Stewart [66, section 4.7 and corollary 2.6], that $\mathrm{im}(U_1^{(i)})\to\mathrm{im}(U_1)$ in the gap topology if $i\to\infty$, i.e., $\mathcal{U}_{[k,l]}^{(i)}\to\mathcal{U}_{[k,l]}$ if $i\to\infty$. We also refer to Davis and Kahan [10, section 2 (the $\sin 2\theta$ theorem) and theorem 8.2].

(iii) For $A_0>0$ the mapping $A\to A^{-1}$ is continuous in A_0, so it suffices to show that $A\to A^{1/2}$ is continuous in A_0. If $A=U\Sigma U^T>0$ with $UU^T=U^TU=I_n$ and $\Sigma=\mathrm{diag}(\sigma_1,\dots,\sigma_n)$ then $A^{1/2}=U\Sigma^{1/2}U^T$. According to (i) the mapping $A\to\Sigma$ is continuous, and by applying (ii) and lemma 5–10–1 for the eigenspaces of A_0 corresponding to distinct eigenvalues it follows that there exists a choice for U such that $A\to U$ is continuous in A_0. Hence $A\to A^{1/2}$ is continuous in A_0. ■

The two foregoing lemmas are used to prove the theorem.

Notation. Let $\tilde{w}=w_r|_{\mathcal{T}}$ for a realization w_r of a stochastic process w in the class $\mathbb{G}^c_{c_{tol}}$ or $\mathbb{G}_{(\varepsilon_{tol},\bar{d})}$ and let $A^P_{c_{tol}}(w)=\mathcal{B}_c\in\mathbb{B}$ and $A^P_{\varepsilon_{tol}}(w)=\mathcal{B}_\varepsilon\in\mathbb{B}$ be as defined in definition 5–7, cf. proposition 5–9. Let $V^c_t:=v_t(L^c_t)$ and $V^\varepsilon_t:=v_t(L^\varepsilon_t)$, where L^c_t

and L_t^ε are the predictive spaces in (CPF) corresponding to $\mathcal{B}_c$ and $\mathcal{B}_\varepsilon$ respectively, as defined in section II.3.2.6. Let $e(\mathcal{B}_c)$ and $e(\mathcal{B}_\varepsilon)$ denote the tightest equation structures of $\mathcal{B}_c$ and $\mathcal{B}_\varepsilon$ respectively. We will show that for $T:=\#(\mathcal{T})$ sufficiently large, a.s. $P^P_{c_{tol}}(\tilde{w})=:\mathcal{B}_c(\mathcal{T})$ and $P^P_{(\varepsilon_{tol},\bar{d})}(\tilde{w})=:\mathcal{B}_\varepsilon(\mathcal{T})$ are singletons. We denote the corresponding predictive spaces by $V_t^c(\mathcal{T})$ and $V_t^\varepsilon(\mathcal{T})$ and the corresponding tightest equation structures by $e^c(\mathcal{T})$ and $e^\varepsilon(\mathcal{T})$ respectively. Further we use the notation of the algorithms in sections 4.3.1 and 4.3.2 and we use a $*$ to indicate symbols corresponding to these algorithms when applied to the process w, cf. proposition 5–9. Finally the condition that $\#(\mathcal{T})$ should be sufficiently large is denoted by $T\to\infty$. □

We now prove the theorem by induction and first consider $P^P_{c_{tol}}$.

Let $w\in\mathbb{G}^c_{c_{tol}}$. We will prove by induction that a.s. $e^c(\mathcal{T})\to e(\mathcal{B}_c)$ and $V_t^c(\mathcal{T})\to V_t^c$ in the Grassmannian topology. By choosing bases $M^{(t)}(\mathcal{T})$ in $V_t^c(\mathcal{T})$ which converge to bases $M^{(t)}$ of V_t^c and defining $R(\mathcal{T}):=\text{col}(v_t^{-1}(M^{(t)}(\mathcal{T}));\ t\in\mathbb{Z}_+)$ and $R:=\text{col}(v_t^{-1}(M^{(t)});\ t\in\mathbb{Z}_+)$ we get $\mathcal{B}_c(\mathcal{T})=\mathcal{B}(R(\mathcal{T}))$, $\mathcal{B}_c=\mathcal{B}(R)$, while $R(\mathcal{T})\to R$ in Euclidean sense where R is bilaterally row proper, cf. proposition II.3–8, definition II.3–15 and assumption 4–7(iv). Hence a.s. $\mathcal{B}_c(\mathcal{T})\to\mathcal{B}_c$, cf. section 5.3.2, which shows consistency of $P^P_{c_{tol}}$.

It remains to show that a.s. $e^c(\mathcal{T})\to e(\mathcal{B}_c)$ and $V_t^c(\mathcal{T})\xrightarrow{(G)}V_t^c$. Consider the algorithm of section 4.3.1. Note that c_{tol} is sensible for $T\to\infty$, i.e., assumption 4–7(i) is then satisfied. As $w\in\mathbb{G}^c_{c_{tol}}\subset\mathbb{G}$ it follows from assumption 5–6 that a.s. $S(\tilde{w},t)\to S(w,t)=:S^*(t)$ for all $t\in\mathbb{Z}_+$ if $T\to\infty$. So assume henceforth that w_r satisfies assumption 5–6(ii), then it remains to show that $\{S(\tilde{w},t)\to S^*(t);\ t\in\mathbb{Z}_+\}\Rightarrow\{e^c(\mathcal{T})\to e(\mathcal{B}_c)$ and $V_t^c(\mathcal{T})\xrightarrow{(G)}V_t^c$ for all $t\in\mathbb{Z}_+$ if $T\to\infty\}$.

First consider step 0 of the algorithm of section 4.3.1. As $w\in\mathbb{G}_{c_{tol}}$ it follows from definition 5–8 and assumption 4–7(iii) that $\sigma^{*(0)}_{q-e_0^{tol}}>\sigma^{*(0)}_{q-e_0^{tol}+1}$. As $S(\tilde{w},0)\to S^*(0)$ it follows from lemma 5–10–2(i) that assumption 4–7(iii) for $t=0$ is satisfied for $T\to\infty$. It then follows from step 2.2 and lemma 5–10–2(ii) that $e_0^c(\mathcal{T})\to e_0(\mathcal{B}_c)$ and $g(V_0^c(\mathcal{T}),V_0^c)\to 0$, and from lemma 5–10–1 that hence $V_0^c(\mathcal{T})\xrightarrow{(G)}V_0^c$. It follows from Stewart [66, theorem 2.2] (cf. lemma IV.5–2), that the projection operators $P_1(\mathcal{T})$ and P_1^* of step 3.0 for $t=1$ satisfy $\|P_1(\mathcal{T})-P_1^*\|\to 0$. Note that the dimensions of $P_1(\mathcal{T})$ are equal to those of P_1^* if $e_0^c(\mathcal{T})=e_0(\mathcal{B}_c)$, i.e., for $T\to\infty$.

Next suppose that for some $t\leq\max\{\tau;\ e_\tau(\mathcal{B}_c)\neq 0\}=:t^*$ it is proved that for all $k\in[0,t-1]$ $e_k^c(\mathcal{T}) \to e_k(\mathcal{B}_c)$, $V_k^c(\mathcal{T}) \xrightarrow{(G)} V_k^c$, and $P_{k+1}(\mathcal{T}) \to P_{k+1}^*$. As $S(\tilde{w},t) \to S^*(t)$ there holds $P_t(\mathcal{T})S(\tilde{w},t)P_t(\mathcal{T})^T \to P_t^*S^*(t)P_t^{*T}$. It follows from definition 5-8 and assumption 4-7(*ii*) for w that $S_-^{(t)}$ and $S_+^{(t)}$ as defined in step 3.1 for $\tilde{w}$ have full rank if $T \to \infty$. Let $C_t(\mathcal{T}):=(S_-^{(t)})^{-1/2}S_{-+}^{(t)}(S_+^{(t)})^{-1/2}$ and let C_t^* be defined analogously in terms of S^*. It follows from lemma 5-10-2(*iii*) that $C_t(\mathcal{T}) \to C_t^*$. By considering $C_t(\mathcal{T})C_t^T(\mathcal{T})$ it follows from lemma 5-10-2(*i*) and assumption 4-7(*iii*) for w that this assumption also is satisfied for $\tilde{w}$ if $T \to \infty$, hence $e_t^c(\mathcal{T}) \to e_t(\mathcal{B}_c)$ for $T \to \infty$. Applying lemma 5-10-2(*i*) and (*ii*) to $C_t(\mathcal{T})C_t^T(\mathcal{T})$ and $C_t^T(\mathcal{T})C_t(\mathcal{T})$ it follows that in step 3.2 $g(V_t^c(\mathcal{T}),V_t^c) \to 0$ and hence $V_t(\mathcal{T}) \xrightarrow{(G)} V_t^c$ if $T \to \infty$. Moreover assumption 4-7(*iv*) for $\tilde{w}$ is implied by that for w if $T \to \infty$ and hence the laws identified at step t for $\tilde{w}$ then are really t-th order laws, i.e., $V_k^c(\mathcal{T})$ then remains unchanged for $k\leq t-1$. As now $g(V_k^c(\mathcal{T}),V_k^c) \to 0$ for all $k\leq t$ if $T \to \infty$ it follows that in step 3.0 for $t+1$ $\|P_{t+1}(\mathcal{T})-P_{t+1}^*\| \to 0$, cf. lemma IV.5-2. This concludes the inductive part and shows that for $t\leq t^*$ $e_t^c(\mathcal{T}) \to e_t(\mathcal{B}_c)$ and $V_t^c(\mathcal{T}) \xrightarrow{(G)} V_t^c$ if $T \to \infty$.

Finally consider orders $t>t^*$. As $e_t(\mathcal{B}_c)=0$ for $t>t^*$, the fact that $e_t^c(\mathcal{T})=e_t(\mathcal{B}_c)$ for $t\leq t^*$ and $T \to \infty$ implies that for $\tilde{w}$ it is, for the given c_{tol}, allowable not to accept any law of order $t>t^*$, for $T \to \infty$. Moreover, for $t\leq\bar{d}(\mathcal{T})$ $A:=\{\tilde{w}\in(\mathbb{R}^q)^{\mathcal{T}};\ \det(S(\tilde{w},t))=0\}$ is a proper algebraic variety and hence has Lebesgue measure zero, cf. Federer [12, section 2.6.5]. As $w\in\mathbb{G}_{c_{tol}}^c$ the continuity of w implies that $S(\tilde{w},t)>0$ a.s. Then also $P_tS(\tilde{w},t)P_t^T>0$ a.s., and definition 3-9 implies that hence $e_t^c(\mathcal{T})=0$ a.s. for all $t\in[t^*+1,\bar{d}(\mathcal{T})]$. This shows that $V_t^c(\mathcal{T}) \xrightarrow{(G)} V_t^c$, $t\in\mathbb{Z}_+$, if $T \to \infty$, as desired.

The consistency of $P^P_{(\varepsilon_{tol},\bar{d})}$ on $\mathbb{G}_{(\varepsilon_{tol},\bar{d})}$ is proved in a completely analogous way by using definition 5-8, the first remark in section 5.3.5, assumption 4-9, and the algorithm of section 4.3.2. Note that $\sigma_{q-e_0}^{*(0)}>(\bar{\varepsilon}_0^{tol})^2$, which implies that for $\tilde{w}$ also $\sigma_{q-e_0}^{(0)}>(\bar{\varepsilon}_0^{tol})^2$ a.s. for $T \to \infty$, and in this case $e_0^\varepsilon(\mathcal{T})=e_0(\mathcal{B}_\varepsilon)$. Similar arguments hold true for steps $t\leq t^*$. That $V_t^\varepsilon(\mathcal{T}) \xrightarrow{(G)} V_t^\varepsilon$ a.s. for $t\leq t^*$ follows in a way analogous to the proof of $V_t^c(\mathcal{T}) \xrightarrow{(G)} V_t^c$. Further note that $e_t^\varepsilon(\mathcal{T})=e_t(\mathcal{B}_\varepsilon)=0$ for all $t>\bar{d}$. Finally we consider $t\in[t^*+1,\bar{d}]$. Definitions 3-12 and 5-8 and the fact that $e_t(\mathcal{B}_\varepsilon)=0$ imply that $1-(\sigma_1^{*(t)})^2>(\bar{\varepsilon}_t^{tol})^2$, and it follows by induction that hence $1-(\sigma_1^{(t)})^2>(\bar{\varepsilon}_t^{tol})^2$ and $e_t^\varepsilon(\mathcal{T})=0$ a.s. for all $t\in[t^*+1,\bar{d}]$ if $T \to \infty$. It follows that $V_t^\varepsilon(\mathcal{T}) \xrightarrow{(G)} V_t^\varepsilon$ a.s. for all $t\in\mathbb{Z}_+$ if $T \to \infty$, as desired. Note that now continuity of w is not required.

This conclues the proof of theorem 5–10. ■

Proof of theorem 5-11

Let $\tilde{w}_0 \in \Omega^P_{c_{tol}}$ and $\mathcal{B}_0 := P^P_{c_{tol}}(\tilde{w}_0)$, $t^* := \max\{\tau;\ e_\tau(\mathcal{B}_0) \neq 0\}$. Let $\tilde{w}_k \to \tilde{w}_0$ if $k \to \infty$, hence $S(\tilde{w}_k,t) \to S(\tilde{w}_0,t)$ for all $t \leq \bar{d}(\mathcal{T})$ if $k \to \infty$. In the proof of theorem 5–10 let $\tilde{w}_0, \tilde{w}_k$, $S(\tilde{w}_0,t)$ and $S(\tilde{w}_k,t)$ play the role of w, $\tilde{w} = w_r|_{\mathcal{T}}$, $S^*(t)$ and $S(\tilde{w},t)$ respectively. As for $\tilde{w}_0$ assumption 4–7(*ii*), (*iii*) and (*iv*) are satisfied it follows directly from the proof of theorem 5–10 that these assumptions also are satisfied for $\tilde{w}_k$ if $k \to \infty$, for all $t \leq t^*$. Definition 3–9 implies that $P^*_t S(\tilde{w}_0,t) P^{*T}_t > 0$ for all $t \in [t^*+1, \bar{d}(\mathcal{T})]$. From this it follows by induction that $P_t S(\tilde{w}_k,t) P^T_t > 0$ if $k \to \infty$. In this case $e_t(P^P_{c_{tol}}(\tilde{w}_k)) = e_t(\mathcal{B}_0)$ for all $t \leq \bar{d}(\mathcal{T})$ and $\tilde{w}_k \in \Omega^P_{c_{tol}}$ if $k \to \infty$. This shows that $\Omega^P_{c_{tol}}$ is open. The proof of theorem 5–10 for $t \leq t^*$ moreover implies that $P^P_{c_{tol}}(\tilde{w}_k) \to P^P_{c_{tol}}(\tilde{w}_0)$ if $k \to \infty$, which proves the continuity of $P^P_{c_{tol}}$ on $\Omega^P_{c_{tol}}$.

That $\Omega^P_{\varepsilon_{tol}}$ is open and that on this set $P^P_{\varepsilon_{tol}}$ and $\bar{\bar{P}}^P_{\varepsilon_{tol}}$ are continuous also follows directly along the lines of the proof of theorem 5–10 by taking $\bar{d} := \bar{d}(\mathcal{T})$. Finally note that the assumptions on $\Omega^P_{\varepsilon_{tol}}$ imply that none of the restrictions $\bar{\varepsilon}^{tol}_t$ is critical. Hence $P^P_{\varepsilon_{tol}} = \bar{\bar{P}}^P_{\varepsilon_{tol}}$ on $\Omega^P_{\varepsilon_{tol}}$. ■

Proof of theorem 5-12

Let $\tilde{w}_0 \in \Omega^D_{c_{tol}}$, then $P^D_{c_{tol}}(\tilde{w}_0)$ is a singleton, see theorem 4–4(*i*). Let $\tilde{w}_k \to \tilde{w}_0$ if $k \to \infty$, hence $S(\tilde{w}_k,t) \to S(\tilde{w}_0,t)$ for all $t \leq \bar{d}(\mathcal{T})$ if $k \to \infty$. From the definition of $\Omega^D_{c_{tol}}$ it follows along the lines of the proof of theorems 5–10 and 5–11 that $S(\tilde{w}_k,t)$ also satisfies assumption 4–3(*ii*) and (*iii*) for all $t \leq \bar{d}(\mathcal{T})$ if $k \to \infty$, and hence that $\Omega^D_{c_{tol}}$ is open. That $P^D_{c_{tol}}(\tilde{w}_k) \to P^D_{c_{tol}}(\tilde{w}_0)$ if $k \to \infty$ also follows directly along the lines of the proof of theorem 5–10.

The results for $\Omega^D_{\varepsilon_{tol}}$, $P^D_{\varepsilon_{tol}}$, and $\bar{\bar{P}}^D_{\varepsilon_{tol}}$ follow in a completely similar way. As none of the restrictions $\bar{\varepsilon}^{tol}_t$ is critical on $\Omega^D_{\varepsilon_{tol}}$ it follows that $P^D_{\varepsilon_{tol}}(\tilde{w}) = \bar{\bar{P}}^D_{\varepsilon_{tol}}(\tilde{w})$ for $\tilde{w} \in \Omega^D_{\varepsilon_{tol}}$. ■

REFERENCES

[1] Akaike, H., Canonical correlation analysis of time series and the use of an information criterion, in Mehra, R.K., and D.G. Lainiotis (eds.), *System identification: advances and case studies*, Academic Press, New York, 1976, pp. 27–96.

[2] Akhiezer, N.I., and I.M. Glazman, *Theory of linear operators in Hilbert space*, volume I, Frederick Ungar, New York, 1961.

[3] Anderson, B.D.O., and J.B. Moore, *Optimal filtering*, Prentice-Hall, Englewood Cliffs, New Jersey, 1979.

[4] Anderson, T.W., *The statistical analysis of time series*, Wiley, New York, 1971.

[5] Box, G.E.P., and G.M. Jenkins, *Time series analysis, forecasting and control*, Holden-Day, San Francisco, 1970.

[6] Brillinger, D.R., *Time series analysis, data analysis and theory*, Holt, Rinehart and Winston, New York, 1975.

[7] Caines, P.E., On the scientific method and the foundations of system identification, in Byrnes, C.I., and A. Lindquist (eds.), *Modelling, identification and robust control*, North-Holland, Amsterdam, 1986, pp. 563–580.

[8] Chen, C.T., *Linear system theory and design*, Holt, Rinehart and Winston, New York, 1984.

[9] Corrêa, G.O., and K. Glover, Pseudo-canonical forms, identifiable parametrizations and simple parameter estimation for linear multivariable systems, *Automatica* 20 (4), 1984, pp. 429–452.

[10] Davis, C., and W.M. Kahan, The rotation of eigenvectors by a perturbation III, *SIAM journal on numerical analysis* 7 (1), 1970, pp. 1–46.

[11] Davis, M.H.A., and R.B. Vinter, *Stochastic modelling and control*, Chapman and Hall, London, 1985.

[12] Federer, H., *Geometric measure theory*, Springer, Berlin, 1969.

[13] Finesso, L., and G. Picci, Linear statistical models and stochastic realization theory, in *Lecture notes in control and information sciences* 62 part 1, Springer, Berlin, 1984, pp. 445–470.

[14] Fomby, T.B., R.C. Hill and S.R. Johnson, *Advanced econometric methods*, Springer, New York, 1984.

[15] Gantmacher, F.R., *The theory of matrices*, volume I, Chelsea, New York, 1959.

[16] Gevers, M., and V. Wertz, Uniquely identifiable state-space and ARMA parametrizations for multivariable linear systems, *Automatica* 20 (3), 1984, pp. 333–347.

[17] Glover, K., All optimal Hankel norm approximations of linear multivariable systems and their L^∞ error bounds, *International journal of control* 39 (6), 1984, pp. 1115–1193.

[18] Glover, K., and J.C. Willems, Parametrizations of linear dynamical systems, canonical forms and identifiability, *IEEE transactions on automatic control* AC-19 (6), 1974, pp. 640–646.

[19] Golub, G.H., and C.F. Van Loan, An analysis of the total least squares problem, *SIAM journal on numerical analysis* 17 (6), 1980, pp. 883–893.

[20] Golub, G.H., and C.F. Van Loan, *Matrix computations*, Johns Hopkins University Press, Baltimore, 1983.

[21] Guidorzi, R., Invariants and canonical forms for systems structural and parametric identification, *Automatica* 17 (1), 1981, pp. 117–133.

[22] Hannan, E.J., *Multiple time series*, Wiley, New York, 1970.

[23] Hannan, E.J., and M. Deistler, *The statistical theory of linear systems*, Wiley, New York, 1988.

[24] Hannan, E.J., and L. Kavalieris, Multivariate linear time series models, *Advances in applied probability* 16 (3), 1984, pp. 492–561.

[25] Hazewinkel, M., and R.E. Kalman, On invariants, canonical forms and moduli for linear, constant, finite dimensional, dynamical systems, in *Lecture notes in economics and mathematical systems* 131, Springer, Berlin, 1976, pp. 48–60.

[26] Heij, C., Exact modelling of a finite data sequence, *Proceedings of 25th IEEE conference on decision and control*, Athens, 1986, pp. 1743–1744.

[27] Heij, C., Approximate modelling of deterministic systems, in Curtain, R.F. (ed.), *Modelling, robustness and sensitivity reduction in control systems*, NATO ASI series, Springer, Berlin, 1987, pp. 271–283.

[28] Heij, C., Exact modelling of a finite time series, *SIAM journal on control and optimization* 26 (1), 1988, pp. 83–111.

[29] Heij, C., and J.C. Willems, Consistency analysis of approximate modelling procedures, in Byrnes, C.I., C.F. Martin and R.E. Saeks (eds), *Linear circuits, systems and signal processing : theory and application*, North-Holland, Amsterdam, 1988, pp. 445–456.

[30] Heij, C., and J.C. Willems, A deterministic approach to approximate modelling, in Willems, J.C. (ed.), *From data to model*, Springer, Heidelberg. To appear.

[31] Hinrichsen, D., and J.C. Willems (eds.), Special issue on parametrization problems, *IMA journal of mathematical control and information* 3 (2&3), 1986, pp. 59–254.

[32] Jayant, N.S., and P. Noll, *Digital coding of waveforms*, Prentice-Hall, Englewood Cliffs, New Jersey, 1984.

[33] Kailath, T., *Linear systems*, Prentice-Hall, Englewood Cliffs, New Jersey, 1980.

[34] Kalman, R.E., On minimal partial realizations of a linear input/output map, in Kalman, R.E., and N. De Claris (eds.), *Aspects of network and system theory*, Holt, Rinehart and Winston, New York, 1971, pp. 385–407.

[35] Kalman, R.E., Identification from real data, in Hazewinkel, M., and A.H.G. Rinnooy Kan (eds.), *Current developments in the interface: economics, econometrics, mathematics*, Reidel, Dordrecht, 1982, pp. 161–196.

[36] Kalman, R.E., System identification from noisy data, in Bednarek, A.R., and L. Cesari (eds.), *Dynamical systems II*, Academic Press, New York, 1982, pp. 135–164.

[37] Kalman, R.E., Identifiability and modelling in econometrics, in Krishnaiah, P.R. (ed.), *Developments in statistics*, volume 4, Academic Press, New York, 1983, pp. 97–136.

[38] Kalman, R.E., and R.S. Bucy, New results in linear filtering and prediction theory, *Transactions ASME, Journal of basic engineering*, series 83D, 1961, pp. 95–108.

[39] Kalman, R.E., P.L. Falb and M.A. Arbib, *Topics in mathematical system theory*, McGraw-Hill, New York, 1969.

[40] Kato, T., *Perturbation theory for linear operators*, Springer, Berlin, 1966.

[41] Kendall, M.G., and A. Stuart, *The advanced theory of statistics*, 3 volumes, Griffin, London, 1958, 1961, 1966.

[42] Koopmans,T.C. (ed.), *Statistical inference in dynamic economic models*, Cowles commission for research in economics, monograph no. 10, Wiley, New York, 1950.

[43] Kullback, S., *Information theory and statistics*, Wiley, New York, 1959.

[44] Kumar, P.R., and P. Varaiya, *Stochastic systems: estimation, identification and adaptive control*, Prentice-Hall, Englewood Cliffs, New Jersey, 1986.

[45] Lax, P.D., and R.S. Phillips, *Scattering theory*, Academic Press, New York, 1967.

[46] Lindquist, A., and M. Pavon, On the structure of state-space models for discrete-time stochastic vector processes, *IEEE transactions on automatic control* AC-29 (5), 1984, pp. 418-432.

[47] Lindquist, A., and G. Picci, Realization theory for multivariate stationary gaussian processes, *SIAM journal on control and optimization* 23 (6), 1985, pp. 809-857.

[48] Ljung, L., Convergence analysis of parametric identification methods, *IEEE transactions on automatic control* AC-23 (5), 1978, pp. 770-783.

[49] Ljung, L., A non-probabilistic framework for signal spectra, *Proceedings of 24th IEEE conference on decision and control*, Fort Lauderdale, 1985, pp. 1056-1060.

[50] Ljung, L., *System identification: theory for the user*, Prentice-Hall, Englewood Cliffs, New Jersey, 1987.

[51] Ljung, L., and P.E. Caines, Asymptotic normality of prediction error estimators for approximate system models, *Stochastics* 3 (1), 1979, pp. 29-46.

[52] Ljung, L., and T. Söderström, *Theory and practice of recursive identification*, MIT Press, Cambridge, Mass., 1983.

[53] Malinvaud, E., *Statistical methods of econometrics*, North-Holland, Amsterdam, 1970.

[54] Moore, B.C., Principal component analysis in linear systems: controllability, observability, and model reduction, *IEEE transactions on automatic control* AC-26 (1), 1981, pp. 17-32.

[55] Nieuwenhuis, J.W., and J.C. Willems, Continuity of dynamical systems: a system theoretic approach, *Mathematics of control, signals, and systems* 1(2), 1988, pp. 147-165.

[56] Northcott, D.G., *Lessons on rings, modules and multiplicities*, Cambridge University Press, London, 1968.

[57] Payne, H.J., and L.M. Silverman, On the discrete time algebraic Riccati equation, *IEEE transactions on automatic control* AC-18 (3), 1973, pp. 226-234.

[58] Pernebo, L., and L.M. Silverman, Model reduction via balanced state space representations, *IEEE transactions on automatic control* AC-27 (2), 1982, pp. 382-387.

[59] Rissanen, J., Modeling by shortest data description, *Automatica* 14 (5), 1978, pp. 465-471.

[60] Rissanen, J., Stochastic complexity and modeling, *The annals of*

statistics 14 (3), 1986, pp. 1080–1100.
[61] Rosenbrock, H.H., *State-space and multivariable theory*, Wiley, New York, 1970.
[62] Shibata, R., Asymptotically efficient selection of the order of the model for estimating parameters of a linear process, *The annals of statistics* 8 (1), 1980, pp. 147–164.
[63] Silverman, L.M., Realization of linear dynamical systems, *IEEE transactions on automatic control* AC-16 (6), 1971, pp. 554–567.
[64] Slepian, D. (ed.), Key papers in the development of information theory, *IEEE press selected reprint series*, IEEE, New York, 1974.
[65] Sorensen, H.W., Kalman filtering: theory and application, *IEEE press selected reprint series*, IEEE, New York, 1985.
[66] Stewart, G.W., Error and perturbation bounds for subspaces associated with certain eigenvalue problems, *SIAM review* 15 (4), 1973, pp. 727–764.
[67] Stewart, G.W., *Introduction to matrix computations*, Academic Press, New York, 1973.
[68] Tether, A.J., Construction of minimal linear state-variable models from finite input-output data, *IEEE transactions on automatic control* AC-15 (4), 1970, pp. 427–436.
[69] Theil, H., *Principles of econometrics*, Wiley, New York, 1971.
[70] Willems, J.C., Least squares stationary optimal control and the algebraic Riccati equation, *IEEE transactions on automatic control* AC-16 (6), 1971, pp. 621–634.
[71] Willems, J.C., System theoretic models for the analysis of physical systems, *Ricerche di automatica* 10 (2), 1979, pp. 71–106.
[72] Willems, J.C., Input-output and state-space representations of finite-dimensional linear time-invariant systems, *Linear algebra and its applications* 50 (1), 1983, pp. 581–608.
[73] Willems, J.C., From time series to linear system, part I: Finite dimensional linear time invariant systems; part II: Exact modelling; part III: Approximate modelling, *Automatica* 22 (5), 1986, pp. 561–580; 22 (6), 1986, pp. 675–694; 23 (1), 1987, pp. 87–115.
[74] Willems, J.C., Models for dynamics, in Kirchgraber, U., and H.O. Walther (eds.), *Dynamics reported*, volume 2, Wiley and Teubner, 1989, pp. 171–269.
[75] Willems, J.C., and C. Heij, l_2-systems and their scattering representation, in Bart, H., I. Gohberg and M.A. Kaashoek (eds.), *Operator theory and systems*, Operator theory: advances and applications, volume 19, Birkhäuser, Basel, 1986, pp. 443–448.
[76] Willems, J.C., and C. Heij, Scattering theory and approximation of linear systems, in Byrnes, C.I., and A. Lindquist (eds.) *Modelling, identification and robust control*, North-Holland, Amsterdam, 1986, pp. 397–411.
[77] Wolovich, W.A., *Linear multivariable systems*, Springer, New York, 1974.

Symbol index

This index contains symbols which are used more than locally in this monograph. The other symbols are explained in the notation paragraphs in the text. The numbers indicate the pages where the symbols are defined.

procedures

miscellanea

Subject index

Lecture Notes in Control and Information Sciences

Edited by M. Thoma and A. Wyner

Vol. 81: Stochastic Optimization
Proceedings of the International Conference,
Kiew, 1984
Edited by I. Arkin, A. Shiraev, R. Wets
X, 754 pages, 1986.

Vol. 82: Analysis and Algorithms
of Optimization Problems
Edited by K. Malanowski, K. Mizukami
VIII, 240 pages, 1986.

Vol. 83: Analysis and Optimization
of Systems
Proceedings of the Seventh International
Conference of Analysis and Optimization
of Systems
Antiba, June 26-27, 1986
Edited by A. Bensoussan, J. L. Lions
XVI, 901 pages, 1986.

Vol. 84: System Modelling
and Optimization
Proceedings of the 12th IFIP Conference
Budapest, Hungary, September 2–6, 1985
Edited by A. Prékopa, J. Szelezsán, B. Strazicky
XII, 1046 pages, 1986.

Vol. 85: Stochastic Processes
in Underwater Acoustics
Edited by Charles R. Baker
V, 205 pages, 1986.

Vol. 86: Time Series and
Linear Systems
Edited by Sergio Bittanti
XVII, 243 pages, 1986.

Vol. 87: Recent Advances in
System Modelling and
Optimization
Proceedings of the IFIP-WG 7/1
Working Conference
Santiago, Chile, August 27-31, 1984
Edited by L. Contesse, R. Correa, A. Weintraub
IV, 199 pages, 1987.

Vol. 88: Bruce A. Francis
A Course in H_∞ Control Theory
XI, 156 pages, 1987.

Vol. 88: Bruce A. Francis
A Course in H_∞ Control Theory
X, 150 pages, 1987.
Corrected - 1st printing 1987

Vol. 89: G. K. H. Pang/A. G. J. McFarlane
An Expert System Approach to
Computer-Aided Design
of Multivariable Systems
XII, 223 pages, 1987.

Vol. 90: Singular Perturbations
and Asymptotic Analysis
in Control Systems
Edited by P. Kokotovic,
A. Bensoussan, G. Blankenship
VI, 419 pages, 1987.

Vol. 91 Stochastic Modelling
and Filtering
Proceedings of the IFIP-WG 7/1
Working Conference
Rome, Italy, Decembre 10-14, 1984
Edited by A. Germani
IV, 209 pages, 1987.

Vol. 92: L. T. Grujić, A. A. Martynyuk,
M. Ribbens-Pavella
Large-Scale Systems Stability Under
Structural and Singular Perturbations
XV, 366 pages, 1987.

Vol. 93: K. Malanowski
Stability of Solutions to Convex
Problems of Optimization
IX, 137 pages, 1987.

Vol. 94: H. Krishna
Computational Complexity
of Bilinear Forms
Algebraic Coding Theory and
Applications to Digital
Communication Systems
XVIII, 166 pages, 1987.

Vol. 95: Optimal Control
Proceedings of the Conference on
Optimal Control and Variational Calculus
Oberwolfach, West-Germany, June 15-21, 1986
Edited by R. Bulirsch, A. Miele, J. Stoer
and K. H. Well
XII, 321 pages, 1987.

Vol. 96: H. J. Engelbert/W. Schmidt
Stochastic Differential Systems
Proceedings of the IFIP-WG 7/1
Working Conference
Eisenach, GDR, April 6-13, 1986
XII, 381 pages, 1987.

Lecture Notes in Control and Information Sciences

Edited by M. Thoma and A. Wyner

Vol. 97: I. Lasiecka/R. Triggiani (Eds.)
Control Problems for Systems
Described by Partial Differential Equations
and Applications
Proceedings of the IFIP-WG 7.2
Working Conference
Gainesville, Florida, February 3-6, 1986
VIII, 400 pages, 1987.

Vol. 98: A. Aloneftis
Stochastic Adaptive Control
Results and Simulation
XII, 120 pages, 1987.

Vol. 99: S. P. Bhattacharyya
Robust Stabilization Against
Structured Perturbations
IX, 172 pages, 1987.

Vol. 100: J. P. Zolésio (Editor)
Boundary Control and Boundary Variations
Proceedings of the IFIP WG 7.2 Conference
Nice, France, June 10-13, 1987
IV, 398 pages, 1988.

Vol. 101: P. E. Crouch,
A. J. van der Schaft
Variational and Hamiltonian
Control Systems
IV, 121 pages, 1987.

Vol. 102: F. Kappel, K. Kunisch,
W. Schappacher (Eds.)
Distributed Parameter Systems
Proceedings of the 3rd International Conference
Vorau, Styria, July 6-12, 1986
VII, 343 pages, 1987.

Vol. 103: P. Varaiya, A. B. Kurzhanski (Eds.)
Discrete Event Systems:
Models and Applications
IIASA Conference
Sopron, Hungary, August 3-7, 1987
IX, 282 pages, 1988.

Vol. 104: J. S. Freudenberg/D. P. Looze
Frequency Domain Properties of Scalar
and Multivariable Feedback Systems
VIII, 281 pages, 1988.

Vol. 105: Ch. I. Byrnes/A. Kurzhanski (Eds.)
Modelling and Adaptive Control
Proceedings of the IIASA Conference
Sopron, Hungary, July 1986
V, 379 pages, 1988.

Vol. 106: R. R. Mohler (Editor)
Nonlinear Time Series and
Signal Processing
V, 143 pages. 1988.

Vol. 107: Y. T. Tsay, L.-S. Shieh, St. Barnett
Structural Analysis and Design
of Multivariable Systems
An Algebraic Approach
VIII, 208 pages, 1988.

Vol. 108: K. J. Reinschke
Multivariable Control
A Graph-theoretic Approach
274 pages, 1988.

Vol. 109: M. Vukobratović/R. Stojić
Modern Aircraft Flight Control
VI, 288 pages, 1988.

Vol. 110: In preparation

Vol. 111: A. Bensoussan, J. L. Lions (Eds.)
Analysis and Optimization
of Systems
XIV, 1175 pages, 1988.

Vol. 112: Vojislav Kecman
State-Space Models of Lumped
and Distributed Systems
IX, 280 pages, 1988

Vol. 113: M. Iri, K. Yajima (Eds.)
System Modelling and Optimization
Proceedings of the 13th IFIP Conference
Tokyo, Japan, Aug. 31 – Sept. 4, 1987
IX, 787 pages, 1988.

Vol. 114: A. Bermúdez (Editor)
Control of Partial Differential Equations
Proceedings of the IFIP WG 7.2
Working Conference
Santiago de Compostela, Spain, July 6-9, 1987
IX, 318 pages, 1989

Vol. 115: H.J. Zwart
Geometric Theory for Infinite
Dimensional Systems
VIII, 156 pages, 1989.

Vol. 116: M.D. Mesarovic, Y. Takahara
Abstract Systems Theory
VIII, 439 pages, 1989

Lecture Notes in Control and Information Sciences

Edited by M. Thoma and A. Wyner

Vol. 117: K.J. Hunt
Stochastic Optimal Control Theory
with Application in Self-Tuning Control
X, 308 pages, 1989.

Vol. 118: L. Dai
Singular Control Systems
IX, 332 pages, 1989

Vol. 119: T. Başar, P. Bernhard
Differential Games and Applications
VII, 201 pages, 1989

Vol. 120: L. Trave, A. Titli, A. M. Tarras
Large Scale Systems:
Decentralization, Structure Constraints
and Fixed Modes
XIV, 384 pages, 1989

Vol. 121: A. Blaquière (Editor)
Modeling and Control of Systems
in Engineering, Quantum Mechanics,
Economics and Biosciences
Proceedings of the Bellman Continuum
Workshop 1988, June 13-14, Sophia Antipolis, France
XXVI, 519 pages, 1989

Vol. 122: J. Descusse, M. Fliess, A. Isidori,
D. Leborgne (Eds.)
New Trends in Nonlinear Control Theory
Proceedings of an International
Conference on Nonlinear Systems,
Nantes, France, June 13-17, 1988
VIII, 528 pages, 1989

Vol. 123: C. W. de Silva, A. G. J. MacFarlane
Knowledge-Based Control with
Application to Robots
X, 196 pages, 1989

Vol. 124: A. A. Bahnasawi, M. S. Mahmoud
Control of Partially-Known
Dynamical Systems
XI, 228 pages, 1989

Vol. 125: J. Simon (Ed.)
Control of Boundaries and Stabilization
Proceedings of the IFIP WG 7.2 Conference
Clermont Ferrand, France, June 20-23, 1988
IX, 266 pages, 1989

Vol. 126: N. Christopeit, K. Helmes
M. Kohlmann (Eds.)
Stochastic Differential Systems
Proceedings of the 4th Bad Honnef Conference
June 20-24, 1988
IX, 342 pages, 1989

Vol.127: C. Heij
Deterministic Identification
of Dynamical Systems
VI, 292 pages, 1989